T0269136

Molecular breeding for the genetic improvement of forage crops and turf

Molecular breeding for the genetic improvement of forage crops and turf

Proceedings of the 4th international symposium on the molecular breeding of forage and turf, a satellite workshop of the XXth International Grassland Congress, July 2005, Aberystwyth, Wales

edited by:

M.O. Humphreys

Wageningen Academic
P u b l i s h e r s

Subject headings:
DNA markers and QTL analysis
Genomics and gene discovery
Genetic diversity

ISBN 9076998736

First published, 2005

Wageningen Academic Publishers
The Netherlands, 2005

IGER local Scientific Committee

Joint Chairs
Mervyn Humphreys
Chris Pollock

Members:
Ian King
Iain Donnison
Helen Ougham
Phillip Morris
Mike Humphreys
Michael Abberton
Pete Wilkins
Sid Thomas

IGER local Administration Committee

Mervyn Humphreys
Ann Davies
Liz Griffiss-White

Molecular Breeding of Forage and Turf International Organising Committee

Mervyn Humphreys
German Spangenberg
Reed Barker
Andy Hopkins
Odd Arne Rognli
Hitoshi Nakagawa

Acknowledgements

Financial support was gratefully received from the following organisations for this publication:

Department for Environment Food and Rural Affairs

The Samuel Roberts Noble Foundation

Foreword

Grassland covers 26% of the world's total land area. It produces feed for livestock; maintains soil fertility; protects and conserves soil and water resources; creates a habitat for wildlife; and provides recreational space for sport and leisure while contributing to the general landscape. Managed grassland systems can provide tourism, recreation and environmental benefits at the same time as maintaining sustainable economic outputs.

Forage breeding programmes have produced improvements in both yield and quality. However, to reduce pollution risks from agriculture and enhance environmental quality, resources such as nutrients and water must be used more efficiently. With a widening range of traits, techniques for more accurate, rapid and non-invasive phenotyping and genotyping become increasingly important. The large amounts of data involved require good bioinformatics support. Data of various kinds must be integrated from an increasingly wide range of sources such as genetic resource and mapping information for plant populations through to the transcriptome and metabolome of individual tissues. The merging of data from disparate sources and multivariate data-mining across datasets can reveal novel information concerning the biology of complex systems.

Application of molecular breeding technologies to forage breeding offers potential for faster and more targeted cultivar development. New approaches based on biotechnology are becoming more accessible; including functional genomics leading to marker assisted introgression and selection. Forage species are genetically diverse in comparison to other crops due to an outcrossing mating system, high potential for hybridisation between related species and the absence of genetic bottlenecks caused by domestication. Conservation of this biodiversity across a range of geographical and ecological niches provides a rich resource for allele mining to facilitate response to future challenges, such as changes in climate and land-use. Technologies for monitoring and manipulating gene expression provide new research tools and opportunities in molecular breeding but important questions concerning risk assessment are raised with regard to transgenics.

This book encompasses a broad range of topics associated with policy, strategy, research and development in 8 keynote papers and 146 shorter contributions presented at the 4th International Symposium on the Molecular Breeding of Forage and Turf held in Aberystwyth, Wales, UK during 3rd – 7th July 2005. Previous MBFT Symposia have been held in Japan in 1998, Australia in 2000 and the USA in 2003. On this occasion the MBFT Symposium was held as a Satellite Workshop of the XX International Grassland Congress that took place at University College Dublin, Ireland from 26th June to 1st July 2005. This provided a welcome opportunity for molecular breeding activities to be drawn to the attention of the wider global grassland community and encourage broader interactions.

An up-to-date account of progress and potential in the genetic improvement of grassland is provided in the book which demonstrates how recent advances in molecular techniques are being used to develop breeding objectives and strategies. State-of-the-art molecular techniques and resources are described that encompass a unique range of expertise in genetic mapping, trait dissection, comparative genomics, bioinformatics, gene discovery and risk assessment. Examples of work in progress or recently completed are provided from across the world. The book has broad educational value and should interest grassland users and policy makers as well as plant geneticists and breeders.

My thanks go to the members of the MBFT International Committee, the XX International Grassland Congress Scientific and Workshop Committees, the local IGER Scientific and Administration Committees, the many reviewers of the submitted papers and, of course, the authors themselves. This publication was made possible through the vital and generous support of the Department for Environment Food and Rural Affairs in the UK and the Samuel Roberts Noble Foundation in the USA.

Mervyn Humphreys

May 2005

Table of contents

Section 4: Genomics, model species, gene discovery and functional analysis   **163**

Section 6: Genetics and breeding for symbiosis — 209

Section 7: Transgenics for research and breeding including risk assessment 219

Keynote presentations

Objectives and benefits of molecular breeding in forage species

T. Lübberstedt

Danish Institute of Agricultural Sciences, Research Centre Flakkebjerg, 4200 Slagelse, Denmark, Email: Thomas.luebberstedt@agrsci.dk

Key points

1. The amount of resources and information provided by forage crop genomic programs has dramatically increased during the past few years.
2. Trait-based forward genetic procedures such as mapping and expression profiling have successfully provided new candidate genes or genome regions affecting forage quality. Respective information can easily be transferred across related forage species.
3. Since several genes in major biochemical pathways related to forage traits have been isolated, gene-based reverse genetic approaches (transformation, association studies) are promising.
4. Most genetic experiments are conducted under simplified "artificial" conditions such as on single-spaced plants. Therefore, transferability of respective genetic information to breeding practice needs to be demonstrated.

Keywords: forage genomics, forward genetics, reverse genetics, digestibility

Introduction

The major goals in forage production are (i) to maximize dry matter yield, and (ii) to achieve a high level of forage quality. Since both goals might be negatively correlated, the ultimate goal is to produce a maximum yield of metabolizable energy. Several factors need to be considered when determining forage quality, such as the fed animal species, the forage plant species and cropping system, and the method(s) used for forage quality evaluation that determine the feed - animal interaction. In addition, the optimum diet of animals depends on the product, such as beef or milk for cattle. Ruminants have a much better capability to digest fibrous carbohydrates compared to monogastrics and to convert poor quality protein and nonprotein nitrogen sources (Van Soest, 1974). Furthermore, intake and digestion by animals depends on forage properties such as its dry matter concentration, particle size, and the ensiling process.

In most forage grasses and legumes, the above-ground parts are harvested before or during flowering. An exception is forage maize, which is harvested after seed-set. Dry matter yield can be easily determined, whereas quality evaluation is rather difficult. The direct approach to evaluate the quality of a given forage crop is the conduct of animal feeding trials to maximize the yield of the intended product such as milk or beef. Different parameters were developed for these traits such as Mega Joule "Net-Energy-Lactation" ((MJ) NEL) (Groß, 1979, Weißbach, 1993) for milk production and "Kilo Starch-Units" (kStE) (Zimmer *et al.*, 1980) for beef production and "Metabolizable Energy" (ME) (Menke and Huss, 1987), all reflecting the energy density (J/kg) of forage dry matter (Boberfeld, 1986). Nevertheless, this direct approach of quality evaluation has a number of limitations. Animal trials are rather time-consuming, laborious, and consequently expensive. Hence, it is not possible to handle thousands of plant genotypes, as required in breeding programs. Furthermore, feeding trials depend on a number of additional factors, such as the animal species and genotypes employed, and the mode of feeding forage genotypes, impairing the generalization of the results. In consequence, a number of indirect biological, chemical, and physical methods for quality evaluation have been developed. Furthermore, prediction of the breeding value of forage plant genotypes based on molecular markers would be highly desirable.

Biological methods for quality evaluation can be subdivided into field, *in vitro*, and enzymatic methods. Field methods score the expected nutritive value of plant communities (Boberfeld, 1986) based on the species composition. In case of forage maize, the proportion of ears in total dry matter has been used for quality evaluation (Zscheischler, 1990). A widely used *in vitro* rumen digestion analysis was developed by Tilley and Terry (1963) using a two-step procedure - first rumen liquor and subsequently peptic hydrochloric acid to estimate the *in vitro* digestibility of organic matter. Another *in vitro* test employing rumen liquor determines gas production, protein and fat content to estimate NEL or StE (Menke and Steingass, 1987). Enzymatic methods use cellulase together with peptic hydrochloric acid to estimate NEL and StE (Kirchgessner and Kellner, 1981).

Since digestibility is mainly limited by poorly digestible cell wall components, chemical methods for forage quality evaluation focus on the breakdown and characterization of cell wall fractions within the organic matter. Using detergents Van Soest (1974) separated cell complexes into soluble cell content and insoluble "neutral detergent fibre" (NDF) representing mainly the cell wall fraction. By acidic detergents further fractionation into a lignin ("acid detergent lignin": ADL) and cellulose fraction (ADF-ADL; ADF: "acid detergent fibre") is possible. ADF values can be converted into NEL and StE estimates by convenient equations (Kirchgessner and Kellner, 1981).

All above mentioned approaches are too laborious for routine quality evaluation of large numbers as required in plant breeding. Near-infrared reflectance spectroscopy (NIRS) (Norris *et al.*, 1976) helps to overcome this limitation. By this method, large sample numbers can be investigated with low effort. Infrared spectra of ground materials (1400 to 2600 nm) can be employed to estimate a number of quality parameters, if suitable calibrations exist based on animal trials, biological or chemical methods. However, with the availability of an increasing number of both tools and knowledge at the genome and gene level, the prospects for efficient breeding strategies based on genotypic rather than on phenotypic information have rapidly changed during the past few years.

The most important forage species belong to two families, the monocot grasses (Gramineae) and the dicot legumes (Leguminosae). Both families include more than 500 genera with annual, biennial, and perennial species. Within these families, the most important agronomic species include *Lolium perenne, Lolium multiflorum, Festuca arundinacea, Festuca pratensis, Trifolium repens, Trifolium pratense,* and *Medicago sativa*. These forage crop species also receive most attention with regard to the development and application of molecular tools, in addition to the "forage model species" *Lotus corniculatus* and *Medicago truncatula*. The rapid development of molecular genetic tools in these species is well documented in the past volumes of "Developments in Plant Breeding" devoted to forage crops (Spangenberg, 2001; Hopkins *et al.*, 2003).

The objectives of this paper are to describe the current status of (i) plant genomics activities in forage crops, (ii) known biochemical pathways and respective genes affecting forage quality with regard to reverse genetic approaches, (iii) activities on forward genetic approaches with regard to forage quality, and (iv) prospects and limitations on the implementation of molecular tools into forage crop breeding.

Forage crop genomics: tools

The ambition of plant genomics (Lander, 1996) is to provide structural information on whole genomes and in multi-parallel experimental approaches to (i) achieve a holistic view on biological processes, (ii) to accumulate information across experiments and species in order to investigate the function and interaction of genes, and, (iii), to transfer information to crops by transgenic approaches or by creating "designer" plants (Pelemann and van der Voort, 2003) based on functional DNA markers (Andersen and Lübberstedt, 2003).

For major crops more than 300.000 gene-derived EST (expressed sequence tag) sequences per species have been generated (http://www.ncbi.nlm.nih.gov/dbEST/dbEST_summary.html). Complete plant genomes have been sequenced so far for the model species *Arabidopsis thaliana* and rice (The *Arabidopsis* Genome Initiative 2000; Goff *et al.*, 2002; Yu *et al.*, 2002). More than 100,000 ESTs each have been generated for the model legumes *L. corniculatus* and *M. truncatula* (http://www.ncbi.nlm.nih.gov/dbEST/dbEST_summary.html). For forage crops, only a limited number of ESTs has been released to public databases (5,800 ESTs for *L. multiflorum*, 6,500 ESTs for *M. sativa*). However, substantial numbers of so far non-released ESTs (>10,000) have been or are currently generated for *L. perenne* (Sawbridge *et al.*, 2003; Asp *et al.*, 2003), *F. arundinacea* (Zhang and Mian, 2003), and *T. repens* (Spangenberg *et al.*, 2003). Furthermore, a comprehenseive collection of "gene thresher" genomic sequences has been produced for *L. perenne* (http://www.vialactia.co.nz/news/newsitem.asp?id=61), and provided to Cold Spring Harbour Laboratory for annotation. Finally, BAC libraries have been reported for *L. perenne*, *F. arundinacea*, and *T. repens* (Farrar *et al.*, 2005; Spangenberg *et al.*, 2003; Donnison *et al.*, 2002).

To improve complex traits such as forage quality, this structural genomic information has two major implications. Firstly, for those species with comprehensive genome sequence information available, it has become feasible to apply the 'forward genetic' approach of map-based gene isolation, as compared to previous attempts where it was basically impossible to go beyond mapped QTL. Secondly there is the possibility of information transfer across related species due to the evolutionary conserved gene order in chromosome blocks or even chromosomes (Devos and Gale, 2000). This approach of using syntenic relationships to identify relevant genes in forage crops is extremely promising due to the close relationships existing within the grass and legume families. Close syntenic relationships among different grass (Alm *et al.*, 2003) and legume (VandenBosch and Stacey, 2003) species have been demonstrated. The concept for identifying orthologous sequences has meanwhile successfully been used in forage grasses (e.g., Armstead *et al.*, 2004; Jensen *et al.*, 2004).

In the frame of "functional genomics", efficient tools for multi-parallel and rapid testing of gene function including microarray-based expression profiling (Aharoni and Vorst, 2002), comprehensive mutant collections and virus induced gene silencing (VIGS) (Constantin *et al.*, 2004) have been developed for plants. For the model species *A. thaliana*, the ambition is to characterize the function of all genes of this species until 2010. Comprehensive functional genomics projects are also underway for *L. corniculatus* and *M. truncatula* (VandenBosch and Stacey, 2003). The term "function" relates to some basic characteristics of genes (e.g. mutant phenotype, biochemical properties, expression pattern of selected genotypes). Functional genomics will provide new candidate genes at high speed for several traits due to better understanding of their biochemical role. Genes of interest can, in principle, be identified for any forage crop by exploiting information based on sequence homology or conserved map position provided from model species. Recently, an increasing number of genomic tools have been developed for the major forage crops. These include gene-derived markers (e.g., Faville *et al.*, 2004; Lübberstedt *et al.*, 2003; Sledge *et al.*, 2003; Saha *et al.*, 2004) and microarrays (Spangenberg *et al.*, 2001, 2003). Moreover, transformation for *in vivo* validation of gene function is established for the major forage crops (Spangenberg *et al.*, 2001) and, more recently, VIGS has been established for legume species (Constantin *et al.*, 2004).

Traits: forward genetics

Despite rapid progress in the last decade in generating sequences and tools in plant genomics, the function of more than 90% of all genes is still unknown even for the model species *A. thaliana*. Thus, reverse genetics approaches can currently only include a minor fraction of all possible genes. Forward genetic approaches are based on traits of interest (e.g. forage quality characters), and genome regions or genes are associated with trait variation. Quantitative trait loci (QTL) mapping combines conventional "black box" quantitative genetics with a marker gene-based approach. Several QTL mapping studies for forage traits have been conducted, e.g., in maize (Lübberstedt *et al.*, 1997; 1998; Barriere *et al.*, 2003), and ryegrass (Cogan *et al.*, 2004). Combined with the availability of BAC libraries, map-based gene isolation has generally become possible. Another option is the comparison of QTL with candidate gene locations in order to identify the most promising known candidate genes (Barriere *et al.*, 2003, Cogan *et al.*, 2004). These map-based approaches can be extended to introgression lines as shown for ryegrass and *Festuca* in the EU project SAGES (http://www.iger.bbsrc.ac.uk/SAGES/). This approach is promising in forage crops since it makes use of synteny both to exploit genomic tools across related species, and gene materials to broaden genetic variation.

Additional genomic tools have already been employed in forward genetic approaches to monitor genes associated with forage traits. In maize, publicly available microarrays have been used to compare expression profiles of bm isogenic lines or extremes from QTL mapping populations (Lübberstedt *et al.*, 2004). Since several hundred genes were found to be significantly differentially expressed between these genetic contrasts in relation to cell wall digestibility, the next crucial step is to identify the most relevant candidates by comparison of expression profiling experiments across different genetic contrasts, or by comparison of map positions of differentially expressed genes with QTL locations. For ryegrass and white clover, programs for forage plant gene discovery based on expression profiling have been announced (Spangenberg *et al.*, 2001; http://www.dafgri.dk). It can be foreseen that VIGS or TILLING will be useful for the identification of additional genes affecting forage quality.

Pathways and genes: reverse genetics

Prerequisite for reverse genetics approaches is the availability of proven or "qualified" candidate genes affecting traits. Forage quality is determined by cell wall properties or cell content composition. The digestibility of cell walls mainly in grasses but also in legumes is often limited by the content and composition of the lignin fraction. With respect to cell content, water soluble carbohydrates such as fructans are most relevant in grasses, whereas proteins and tannins are major determinants of legume quality.

The lignin biosynthetic pathway is well characterized (Boudet *et al.*, 1995), especially the common phenylpropanoid pathway starting with the deamination of phenylalanine and providing hydroxycinnamoyl CoAs. The enzymes involved in the common phenylpropanoid pathway are phenylalanine ammonia-lyase (PAL), cinnamate hydroxylase (C4H), coumarate hydroxylase (C3H), caffeic O-methyltransferases (COMT), ferulate hydroxylase (FA5H), and hydroxycinnamate CoA ligases (4CL). In total 34 genes have been identified in the *A. thaliana* genome coding for enzymes in the monlignol biosynthesis (Raes *et al.*, 2003). The end products of this common pathway, the hydroxycinnamoyl CoAs, are the precursors of the major classes of phenolic compounds which accumulate in plant tissues, e.g. flavonoids, stilbenes, phenolamides as well as lignins. Subsequently, cinnamoyl CoA reductase (CCR) and cinnamyl alcohol dehydrogenase (CAD) are specifically involved in biosynthesis of the lignin monomers p-Coumaryl, Coniferyl, and Sinapyl alcohol. In maize, genes for COMT (Collazo *et al.*, 1992) and CAD (Halpin *et al.*, 1998) have been isolated. Defect alleles of both genes have been shown to correspond to brown midrib mutations (COMT: bm3; CAD: bm1) known for long time (Barriere and Argillier, 1993). Several independent studies on bm1 and bm3 have proved already the concept of increasing silage quality by altering the lignin biosynthetic pathway (Barriere and Argillier, 1993) and reducing lignin content. However, the application of bm mutants in plant breeding has been hampered so far by their strong negative pleiotropic effects on yield characters and lodging. Therefore, one current target for improving feed quality is the identification of optimal alleles both for quality but also agronomic performance at well characterized genes such as COMT (Fontaine *et al.*, 2003, Lübberstedt *et al.*, 2004). Another target is the identification and evaluation of additional promising candidate genes affecting lignification and cell wall formation.

Lignins exhibit a high degree of structural variability depending on the relative proportion of three monolignols, different types of interunit linkages, and the occurrence of non conventional lignin units within the polymer (Boudet and Grima-Pettenati, 1996). Polymerization of monolignols involves peroxidases and laccases in an oxidation step but is generally not well understood (Boudet *et al.*, 1995). Laccase was the first enzyme shown to be able to polymerize lignin monomers in vitro (Freudenberg *et al.*, 1958). Several studies indicated that laccase and laccase-like activities are closely correlated with lignin deposition in developing xylem (Davin *et al.*, 1992). Nersissian and Shipp (2002) identified 19 laccases in the genome of *A. thaliana*. In case of peroxidases, association between allelic variation at one Prx locus with forage digestibility in maize has been demonstrated (Guillet-Claude *et al.*, 2004). Thus, a minimum of 70-80 candidate genes code for enzymes that are directly involved in lignin formation, based on the small *A. thaliana* genome. Furthermore, gene families involved in cellulose or hemicellulose formation might also affect cell wall digestibility (Barriere *et al.*, 2003, Ralph *et al.*, 2004). In conclusion, the challenge is not in finding candidate genes but in the identification of the most promising genes among those candidates for transgenic or marker-based approaches.

The digestibility of grasses as a general trait becomes markedly reduced during the course of the growing season. This reduction is largely caused by an increase in the content of poorly digestible cell wall structural components. In parallel, there is a decrease in the content of soluble carbohydrates – "sugars". Varieties of ryegrass with a high stable level of carbohydrates in the form of fructans have been shown to retain a high degree of digestibility throughout the growing season. Poorly digestible structural components create an imbalance between carbohydrate and protein levels during ruminant fermentation, leading to a loss of nitrogen (ammonia) to environment. Grass varieties with an increased level of soluble carbohydrates will lead to a more efficient uptake of proteins in ruminants, and thus, more efficient milk and meat production. Fructans are polymers of fructose, and have a general structure of a glucose linked to multiple fructose units (polyfructosylsucroses). In contrast to the uniform structure of bacterial fructans, plant fructans represent five major classes of structural distinct fructans according to the linkages between the fructose units; inulins, levans, inulin neoseries, levan neoseries, and mixed type levans. It is now known that four out of five different fructosyltransferases (FT), each with their own specificity, are needed to synthesize the wide variety of fructans found in plants; 1-SST (1-sucrose:sucrose fructosyltransferase), 1-FFT (1-fructan:fructan fructosyltransferase), 6-FFT (6-fructan:fructan fructosyltransferase), 6-SFT (Sucrose fructosyl 6-transferase) and 6G-FFT (fructan:fructan 6G-fructosyltransferase). Other enzymes that are involved in fructan degradation are fructanhydrolases (FH), and invertases. Fructanhydrolase is a beta-fructofuranosidase and can uncouple fructose units from fructans with sucrose as an end product. Invertase, which is active in the vacuole, cleaves one sucrose molecule into glucose and fructose. It's also capable of cleaving fructose molecules from smaller fructans. This fructan hydrolysing activity decreases with a higher degree of polymerization of the fructan. The type of fructans and FTs varies among different monocot species. Levan type fructans are abundant in *Triticum*, *Hordeum* and *Bromus*, whereas inulin types are characteristic to e.g. *Lolium* species. Interestingly, bifurcose (a product of 6-SFT and precursor to the levan type and one of the routes to the levan neoseries) has not been found in *Lolium* species. According to this, four enzymes would be necessary to account for the synthesis of the fructans identified in *L. perenne*, namely 1-SST, 1-FFT, 6G-FFT and 6FFT (Parvis *et al.*, 2001). Genes coding respective enzymes in forage crops have been isolated recently (Gallagher *et al.*, 2004; Chalmers *et al.*, 2003).

Forage legumes are highly digestible as compared to forage grasses. However, proteolysis and microbial deamination might lead to protein loss in the rumen, not fully compensated by post-ruminal absorption (Robbins *et al.*, 2002). High digestion rates may result in protein foaming and rumen pasture bloat as a digestive disorder (Gruber *et al.*, 2001). Moreover, amino acid composition of the protein fraction determines it's nutritive value (Spangenberg *et al.*, 2001). A major role for reducing the high digestion rates has been assigned to condensed tannins (Gruber *et al.*, 2001). Condensed tannins are polymeric flavonoids with protein-precipitating properties. Whereas high amounts of condensed tannins are detrimental to ruminant digestion, moderate levels (2-3% of dry matter) improve forage legume quality by reducing ruminal digestion rates and avoidance of protein foaming, and thus lead to higher rates of protein conversion into animal products (Robbins *et al.*, 2002). Highly nutritious species such as white clover and alfalfa have a low level of endogeneous tannins as compared to tanniferous forages like Lotus. The initial steps in condensed tannin biosynthesis belong to the general flavonoid pathway and include enzymes like chalcone synthase. Whereas genes coding for enzymes of the general flavonoid have been isolated for long time, the first genes coding for enzymes of the condensed tannin specific pathway such as Leucoanthocyanidin reductase have been isolated more recently (Tanner *et al.*, 2003). As for the lignin biosynthesis pathway, regulatory genes coding for transfactors have been envisaged as targets to manipulate tannin content, such as myb- or myc-like genes (Gruber *et al.*, 2001).

Once qualified candidate genes affecting forage quality have been identified and isolated, these genes can be used to (i) create new genetic variation not available in elite germplasm of forage crops, or (ii) monitor and exploit existing genetic variation in a more targeted way. Transgenic approaches have been successfully employed both for improving cell wall digestibility and cell content composition in different forage crops, either by overexpression of novel genes, or by suppression of genes using antisense or RNAi technology (Spangenberg *et al.*, 2001). One of the most obvious examples is the production of condensed tannins in alfalfa or white clover by overexpression of genes from the tannin biosynthesis (Gruber *et al.*, 2001). Another major research area is the down-regulation of genes from the lignin biosynthesis pathway (Ralph *et al.*, 2004). Down-regulation of the maize bm3 orthologue coding for COMT was shown to successfully alter both lignin content and composition in grasses (Chen *et al.*, 2003). However, implementation of transgenic approaches for variety production suffers from a lack of acceptance in several countries, and requires extensive risk evaluation (Wang *et al.*, 2003), which is especially crucial for the mostly outbreeding forage crops endogeneous in the relevant production areas. An alternative, at least for the knock-out approach, is the generation and screening of new genetic variation by TILLING (McCallum *et al.*, 2000). However, public TILLING populations are currently not available for forage crops, but in preparation for maize (http://genome.purdue.edu/maizetilling/), ryegrass (http://www.intl-pag.org/pag/13/abstracts/PAG13_W100.html), and *M. truncatula* (May *et al.*, 2003).

The second important reverse genetics area is the monitoring of allelic variation with a view to the development of functional markers (Andersen and Lübberstedt, 2003). In this context, association studies have recently been adapted to plants from human genetics and proven valuable to identify sequence motifs within genes affecting a trait of interest (Flint-Garcia *et al.*, 2003). Thornsberry *et al.* (2001) identified in a pioneering study nine SNP or INDEL polymorphisms in the maize dwarf 8 gene associated with flowering time. More recently, polymorphisms within CCoAOMT-2 (Guillet-Claude *et al.*, 2004a), a peroxidase (Guillet-Claude *et al.*, 2004b), and COMT (Lübberstedt *et al.*, 2004) were associated with cell wall digestibility in maize. Association studies based on candidate genes are especially promising in species with a generally low linkage disequilibrium (Flint-Garcia *et al.*, 2003), as can be expected for outcrossing forage crops. Studies on systematic allele-sequencing and association studies in ryegrass are currently ongoing in the EU project GRASP (http://www.grasp-euv.dk). However, sequence motifs showing association with the trait of interest need further validation (Andersen and Lübberstedt, 2003) before converting them into robust functional markers.

Molecular breeding: benefits

Independent of the breeding programme, the process of breeding new cultivars includes three phases: I) generation of genetic variation, II) development of genetic components for producing new varieties (such as inbred lines in hybrid breeding), and III) testing of experimental varieties (Becker, 1993). Molecular breeding benefits all three phases, but is also useful in the context of variety registration and protection as well as for the characterization and management of genetic resources. The major approaches provided by genomics are based on transgenes or markers. The predictive value of markers depends on whether they are random DNA markers, gene-derived or functional markers (Andersen and Lübberstedt, 2003).

The major benefit of transgenic approaches is a broadening of genetic variation, especially if respective genes are lacking in the target species (Spangenberg *et al.*, 2001). Markers are useful to establish heterotic groups (Riday and Brummer, 2003), and to assign genotypes or populations to heterotic groups. This topic might become increasingly relevant, if forage crop breeding moves from population or synthetic breeding to hybrid breeding (Riday and Brummer, 2003). Furthermore, markers might assist identification of genetically divergent parent genotypes or populations with a maximum usefulness (Lamkey *et al.*, 1995; Schnell, 1983) to generate better varieties. Finally, recurrent selection programs might benefit from the application of markers ensuring, e.g., a sufficient level of genetic variation over several selection cycles.

Marker-assisted selection (MAS) and backcrossing (MAB) are major applications of molecular markers. MAS is especially promising for traits with low heritability (Lande and Thompson, 1990), whereas MAB allows tracing of favourable alleles, which is especially useful in case of recessive gene action. In case of MAB, including transfer of trangenes across genotypes and populations, markers are useful for background selection (Frisch *et al.*, 2001) to recover the elite parent background efficiently in a short time. For MAS, an increasing number of candidate gene-derived or even functional markers (Andersen and Lübberstedt, 2003) will become available in the near future, as demonstrated for ryegrass recently (Faville *et al.*, 2004). Functional markers will reduce the risk of a Type 3 error in declaring QTL-marker associations (Dudley, 2003), i.e., declaring in case of a significant QTL the wrong marker allele as being linked to a favourable QTL allele. In addition, markers can be employed to predict the performance of components of hybrid or synthetic varieties using BLUP (Bernardo, 2002).

Transgenic approaches might be employed for controlled crosses in view of hybrid seed production but also in the context of controlled flowering of grasses to increase forage quality by reducing the stem fraction. This would, in addition, minimize pollen flow of transgenic plants to natural populations but also pollution with pollen allergens (Spangenberg *et al.*, 2001). Furthermore, transgenic traits or bar-codes can be employed for variety description and protection (Spangenberg *et al.*, 2001). During the final steps of variety production markers can be used to (i) reduce the amount of experimental testing, (ii) confirm hybridity, and (iii) fulfil DUS criteria in variety registration (Tommasini *et al.*, 2001). Finally, gene bank collections might benefit from molecular markers to describe and maintain genetic resources, as well as to establish core collections.

Levels of complexity: limitations

Forage crop breeding is characterized by several layers of complexity. At the trait level, direct evaluation of feed quality is too costly to be performed for several genotypes or populations as required in breeding programs. Therefore, indirect biological and chemical methods have been established, meanwhile substituted by more rapid physical methods like NIRS, often calibrated to the indirect chemical or biological methods. Genetic markers based on results obtained with, e.g., NIRS, are currently developed and implemented in breeding programs. Thus, the increasing distance between original animal trials and current indirect methods require re-calibration to avoid artefacts based on the indirect methods used.

At the feed or plant level, a given variety is often only part of the animal diet fed together with minerals, additives etc. If used for grazing, a single variety typically is only part of a mixture between different grasses and legumes. Forages are grown in swards and not at single-spaced plant level as a number of genetic experiments, with quite variable cultivation regimes (e.g., regarding the number of cuts). Further aspects adding to the complexity are different ploidy levels within crops such as ryegrass, and symbiosis with endophytes or root nodule bacteria. In contrast, many genetic experiments are performed under simplified conditions to establish sound phenotype – genotype associations, preferably (i) at the diploid level, (ii) in monoculture, (iii) at per se level, and (iv) for single spaced plants. Therefore, a crucial question is, to what extent are results obtained in "artificial" experimental situations transferable and, thus, valuable to operational breeding programmes?

For some forage crops such as alfalfa, commercial varieties are mainly tetraploid although diploids also exist. In this case, QTL mapping in diploids is much more straightforward. However, QTL detected at diploid level might not be functional at the autopolyploid level. A well known example for differences in gene action at diploid and tetraploid level is the presence of gametophytic self-incompatibility in diploid potatoes, whereas autotetraploid potatoes generally are self fertile (Becker, 1993). Furthermore, autopolyploids generally have enlarged cells and vegetative organs as compared to diploid forms (Becker, 1993). This implies that tetraploid performance can only partly be predicted based on "diploid information". Similarly, prediction of genotype or family performance to be grown in swards based on information obtained at the single-spaced plant level might be poor. Posselt (1984) reported a generally lower heterosis for agronomic traits for ryegrass in swards as compared to spaced plants. Furthermore, low correlations were found in ryegrass for seed yield components evaluated in plots versus single plants (Elgersma, 1990). Thus, depending on the trait of interest, the mode of testing genotype or family performance is essential with regard to the transferability of information for breeding of superior varieties under practical conditions.

Genome regions increasing GCA within a given synthetic are of highest priority for synthetic breeding. Hence, evaluation of testcross rather than per se performance (after cloning of mapped genotypes) will be preferable. In hybrid breeding, per se performance of inbreds is of minor interest compared to that of hybrid performance. In an experiment on mapping of QTL for forage traits in maize, four segregating populations were established within the flint heterotic pool and evaluated for forage traits after testcrossing to elite dent tester inbreds at the hybrid level (Lübberstedt *et al.*, 1997a, b, 1998). The predictive value of QTL was evaluated by comparing QTL results across testers within one population or across populations using the same tester. The three small validation populations had zero, one, or both parent lines in common with the large calibration population. Generally, the number of common QTL across populations increased with the genetic similarity of mapping populations. Almost all QTL detected in the small independent sample were also detected in the calibration population, both derived from the same cross. For unrelated mapping populations, about 70% of the detected QTL were specific to each population. However, consistency of QTL across populations as well as testers was highly trait-dependent. In conclusion, QTL or genes identified in different populations or test systems (like plots versus single spaced plants) need re-evaluation in breeding populations under practical relevant conditions.

Conclusions and perspective

During the past decade, the availability of genomic tools and information in major forage crops has dramatically increased. Numerous genes have been isolated and characterized in biochemical pathways relevant to forage quality. First studies demonstrated the usefulness of these new tools and information in both reverse and forward genetic approaches for more efficient breeding of forage crop varieties. Besides continued development of genomic tools and their application in basic research, the next challenge will be the implementation of these resources in experimental procedures delivering relevant information for practical breeding. Besides implementation of experimental approaches such as haplotype of association mapping (Flint-Garcia *et al.*, 2003) in forage crops, this will require phenotypic testing close to agronomic practice.

References

Aharoni, A. & O. Vorst (2002). DNA microarrays for functional plant genomics. *Plant Molecular Biology*, 48, 99-118.

Alm, V., C. Fang, C.S. Busso, K.M. Devos, K. Vollan, Z. Grieg & O.A. Rognli (2003). A linkage map of meadow fescue and comparative mapping with other Poaceae species. *Theoretical and Applied Genetics*, 108, 25-40.

Andersen J.R. & T. Lübberstedt, (2003). Functional markers in plants. *Trends in Plant Science*, 8, 554-560.

Armstead IP, L.B. Turner, M. Farrell, L. Skot, P. Gomez, T. Montoya, I.S. Donnison, I.P. King & M.O. Humphreys (2004). Synteny between a major heading-date QTL in perennial ryegrass (Lolium perenne L.) and the Hd3 heading-date locus in rice. *Theoretical And Applied Genetics*, 108, 822-828.

Asp, T., C.H Andersen, T. Didion, P.B. Holm, K.K. Nielsen, & T. Lübberstedt (2003). EST development for nutrient utilisation and forage quality of ryegrass, In: 25th EUCARPIA fodder crops and amenity grasses section meeting, Brünn, 98

Barrière, Y. & O. Argillier (1993). Brown-midrib genes of maize: a review. *Agronomie*, 13, 865-876.

Barrière, Y., C. Guillet, D. Goffner & M. Pichon (2003). Genetic variation and breeding strategies for improved cell wall digestibility in annual forage crops. A review. *Animal Research*, 52, 193-228.

Becker, H. (1993). Pflanzenzüchtung, Ulmer Verlag, Stuttgart.

Bernardo, R.(2002). Breeding for quantitative traits in plants. Stemma press, Woodburry, MN

Boberfeld, W.O.V. (1986). Grünlandlehre, Ulmer Verlag, Stuttgart.

Boudet, A.M., & J. Grima-Pettenati (1996). Lignin genetic engineering. *Molecular Breeding,* 2, 25-39.

Boudet, A.M., C. Lapierre & J. Grima-Pettenati (1995). Biochemistry and molecular biology of lignification. *New Phytol.*, 129, 203-236.

Chalmers, J., X. Johnson, A. Lidgett & G.C. Spangenberg (2003). Isolation and characterization of a sucrose:sucrose 1-fructosyltransferase gene from perennial ryegrass. *J Plant Physiology* 160: 1385-1391

Chen, L., C.K. Auh, P. Dowling, J. Bell & Z.Y. Wang (2003). Improving forage quality of tall fescue by genetic manipulation of lignin bioynthesis. In: Hopkins A., Z.-Y. Wang., R. Mian , M. Sledge & R.E. Barker (eds.) Molecular breeding of forage and turf. - Developments in Plant Breeding, Kluwer Academic Publishing, Dordrecht, 181-188.

Cogan, N.O.I., K.F. Smith, T. Yamada, M.G. Francki, A.C. Vecchies, E.S. Jones, G.C. Spangenberg & J.W. Forster (2004). QTL analysis and comparative genomics of herbage quality traits in perennial ryegrass. *c* (in press, available online)

Collazo, P., L. Montoliu, P. Puigdomenech & J. Rigau, (1992). Structure and expression of the lignin O-methyltransferase gene from Zea mays L. *Plant Molecular Biology*, 20, 857-867.

Constantin, GD, B.N. Krath, S.A. MacFarlane, M. Nicolaisen, I.E. Johansen & O.S. Lund (2004). Virus-induced gene silencing as a tool for functional genomics in a legume species. *Plant J ,* 40, 622-632.

Davin, L.B., D.L. Bedgar, T. Katayama & N.G. Lewis (1992). On the stereoselective synthesis of (+)-pinoresinol in Forsythia suspensa from its achiral precursor, coniferyl alcohol. *Phytochemistry*, 31, 3869-3874.

Devos, K.M. & M.D. Gale (2000). Genome relationships: The grass model in current research. *Plant Cell ,* 12, 637-646.

Donnison, I., D.O. O'Sullivan, A. Thomas, P. Canter, H. Thomas, K. Edwards, H.M. Thomas & I. King (2002). Construction of a *Festuca pratensis* BAC library for map based cloning in *Festulolium* introgression lines. Plant, Animal & Microbe Genomes, X Conference, San Diego.

Dudley, J.W. (2003). Population and quantitative genetic aspects of molecular breeding. In: Hopkins A., Z.-Y. Wang, R. Mian, M. Sledge & R.E. Barker (eds.) Molecular breeding of forage and turf. - Developments in Plant Breeding, Kluwer Academic Publishing, Dordrecht, 289-302.

Elgersma, A. (1990). Seed yield related to crop development and to yield components in nine cultivars of perennial ryegrass. *Euphytica*, 49, 141-151.

Farrar, K., I. Thomas, M. Humphreys & I. Donnison I (2005). Construction of a BAC Library for *Lolium perenne* (Perennial Ryegrass). In: Plant Animal Genome Conference XIII abstracts (http://www.intl-pag.org/pag/13/abstracts/PAG13_P052.html)

Faville, M.J., A.C. Vecchies, M. Schreiber, M.C. Drayton, L.J. Hughes, E.S. Jones, K.M. Guthridge, K.F. Smith, T. Sawbridge, G.C. Spangenberg, G.T. Bryan & J.W. Forster (2004). Functionally associated molecular genetic marker map construction in perennial ryegrass. *Theoretical And Applied Genetics* (in press, online available).

Flint-Garcia, M., J.M. Thornsberry & E.S. Buckler (2003). Structure of linkage disequilibrium in plants. *Annu. Rev. Plant Biol.*, 54, 357–74.

Fontaine, A.S., Y. Barrière (2003). Caffeic acid O-methyltransferase allelic polymorphism characterization and analysis in different maize inbred lines. *Mol. Breed.*, 11, 69-75.

Freudenberg, K., J.M. Harkin, M. Rechert & T. Fukuzumi (1958). Die an der Verholzung beteiligten Enzyme. Die Dehydrierung des Subaoinalkohols. Chem. Ber. , 91, 581-590.

Frisch, M. & A.E. Melchinger (2001). Marker-Assisted Backcrossing for Simultaneous Introgression of Two Genes. *Crop Sci.*, 41, 1716.

Gallagher, J.A., A.J. Cairns & C.J. Pollock (2004). Cloning and characterization of a putative fructosyltransferase and two putative invertase genes from the temperate grass *L. temulentum* L. *J. Exp. Bot.*, 55, 557-569.

Goff, S.A. *et al.* (2002). A draft sequence of the rice genome (Oryza sativa L. ssp. japonica). *Science*, 296, 92-100.

Groß, F. (1979). Nährstoffgehalt und Verdaulichkeit von Silomais. 1.Mitt.: Bewertung von Silomais, Das wirtschaftseigene Futter, 25, 215-225.

Gruber, M.Y., H. Ray & L. Blahut-Beatty (2001). Genetic manipulation of condensed tannin synthesis in forage crops. In: Spangenberg, G. (ed.) Molecular breeding of forage crops - Developments in Plant Breeding, Kluwer Academic Publishing, Dordrecht.

Guillet-Claude, C., C. Birolleua-Touchard, D. Manicacci, P.M. Rogowsky, J. Rigau, A. Murigneux, J.-P. Martinant & Y. Barrière (2004). Nucleotide diversity of the ZmPox3 maize peroxidase gene: relationships between a MITE insertion in exon 2 and variation in forage maize digestibility. *BMC Genetics*, 5, 19.

Guillet-Claude, C., C. Birolleua-Touchard, D. Manicacci, M. Fourmann, S. Barraud, V. Carret, J.-P. Martinant & Y. Barrière (2004). Genetic diversity associated with variation in silage corn digestibility for three O-methyltransferase genes involved in lignin biosynthesis. *Theor. Appl. Genet.* (in press, available online).

Halpin, C., K. Holt, J. Chojecki, D. Oliver & B. Chabbert (1998). Brown midrib maize (bm1) – a mutation affecting the cinnamyl alcohol dehydrogenase gene. *Plant J*, 14, 545-553.

Hopkins, A., Z.-Y. Wang, R. Mian, M. Sledge & R.E. Barker (2003). Molecular breeding of forage and turf. In: Developments in Plant Breeding, Kluwer Academic Publishing, Dordrecht.

Humphreys, J.M. & C. Chapple,C. (2002). Rewriting the lignin roadmap. *Curr. Opin. Plant Biol.*, 5, 224-229

Jensen, L.B., J. Andersen, U. Frei, Y. Xing, C. Taylor, P.B. Holm & T. Lübberstedt (2004). QTL mapping of vernalization response in perennial ryegrass reveals cosegregation with an orthologue of wheat VRN1. *Theor. Appl. Genet.*, in press.

Kirchgessner, M. & R.J. Kellner. (1981). Schätzung des energetischen Futterwertes von Grün- und Rauhfutter durch die Cellulasemethode, *Landw. Forschung*, 34, 276-281.

Lamkey, K.R., B.J. Schnicker & A.E. Melchinger (1995). Epistasis in elite maize hybrids and choice of generation for inbred line development. *Crop Sci.*, 35, 1272-1281.

Lande, R. & R. Thompson (1990). Efficiency of Marker-Assisted Selection in the Improvement of Quantitative Traits. *Genetics*, 124, 743-756.

Lander, E.S. (1996). The New Genomics: Global Views of Biology. *Science*, 274, 536-539.

Lübberstedt, T., A.E. Melchinger, C.C. Schön, H.F. Utz & D. Klein (1997a). QTL mapping in testcrosses of European flint lines of maize: I. Comparison of different testers for forage yield traits. *Crop Sci.*, 37, 921-931.

Lübberstedt, T., A.E. Melchinger, D. Klein, H. Degenhardt & C. Paul (1997b). QTL mapping in testcrosses of European flint lines of maize: II. Comparison of different testers for forage quality traits. *Crop Sci.*, 37, 1913-1922.

Lübberstedt, T., A.E. Melchinger, S. Fähr, D. Klein, A. Dally & P. Westhoff (1998). QTL mapping in testcrosses of Flint lines of maize: III. Comparison across populations for forage traits. *Crop Sci.*, 38, 1278-1289.

Lübberstedt, T., B.S. Andreasen, P.B. Holm *et al.* (2003). Development of ryegrass allele-specific (GRASP) markers for sustainable grassland improvement - a new EU Framework V project. *Czech Journal of Genetics and Plant Breeding*, 39, 125-128.

Lübberstedt, T., I. Zein., J.R. Andersen, G. Wenzel, B. Krützfeldt, J. Eder, M. Ouzunova & Chun (2005). Development and application of functional markers in maize. *Euphytica* (submitted).

May, G.D. (2003). From models to crops: integrated Medicago genomics for alfalfa improvement. In: Hopkins, A., Z.-Y. Wang, R. Mian, M. Sledge & R.E. Barker (eds.) Molecular breeding of forage and turf. - Developments in Plant Breeding, Kluwer Academic Publishing, Dordrecht, 325-332.

McCallum, C.M. *et al.* (2000). Targeting induced local lesions IN genomes (TILLING) for plant functional genomics. *Plant Physiol.* 123, 439-442.

Menke, K.H. & H. Steingaß (1987). Schätzung des energetischen Futterwertes aus der in vitro mit Pansensaft bestimmten Gasbildung und der chemischen Analyse, II. Regressionsgleichungen. *Übersichten zur Tierernährung*, 15, 59-93.

Nersissian, A.M. & E.L. Shipp (2002). Blue copper-binding domains. *Adv Prot Chemistry*, 60, 271-340.

Norris, K.H., R.F. Barnes, J.E. Moore & J.S. Shenk. (1976). Predicting forage quality by infrared reflectance spectroscopy. *J. Anim. Sci.*, 43, 889-897.

Pavis, N., J. Boucaud & M. P. Prud'homme (2001) Fructans and fructan-metabolizing enzymes in leaves of *Lolium perenne. New Phytologist*, 150, 97-110.

Peleman, J.D. & J.R. van der Voort (2003). Breeding by design. *Trends In Plant Science*, 8, 330-334.

Posselt, U. (1984). Hybridzüchtung bei Lolium perenne. *Vorträge für Pflanzenzüchtung*, 5, 87-100.

Raes, J., A. Rohde, J.H. Christensen, Y. van de Peer & W. Boerjan (2003). Genome-wide characterisation of the lignification toolbox in Arabidopsis. *Plant. Physiol.*, 133,1051-1071.

Ralph, J., S. Guillaumie, J.H. Grabber, C. Lapierre & Y. Barrière (2004). Genetic and molecular basis of grass cell-wall degradability. III. Towards a forage grass ideotype. *Biologies*, 327, 467-479.

Riday, H. & E.C. Brummer (2003). Dissection of heterosis in alfalfa hybrids. In: Hopkins, A., Z.-Y. Wang, R. Mian, M. Sledge & R.E. Barker (eds.) Molecular breeding of forage and turf. - Developments in Plant Breeding, Kluwer Academic Publishing, Dordrecht, 181-188.

Robbins, M.P., G.G. Allison, A.J.E. Bettany, S.J. Dalton, T.E. Davies & B. Hauck (2002). Biochemical and molecular basis of plant composition determining the degradability of forage for ruminant nutrition. In: Grassland science in Europe - Multifunctional Grassland: Quality Forages, Animal Products and Landscapes. Durand, J.-L., J.-C. Emile, C. Huyghe & G. Lemaire (eds.) Proc. 19th General Meeting of the European Grassland Federation, La Rochelle, France pp. 37-43.

Saha, M.C., M.A.R. Mian, I. Eujayl, J.C. Zwonitzer, L. Wang & G.D. May (2004). Tall fescue EST-SSR markers with transferability across several grass species. *Theor. Appl. Genet.*, 109, 783 – 791.

Sawbridge, T., E.-K. Ong, C. Binnion, M. Emmerling, R. McInnes, K. Meath, N. Nguyen, K. Nunan, M. O'Neill, F. O'Toole, C. Rhodes, J. Simmonds, P. Tian, K. Wearne, T. Webster, A. Winkworth & G. Spangenberg (2003). Generation and analysis of expressed sequence tags in perennial ryegrass. *Plant Sci.*, 165, 1089-1100.

Schnell. F.W. (1983). Probleme der Elternwahl – Ein Überblick. In: Arbeitstagung der Arbeitsgemeinschaft der Saatzuchtleiter in Gumpenstein, Austria. Nov. 22-24 1983, pp. 1 – 11, Press by Bundesanstalt für alpenländische Landwirtchaft, Gunpenstein, Austria

Sledge, M., I. Ray, & M.A.R. Mian (2003). EST-SSRs for genetic mapping in alfalfa. In: Hopkins, A., Z.-Y. Wang, R. Mian, M. Sledge & R.E. Barker (eds.) Molecular breeding of forage and turf. - Developments in Plant Breeding, Kluwer Academic Publishing, Dordrecht, 239-244.

Spangenberg, G. (2001). Molecular breeding of forage crops. In: Developments in Plant Breeding, Kluwer Academic Publishing, Dordrecht

Spangenberg, G., R. Kalla, A. Lidgett, T. Sawbridge, E.K. Ong& U. John (2001). Breeding forage plants in the genome era. In: Spangenberg, G. (ed.) Molecular breeding of forage crops - Developments in Plant Breeding, Kluwer Academic Publishing, Dordrecht.

Spangenberg *et al.* (2003). Integrated resources for pastoral functional genomics: EST collections, BAC libraries, VIGS systems and microarray-based expression profiling in perennial ryegrass (*Lolium perenne*), white clover (*Trifolium repens*) and *Neotyphodium* grass endophytes. Plant & Animal Genomes XI Conference (http://www.intl-pag.org/11/abstracts/W21_W131_XI.html)

Tanner, G.J., K.T. Francki, S. Abrahams, J.M. Watson, P.J. Larkin & A.R. Ashton (2003). Proanthocyanidin biosynthesis in plants. *J. Biol. Chem.*, 278, 31647-31656.

The Arabidopsis Genome Initiative (2000). Analysis of the genome sequence of the flowering plant Arabidopsis thaliana. *Nature*, 408, 796-815.

Thornsberry, J.M. *et al.* (2001). Dwarf8 polymorphisms associate with variation in flowering time. *Nat. Genet.*, 28, 286-289.

Tilley, J.M.A. & R.A. Terry (1963). A two-stage technique for the in vitro digestion of forage crops. *J. Brit. Grassl. Soc.*, 18, 104-111.

Tommasini, L., J. Batley, G.M. Arnold, R.J. Cooke, P. Donini, D. Lee, J.R. Law, C. Lowe, C. Moule, M. Trick & K.J. Edwards (2003). The development of multiplex simple sequence repeat (SSR) markers to complement distinctness, uniformity and stability testing of rape (*Brassica napus* L.) varieties. *Theor. Appl. Genet.*, 106, 1091-1101.

VandenBosch, K.A. & G. Stacey (2003). Summaries of legume genomics projects from around the globe. Community resources for crops and models. *Plant Physiol.*, 131, 840-865.

Van Soest, P.J. (1974). Composition and nutritive value of forages, in: Heath, M.E., D.S. Metcalfe & R.F. Barnes (eds.) Forages (3rd edition), Iowa State University Press, Ames, Iowa, pp. 53-63.

Wang, Z.-Y., A. Hopkins, R. Lawrence, J. Bell & M. Scott (2003). Field evaluation and risk assessment of trangenic tall fescue plants. In: Hopkins, A., Z.-Y. Wang, R. Mian, M. Sledge & R.E. Barker (eds.) Molecular breeding of forage and turf. - Developments in Plant Breeding, Kluwer Academic Publishing, Dordrecht, 367-379.

Weißbach, F. (1993). Bewerten wir die Qualität des Maises richtig ? *Mais*, 21, 162-165.

Yu, J. *et al.* (2002). A draft sequence of the rice genome (Oryza sativa L. ssp. indica). *Science* 296, 79-92.

Zhang, Y., M.A.R. Mian (2003). Functional genomics in forage and turf – present status and future prospects. *African J. Biotech.*, 2, 521-527.

Zimmer, E., H.H. Theune & M. Wermke (1980). Estimation of nutritive value of silage maize by using chemical parameters and in vitro digestibility. In: Pollmer, W.G. & R.H. Phipps (eds.), Improvement of quality traits of maize for grain and silage use, Martinus Nijhoff Publishers, The Hague, Boston, London, pp. 447-465.

Zscheischler, J. (1990). Handbuch Mais, DLG-Verlag, Frankfurt/Main.

Introgression mapping in the grasses

I.P. King[1], J. King[1], I.P. Armstead[1], J.A. Harper[1], L.A. Roberts, H. Thomas[1], H.J. Ougham, R.N. Jones[2], A. Thomas[1], B.J. Moore[1,2], L. Huang[1] and I.S. Donnison[1]

[1]*Molecular and Applied Genetics Team, Institute of Grassland and Environmental Research, Plas Gogerddan, Aberystwyth, SY23 3EB, Wales, United Kingdom,*
e-mail: ian.king@bbsrc.ac.uk
[2]*Institute of Biological Sciences, The University of Wales Aberystwyth, Ceredigion SY23 3DA, Wales, UK*

Key points

1. *Lolium perenne/Festuca pratensis* hybrids and their derivatives provide an ideal system for intergeneric introgression.
2. The *Lolium perenne/Festuca pratensis* system is being exploited to elucidate genome organisation in the grasses, determination of the genetic control of target traits and the isolation of markers for MAS in breeding programmes.
3. The potential of the system as an aid to contig the *Lolium* and *Festuca* genomes and for gene isolation is discussed.

Keywords: introgression mapping, introgression landing, *Lolium perenne/Festuca pratensis*, contig, gene isolation

Introduction

The *Lolium perenne* (*Lp*) and *Festuca pratensis* (*Fp*) hybrids and their derivatives exhibit a unique combination of characteristics, not seen in other plant species, which makes the *Lp/Fp* system an ideal model for intergeneric introgression.
The combination of characters exhibited by *Lp/Fp* hybrids and their derivatives include:
1. A high frequency of recombination between *Lp* and *Fp* chromosomes facilitates the transfer of *F. pratensis* chromosome segments, carrying target genes, into *Lolium.*
2. Although the chromosomes of the two species recombine at high frequency at meiosis, they can be distinguished easily through genomic *in situ* hybridisation (GISH). GISH analysis allows the identification and classification of *Lolium /Fp* introgressions, i.e. confirmation of the introgression of *Fp* segments into *Lolium*, and an estimation of their physical size.
3. Recombination occurs along the entire length of *Lolium/Fp* bivalents permitting the transfer of any *Fp* gene into *Lolium.*
4. A high frequency of marker polymorphism is observed between *Fp* and *Lolium* which aids the mapping of target *Fp* genes on introgressed *Fp* segments.
5. The system also facilitates the rapid identification of markers located on an introgressed *F. pratensis* chromosome segment by the screening of a *Lolium/Fp* introgression together with the parental and hybrid germplasm from which it was derived. Any polymorphic marker present in the *Fp* parent, the *Lolium/Fp* hybrid and the introgression line itself, but not the *Lolium* parents, must be located within the introgressed *Fp* chromosome segment.

The work described in this paper enables the elucidation of gene organization along *Lolium/Fp* chromosomes; the determination of the relationship between gene distribution and recombination, allowing comparisons to be made between genome organisation in a small genome species, i.e. rice, and large genome monocots such as grass, wheat and others; the determination of the genetic control of key traits; and provides a resource for gene isolation via a chromosome landing strategy, i.e. Introgression Landing.

Development of *Lp/Fp* substitution lines

The forage grass *Lolium perenne* (*Lp*) (2n=2x=14) can be readily hybridised with *Festuca pratensis* (*Fp*) (2n=2x=14) to form a 14-chromosome hybrid which exhibits full pairing at metaphase I of meiosis but nearly complete sterility (Lewis, 1966; Jauhar, 1975). However, *LpLpFp* triploids, derived by hybridising synthetic tetraploid *Lp* with diploid *Fp* show both male and female fertility. When these triploids are backcrossed to diploid *Lp* they give rise to BC$_1$ progeny containing 14 chromosomes (Figure 1). Although the majority of the genome of these individuals is derived from the *Lp* parent, over 74% carry one or more *Fp* chromosome segments. Most plants carry one or two introgressed *Fp* chromosome segments (King *et al.*, 1998). Recombination, albeit at varying frequencies, has been observed to occur along the entire length of the chromosomes (I. King *et al.*, 1998; 1999; J. King *et al.*, 2002a). In addition to individuals carrying *Lp/Fp* recombinant chromosomes, 14 chromosome plants carrying 13*Lp* chromosomes and 1 *Fp* chromosome, i.e. monosomic substitutions, have been isolated (Figure 1).

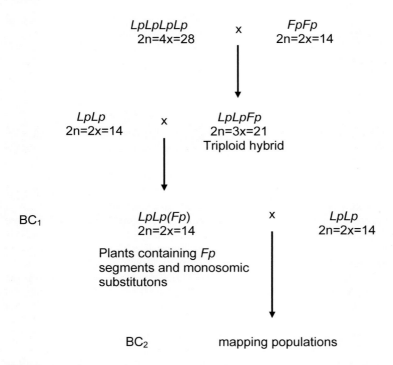

Figure 1 Crossing scheme for producing *Lolium/Festuca* introgressions and monosomic substituions

In order to isolate all seven possible substitution lines 550 BC$_1$ individuals were screened using RFLPs, AFLPs and SSR's (SSR's were derived from IGER's genomic microsatellites and from Vialactia's Gene Thresher library). All seven monosomic substitution lines have now been isolated and microsatellite analysis reveals that, at least at the macro-level, synteny has been maintained between *Lp* and *Fp*.

Genetic and introgression mapping of *Fp* chromosome 3

A single *Lp/Fp* monosomic substitution that carried a *Fp* chromosome homoeologous to the Triticeae group 3 chromosomes and rice chromosome 1 (King *et al.*, 2002a) was selected for further research.

Development of a genetic map of Fp chromosome 3

The *Lp/Fp* group 3 monosomic substitution, which carries a nucleolar organiser region, was backcrossed to diploid *Lp* to produce a BC$_2$ mapping population (Figure 1). This population was analysed with AFLP's using two enzyme combinations (*Eco*R1/*Tru*91 and *Hin*dIII/*Tru*91). The initial AFLP analysis was carried out in order to find markers specific to the *Festuca* chromosome in the monosomic substitution (King *et al.*, 2002a). To do this DNA from the four plants involved in the production of the substitution, i.e. *Fp, Lp* (tetraploid), *Lp* (diploid) and the *Lp/Lp/Fp* triploid hybrid, as well as the monosomic substitution itself, were screened. Markers found to be present in *Fp*, the triploid hybrid and the monosomic substitution, but absent from *Lp* (both the diploid and tetraploid genotypes) were classified as being specific to the *Fp* group 3 chromosome. Thirty five primer pairs generating 104 markers specific to *Fp* chromosome 3 were used to analyse 148 randomly selected plants. To keep scoring errors to a minimum all AFLP primer pairs/genotypes giving rise to singletons and questionable markers [as defined by the mapping programme – JOINMAP 2.0 (Stam, 1993)] were re-run and re-analysed at least once. This gave rise to a final genetic map of 81cM for the chromosome.

In addition to AFLP analysis the ability to distinguish the *Fp* chromosome in the *Lp/Fp* monosomic substitution line at meiosis using GISH enabled us to study the frequency of chiasmata in the *Lp/Fp* bivalent (King *et al.*, 2002a). This analysis allowed us to predict the expected genetic length of the *Fp* chromosome derived from AFLP analysis. The predicted genetic length of the *Festuca* chromosome can be calculated from the mean number of chiasmata scored in the *Fp/Lp* bivalent (µ), visualised with GISH, using the expression 50µcM, For example, a chromosome with an average of two chiasmata is expected to have a genetic length of 50 x 2 = 100cM (Kearsey and Pooni, 1996). The average chiasma frequency for the *Lp/Fp* bivalent was 1.522 on the basis of 347 chiasmata in 228 PMCs. This value provided an estimate of the genetic length of the *Fp* chromosome using the expression 50µcM, i.e. 50 x 1.522 = 76cM. Thus the genetic linkage map of 81cM constructed from the AFLP markers was not significantly different from that predicted by chiasma frequencies, i.e. 76cM. This demonstrates a 1:1 correspondence between chiasma frequency and recombination rate (King *et al.*, 2002a). This result is in contrast to many reports (Nilsson *et al.*, 1993), which have shown large discrepancies between estimated genetic distances based on chiasma frequencies and genetic distances based on the segregation of genetic markers.

Development of an introgression map of Fp chromosome 3

The genetic linkage map of the *Fp* chromosome was used to select 16 plants from the BC$_2$ mapping population for physical mapping (King *et al.*, 2002b). These plants were chosen because they showed a relatively even spread of recombination points along the chromosome and, where possible, recombination points on either side of the centromere and NOR.

Measurements taken of the recombinant chromosomes from mitotic root tip preparations in the BC$_2$ plants were: 1) total length of chromosome; 2) distance of recombination site or sites from both telomeres; 3) position of the NOR site measured from the edge of the NOR nearest to the telomere. Measurements of *Lp*/*Fp* recombinant chromosomes were taken from enlarged projections of at least ten separate chromosomes for each of the BC$_2$ plants used for physical mapping. However, the *Lp* and *Fp* genomes differ in size with the *Fp* genome being 7% larger than the *Lp* genome (Bennett *et al.*, 1982; Hutchinson *et al.*, 1979). Therefore the larger the *Fp* chromosome segment the larger the recombinant chromosome size and it is necessary to use an expansion factor to enable size comparisons of *Fp* segments in different genotypes to be made (King *et al.* 2002b).

The 16 BC$_2$ plants used for the physical mapping involved single crossovers (with the exception of BC$_2$ 83). All the *Fp* segments observed extended from one or other of the telomeres. Thus two series of *Fp* segments were looked at using GISH: the first series increased in size from the telomere of the chromosome arm without the NOR, whilst the second series increased in size from the telomere of the chromosome arm carrying the NOR. This resulted in the *Fp* chromosome being split into 18 segments or bins (BC$_2$ 83 contained two *Fp* segments which could be individually measured and mapped). The physical sites of recombination appeared to occur along the whole length of the chromosome including regions close to the centromere and within the NOR (King *et al.*2002b).

By combining the genetic and physical maps it has been possible to assign each of the AFLPs used to genetically map chromosome 3 to one of the 18 physical bins. This work has allowed comparisons to be made between genetic and physical distance.

Two gaps of greater than 10% of the chromosome arm were observed on the physical map of the *Fp* chromosome. The distribution of recombination sites along the whole length of the chromosome, however, shows that the present physical map has the potential to be broken down into much smaller sections by screening large populations for recombination between two markers that flank a specific region of the genome. A comparison of the physical and genetic maps clearly shows how their inter-relationship varies from one part of the chromosome to another. The two gaps on the physical map do not coincide with gaps on the genetic map. In fact, the density of AFLP markers on the genetic map is such that the largest distance between markers is only 5.9cM and there are only two other gaps of between 4 and 5cM present. Of the two gaps on the physical map, the first (a gap of 15.2% of the chromosome) contains 11 AFLP markers spread over just 1.3cM, while the second (a gap of 13.4% of the chromosome) contains eight AFLP markers spread over 9.1cM.

The distribution of recombination sites along the whole length of the chromosome included those very close to the centromere and within the NOR although not between the two. Thus, although the centromere and NOR both cause a reduction in the frequency of recombination in the region between them (see below), recombination itself does take place within these regions.

Recombination levels were found to vary within, as well as between arms. The highest frequency of recombination occurred at a physical distance of between 12 and 20% from the telomeres. The lowest frequency was found between 45 and 75% of the distance along the chromosome (the region of the chromosome containing both the centromere and the NOR).

Our results show that the centromere was physically mapped at 49.2% of the distance along the chromosome. The frequency of recombination started to increase at a distance of only about 5% from the centromere in the arm without the NOR. In contrast, it remained extremely low in the NOR arm for the whole of the region between the centromere and the NOR and including the NOR itself, but rose sharply after the end of the NOR. However, the peak in the NOR arm was considerably smaller than the major peak in the non-NOR arm. This result strongly suggests that the NOR, as well as the centromere, causes a reduction in the frequency of recombination. Similar evidence for little or no crossing over between the centromere and NOR has been reported for chromosomes 1B and 6B of wheat (Payne *et al.*, 1984; Dvořák and Chen, 1984; Snape *et al.*, 1985), barley chromosomes 6 and 7 (Linde-Laursen, 1979) and rye chromosome 1R (Lawrence and Appels; 1986).

The results obtained from this work are in agreement with data obtained from a range of other species where it has been shown that there is not a consistent relationship between genetic distance in cM and physical distance in base pairs, and that there is variation in this relationship from one part of the genome to another, e.g. Gustafson and Dillé (1992), CHEN and Gustafson (1995), Werner *et al.* (1992), Hohmann *et al.* (1994), Hohmann *et al.* (1995), Delaney *et al.* (1995), Mickelson-Young *et al.* (1995), Gill *et al.* (1996a and b), Künzel *et al.* (2000). Genetically close markers may actually be far apart in terms of base pairs (or vice versa) due to differences in the frequency of recombination along the length of a chromosome. When considering the average length of DNA per unit of recombination, different segments of a chromosome should therefore be considered independently. For chromosome 4 of *Arabidopsis,* the base-pair to cM ratio varied from 30kb to 550kb per cM (Schmidt *et al.*, 1995). In rice 1cM is on average equal to 240kb, although this figure actually varies from 120 to 1000kb per cM (Kurata *et al.*, 1994). In wheat the variation is even more extreme, with 1cM equal to 118kb in regions of high recombination but 22000kb in regions of low recombination (Gill *et al.*, 1996a and b). Regions corresponding to centromeres, and even some telomeres in tomato and potato, show a ten-fold decrease in recombination compared to other regions in the genome (Tanksley *et al.*, 1992). Reduced recombination frequency in pericentric regions is also seen in many species including the grasses, e.g. wheat (Dvořák and Chen, 1984; Snape *et al.*, 1985; Curtis and Lukaszewski, 1991; Gill *et al.*, 1993 and 1996a and b; for review see Gill and Gill, 1994), barley (Leitch and Heslop-Harrison, 1993; Pedersen *et al.*, 1995; Künzel *et al.*, 2000), rye (for review see Heslop-Harrison, 1991; Wang, 1992), *Lolium* (Hayward *et al.*, 1998; Bert *et al.,* 1999). Nucleolar organiser regions (NORs) may also cause a reduction in the frequency of crossing over, e.g. *Allium schoenophrasum* (J.S. Parker pers. com.). Recombination hotspots also occur (Endo and Gill, 1996; Weng *et al.,* 2000; Künzel *et al.*, 2000).

Thus in the *Lp/Fp* work described here genes located on the *Fp* chromosome arm without the NOR will appear genetically much further apart than genes located on the chromosome arm carrying the NOR. However, the genetic distance between the genes on the two chromosome arms will have very little relevance with regard to the physical distance between genes.

Development of an introgression map of the *Lp/Fp* genomes

Festuca chromosome 3 as been used as a prototype for the development of an introgression map of the *Lp*/Fp genomes. Each of the six remaining *Lp/Fp* monosomic substitution lines have been backcrossed to diploid *Lp* to generate BC$_2$ mapping populations. As with the monosomic substitution carrying *Fp* chromosome 3, the *Lp* and *Fp* homoeologous bivalents in the other monosomic substitution lines undergo high frequencies of recombination along the length of the chromosomes during meiosis, resulting in the generation of *Lp/Fp* recombinant chromosomes. The mapping populations derived from each of these seven substitutions are presently being screened with microsatellite markers (500 SSRs;

IGER/Vialatia) which will enable the genetic mapping of each of the seven substituted *Festuca* chromosomes. These genetic marker profiles are being used to identify individuals, derived from each of the seven mapping populations, which carry *Lp/Fp* recombinant chromosomes with different sized and overlapping *Fp* chromosome segments. The physical size and position of the overlapping *Fp* segments in each of the recombinant chromosomes will then be measured using GISH. These data will then be assembled enabling each of the seven *Festuca* chromosomes to be divided into introgression bins, i.e. each of the seven *Festuca* chromosomes will initially be divided into at least 20 bins composed of no more than 5% of the total chromosome length. However, the resolution of the system is such that it is possible to divide the chromosomes into much smaller bins, i.e. at least 0.4% of the total chromosome length.

Exploitation of the *Lolium/Festuca* introgression maps

A programme at IGER has recently been initiated to bin map every 5th rice BAC. The bin mapping of rice BACs requires that a coding region is identified on each BAC that is to be mapped. This sequence is then used to screen other monocot databases. Primers are designed from conserved regions and these are used to amplify the equivalent sequence in *Lp* and *Fp* via PCR. The products are sequenced and SNP markers that discriminate between *Lp* and *Fp* are designed. Once identified, SNPs can be mapped to a specific bin on a specific chromosome (a similar strategy will be applied using orthologous markers from other monocots such as barley). This strategy is proving successful for *Fp* chromosome 3. At the time of writing 24 rice BACs have been mapped to *Fp* chromosome 3.

Bin mapping rice BACs will enable the determination of the organisation of genes in large genome plant species, such as *Fp*, and allow comparisons to be made with the small genome model, rice. The work will allow determination of the distribution of genes along the *Fp* chromosomes. Are genes (I) evenly distributed along the chromosome and separated by large regions of repetitive DNA? or (II) present in clusters surrounded by repetitive DNA? If clustered, how are the clusters themselves organised: are they evenly distributed or are the clusters clustered? In addition, the work will allow the elucidation of the frequency and distribution of recombination relative to gene density, i.e. do peaks in recombination frequency coincide with gene rich areas? Are some genes or clusters of genes located in areas of low recombination and if so has this led to the development of co-adapted gene complexes which confer a selective advantage? A knowledge of the relationship between recombination and the physical location of target genes is of importance since the success of both conventional breeding programmes and gene isolation, via forward genetics, are heavily dependent on the frequency of recombination in the region in which a target gene is located. Our initial results indicate that the majority of rice BACs are located in areas of low recombination.

Introgression mapping will also facilitate the exploitation of information and technology developed in the model plants. The rice genome project has provided the order and sequence of the genes on each of its 12 chromosomes. This data will enable the isolation of genes, via chromosome landing and walking strategies, from large genome monocot species. For example EST based markers that flank a target gene in a large genome crop species are being used to identify the equivalent region in rice. Since the rice genome has been sequenced, every gene between the flanking markers in this species is known. These rice gene sequences can be used to develop additional markers which will then be used to isolate the target gene via chromosome landing approaches in the crop species itself. However, this approach assumes that the gene order in rice and other monocot crop species is the same. The bin mapping of rice BACs will provide an in depth genome wide assessment of the syntenic relationship between rice and *Lp* and *Fp*. In addition, the maintenance of synteny between rice and grass is likely to indicate a similar relationship between other monocots and rice. Thus the *Lp/Fp* introgression mapping described will benefit the whole monocot research community. In addition, strategies are presently being developed where by introgression maps in combination with other technologies will provide a platform to contig the *Lp/Fp* genomes.

The monosomic substitutions and the individuals making up the physical maps of each *F. pratensis* chromosome will also provide a valuable resource for determining the genetic control of target traits and gene isolation. For example, the seven monosomic substitution lines will be screened for a specific trait. Once a *F. pratensis* chromosome has been identified as carrying a gene(s) controlling the trait the relevant genotypes making up the physical map will also be screened. In this way it will be possible to physically map genes that control key traits. Furthermore, physical mapping/bin mapping will identify syntenic regions in rice thus facilitating the isolation of genes responsible for the control of a trait via map based cloning.

A strategy for the isolation of introgressed *Fp* genes

The unique combination of genetic and cytogenetic characteristics exhibited by *Lp/Fp* and also *Lm/Fp* hybrids and their derivatives (the chromosomes undergo a high frequency of homoeologous recombination at meiosis; the chromosomes of the two species can easily be discriminated by GISH; recombination occurs along the entire length of homoeologous bivalents; a high frequency of marker polymorphism is observed between the two species) are being exploited to isolate tightly linked markers to target genes and to provide a springboard for gene isolation.

A *Fp* chromosome segment which carries a mutation of a gene normally required for leaf yellowing during senescence has been introgressed into *Lm* (Thomas *et al.*, 1987 and Thomas *et al.*, 1997). The stay-green character results from a recessive mutation in the gene, and only plants homozygous for the mutation express the stay-green phenotype. Leaf segments of plants homozygous for the mutation remain green, while plants heterozygous or homozygous for the wild type gene turn yellow as chlorophyll is broken down. The lesion in the chlorophyll breakdown pathway in plants homozygous for the green gene appears to result from the inability of plants to break down pheophorbide to red-chlorophyll-catabolite (RCC) because of a deficiency in pheophorbide-a-oxygenase (PaO) activity (Vicentini *et al.*, 1995; Rodoni *et al.*, 1997; Thomas *et al.*, 2001). Thus the stay green phenotype is believed to result from a mutation in the gene responsible for the production of the PaO enzyme or a regulator gene that controls the expression or activation of the gene/ protein (Roca *et al.*, 2004).

A mapping family was generated from the *Lm/Fp* introgression line carrying the stay-green mutation and AFLP's were used to generate a genetic map of the *Fp* chromosome segment (Moore *et al.*, 2005). AFLP analysis was performed, using the restriction enzyme pairs *Hin*dIII/*Tru*91 and *Eco*R1/*Tru*91. Polymorphisms specific to the *Fp* segment were identified by screening the parents, i.e. tetraploid *Lm*, diploid *Fp*, diploid *Lm*, the *Lm/Lm/Fp* triploid hybrid and the selected BC$_1$ genotype carrying a single small *Fp* chromosome segment. Primer pairs which failed to give a *Fp* specific polymorphism or primer pairs which gave a *Fp* specific polymorphism in the *Fp* parent and *Fp* hybrid but were not in the selected BC$_1$ introgression genotype, i.e. those where the *Fp* specific marker lay outside the introgressed *Fp* chromosomes segment, were discarded. Primer pairs which gave a *Fp* specific polymorphism in the *Fp* diploid parent, the *Lm/Lm/Fp* triploid and the selected BC₁ individuals were selected.

Twenty-two selected AFLP primer pairs, giving 28 *Fp* specific polymorphisms (Figure 2a), were used to screen the mapping population. The segregation of the *Fp* specific polymorphisms in the mapping population was analysed using JOINMAP™ 2.0 (Stam, 1993) to generate a genetic map of the *Fp* chromosome segment (Figure 2a and figure 2b). Each of the individuals of the BC₂ mapping population was also test-crossed to a *Lm* genotype homozygous recessive for the green gene (yy) to determine the presence or absence of the stay green gene in each of the individuals of the mapping population. The data for the presence or absence of the stay-green gene were combined with the AFLP data in order that the senescence mutation could be mapped within the introgressed *Fp* chromosome segment. The final genetic distance of the *Fp* chromosome segment between the terminal *Fp*-derived AFLP markers was estimated to be 19.8cM with the stay-green *sid* mutation located at 9.8cM; the closest flanking markers to *sid* were at 0.6cM and 1.3cM.

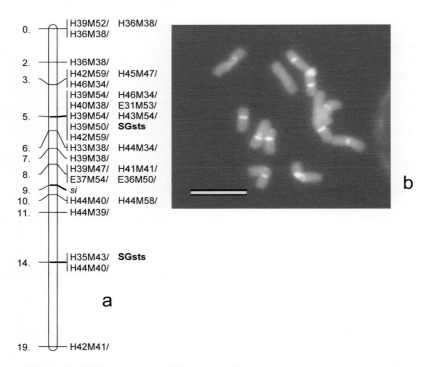

Figure 2a Linkage group of introgressed *F. pratensis* segment containing the stay-green gene (*sid*). Genetic distances are indicated in centiMorgans. The positions of STS markers derived from AFLP/BAC sequencing are indicated in bold-type; the positions of AFLP markers are indicated in normal-type.

Figure 2b BC₁ plant showing the smallest introgressed segment from *F.pratensis* carrying the stay green gene (sid)

Twelve AFLP bands were excised, cloned and sequenced. Primers designed from one of these 12 AFLPs (SG7), which produced a *Fp* specific fragment of 390 bp, immediately distinguished between the *Lolium* and *Festuca* genotypes, the other 11 did not. Most of the internal sequences of the other markers were not polymorphic between the parents and therefore not useful for conversion to STS markers. Therefore a *Fp* BAC library (Donnison *et al.*, 2005) was screened with AFLP-derived primers from another two markers on the other side of the stay-green locus, with the aim of identifying additional sequences suitable for designing new primers. One primer pair identified many BACs indicating that the sequence was present in multiple locations in the genome, but the other (SG2), which generated a *Fp* AFLP fragment of 300 bp, identified two. Given the 2.5x coverage of the BAC library, this fragment was considered to be present as a single copy sequence. BAC DNA of an individual BAC identified from this marker was extracted, digested and re-cloned. End-sequencing of these BAC sub-clones generated 7 Kb of additional sequence. This sequence was used to generate an extra primer pair from a BAC sub-clone fragment which did not show homology to repetitive DNA such as retroelements of other monocot species. This primer pair was tested on *Lolium* and *Festuca* parental DNA and shown to be polymorphic. Primer pairs for this and the other polymorphic marker were then mapped back onto the genetic map and in both cases were found to map precisely to the same position as the original AFLP markers, on either side of the stay-green locus (Figure 2a).

The work described above makes use of *Fp* specific AFLP polymorphisms to map an alien chromosome segment. An alternative strategy will be to derive markers from the sequenced rice genome. Once the region of the rice genome that shows synteny with a *Fp* chromosome segment has been identified, additional markers for the introgressed segment can be developed. This can be achieved by comparing a predicted coding sequence from rice with EST databases from other monocots. Primers can then be developed from regions that show very high conservation. Ninety per cent of such primers have been shown to generate an equivalent sequence in *Lolium* and *Festuca* and a high proportion show polymorphism between the two species. The advantage of this strategy is that it provides large numbers of markers for a specific region of the *Fp* genome that is of interest, as well as possible information on gene function in the model monocot plant species. The potential of isolating *Fp* genes via the use of large numbers of rice markers and high resolution *Lolium*/*Fp* mapping populations is presently being investigated. A large mapping population of over 2000 individuals has been mapped with markers, derived from rice, that closely flank the green gene. This procedure '*Introgression Landing*' has identified a region of the rice genome that carries 30 genes and these will be functionally tested to determine which of these is responsible for the stay-green phenotype. In addition to this, markers that discriminate between *Lp, Lm* and *Fp* have been developed for use in IGER breeding programmes.

Our initial research indicates that *Introgression Landing*, in combination with robust screening procedures for target traits provides a fast and efficient method of isolating closely linked markers to target genes and provides a platform for gene isolation.

Acknowledgements

Julie King is supported by an EMBO Restart Fellowship. Beverley Moore was supported by the Teaching Company Scheme. IGER is sponsored by the BBSRC.

References

Bennett, M.D., J.B Smith & J.S. Heslop-Harrison (1982). Nuclear DNA amounts inangiosperms. *Proc. Royal Soc.* London B., 216, **179**-199.

Bert, P.F., G.Charmet, P. Sourdille, M.D. Hayward & F. Balfourier (1999). A high density molecular map for ryegrass (*Lolium perenne*) using AFLP markers. *Theor. Appl. Genet.*, 99, 445-452.

Chen, J.M. & J.P. Gustafson (1995). Physical mapping of restriction fragment length polymorphisms (RFLPs) in homoeologous group 7 chromosomes of wheat by *in situ* hybridisation. *Heredity*, 75, 225-233.

Curtis, C.A. & A.J. Lukaszewski (1991). Genetic linkage between C-bands and storage protein genes in chromosome 1B of tetraploid wheat. *Theor. Appl. Genet.*, 81, 245-252.

Delaney, D.E., S. Nasuda, T.R. Endo, B.S. Gill & S.H. Hulbert (1995). Cytogenetically based physical maps of the group 3 chromosomes of wheat. *Theor. Appl. Genet.*, 91, 780-782.

Donnison, I.S., D. O'Sullivan, A. Thomas, P. Canter, B.J. Moore, I.P. Armstead, H. Thomas, K. Edwards & I.P. King (2005). Construction of a *Festuca pratensis* BAC library for map based cloning in *Festulolium* substitution lines. *Theor. Appl. Genet.*, in press.

Dvořák, J. & K-C. Chen (1984) Distribution of nonstuctural variation between wheat cultivars along chromosome arm 6Bp: evidence from the linkage map and the physical map of the arm. *Genetics*, 106, 325-333.

Endo, T.R. & B.S. Gill (1996). The deletion stocks of common wheat. *J. Hered.*, 87, 295-307.

Gill, K.S., B.S. Gill & T.R. Endo (1993). A chromosome region specific mapping strategy reveals gene rich telomeric ends in wheat. *Chromosoma*, 102, 374-381.

Gill, K.S. & B.S. Gill (1994). Mapping in the realm of polyploidy: the wheat model. *BioEssays*, 16, 841-846.

Gill, K.S., B.S. Gill, T.R. Endo & E.V. Boyko, (1996a) Identification and high density mapping of gene rich regions in chromosome group 5 of wheat. *Genetics*, 143, 1001-1012.

Gill, K.S., B.S. Gill, T.R. Endo & T. Taylor (1996b). Identification and high-density mapping of the gene-rich regions in chromosome group 1 of wheat. *Genetics*, 144, 1883-1891.

Gustafson, J.P. & J.E. Dillé (1992). Chromosome location of *Oryza sativa* recombination linkage groups. *Proc. Natl. Acad. Sci. USA*, 89, 8646-8650.

Hayward, M.D., J.W. Forster, J.G. Jones, O. Dolstra, C. Evans, (1998). Genetic analysis of *Lolium*. I. Identification of linkage groups and the establishment of a genetic map. *Plant Breed.*, 117, 451-455.

Heslop-Harrison, J.S. (1991). The molecular cytogenetics of plants. *J. Cell Sci.*, 100, 15-21.

Hohmann, U., T.R. Endo, K.S. Gill & B.S. Gill (1994). Comparison of genetic and physical maps of group 7 chromosomes from *Triticum aestivum* L. *Mol. Gen. Genet.*, 245, 644-653.

Hohmann, U., A. Graner, T.R. Endo, B.S. Gill & R.G. Herrmann (1995). Comparison of wheat physical maps with barley linkage maps for group 7 chromosomes. *Theor. Appl. Genet.*, 91, 618-626.

Hutchinson, J., H. Rees & A.G. Seal (1979). An assay of the activity of supplementary DNA in *Lolium. Heredity* 43, 411-421.

Jauhar, P.P. (1975). Chromosome relationships between *Lolium* and *Festuca* (Graminea). *Chromosoma*, 52, 103-121.

Kearsey M.J. & S. Pooni (1996) The genetical analysis of quantitative traits, pp116-119, Publishers, Chapman and Hall, London.

King, I.P., W.G. Morgan, I.P. Armstead, J.A. Harper, M.D. Hayward, A. Bollard, J.V. Nash, J.W. Forster & H.M. Thomas (1998). Introgression mapping in the grasses I. Introgression of *Festuca pratensis* chromosomes and chromosome segments into *Lolium perenne. Heredity*, 81, 462-467.

King, I.P., W.G. Morgan, J.A. Harper & H.M. Thomas, (1999). Introgression mapping in the grasses. II. Meiotic analysis of the *Lolium perenne/Festuca pratensis* triploid hybrid. *Heredity*, 82, 107-112.

King J, L.A. Roberts, M.J. Kearsey, H.M. Thomas, R.N. Jones, L. Huang, I.P. Armstead, W.G. Morgan, & I.P. King (2002a). A demonstration of a 1:1 correspondence between chiasma frequency and recombination using a *Lolium perenne/Festuca pratensis* substitution line. *Genetics*, 161, 315-324.

King J, I.P. Armstead, I.S. Donnison, H.M. Thomas, R.N. Jones, M.J. Kearsey, L.A. Roberts, A. Jones, & I.P. King (2002b). Physical and genetic mapping in the grasses *Lolium perenne* and *Festuca pratensis. Genetics*, 161, 315-324.

Künzel, G., L. Korzun & A. Meister (2000). Cytologically integrated physical restriction fragment length polymorphism maps for the barley genome based on translocation breakpoints. *Genetics*, 154, 397-412.

Kurata, N., Y. Nagamura, K. Yamamoto, Y. Harushima, & N. Sue. (1994). A 300 kilobase interval genetic map of rice including 883 expressed sequences. *Nature Genetics*, 8, 365-372.

Lewis, E.J. (1966). The production and manipulation of new breeding material in *Lolium-Festuca*. In: Hill, AG.G. (ed) Proc. X Int. Grassland Congr., Valtioneuvoston Kirjapaino, Helsinki, pp 688-693.

Lawrence, G.J. & R. Appels (1986). Mapping the nucleolar organiser region, seed protein loci and isozyme loci on chromosome 1R in rye. *Theor. Appl. Genet.*, 71, 742-749.

Leitch, I.J. & J.S. Heslop-Harrison (1993). Physical mapping of 4 sites of 5s rRNA sequences and one site of the α-amylase-2 gene in barley (*Hordeum vulgare*). *Genome*, 36, 517-523.

Linde-Laursen, I. (1979) Giemsa C-banding of barley chromosomes III. Segregation and linkage of C bands on chromosomes 3, 6 and 7. *Hereditas*, 91, 73-77.

Mickelson-Young, L., T.R. Endo & B.S. Gill (1995). A cytogenetic ladder map of the wheat homoeologous group 4 chromosomes. *Theor. Appl. Genet,*. 90, 1007-1011.

Moore, B.J., I.S. Donnison, J.A. Harper, I.P. Armstead, J. King, H. Thomas, R.N. Jones, T.H. Jones, H.M. Thomas, W.G. Morgan, A. Thomas, H.J. Ougham, L. Huang, T. Fentem, L.A. Roberts & I.P. King (2005). Molecular Tagging of a senescence gene by introgression mapping of a stay-green mutation from *Festuca pratensis. New Phytologist*, 165, 801-806.

Nilsson N-O., T. SÄll & B.O. Bengston (1993) Chiasma and recombination data in plants: are they compatible? *Trends Genet.*, 9, 344-348.

Payne, P.I., L.M. Holt, J. Hutchinson & M.D. Bennett, (1984) Development and characterisation of a line of bread wheat, *Triticum aestivum*, which lacks the short arm satellite of chromosome 1B and the *Gli-B1* locus. *Theor. Appl. Genet.* 68, 327-334.

Pedersen, C., H. Giese & I. Linde-Lauresen (1995). Towards an integration of the physical and the genetic chromosome map of barley by *in situ* hybridisation. *Hereditas*, 123, 77-88.

Roca, M., C. James, A. Pružinská, S. Hortensteiner, H. Thomas, & H. Ougham (2004). Analysis of the chlorophyll catabolism pathway in leaves of an introgression senescence mutant of Lolium temulentum. *Phytochemistry*, 65, 1231-1238.

Rodoni S, W. Muhlecker, M. Anderl, B. Krautler, D. Moser, H. Thomas, P. Matile, & S. Hortensteiner (1997). Chlorophyll breakdown in senescent chloroplasts - cleavage of pheophorbide a in two enzymic steps. *Plant Physiology*, 115,669-676.

Schmidt, R., J. West, K. Love, Z. Lenechan, C. Lister (1995). Physical map and organisation of *Arabidopsis thaliana* chromosome 4. *Science*, 270, 480-483.

Snape, J.W., R.B. Flavell, M. O'Dell, W.G. Hughes & P.I. Payne (1985). Intrachromosomal mapping of the nucleolar organiser region relative to three marker loci on chromosome 1B of wheat (*Triticum aestivum*). *Theor. Appl. Genet.,* 69, 263-270.

Stam, P. (1993). Construction of integrated genetic linkage maps by means of a new computer package. JOINMAP. *Plant J.*, 3, 739-744.

Tanksley, S.D., M.W. Ganal, J.P. Prince, M.C. deVicente, & M.W. Bonierbale (1992). High density linkage maps of the tomato and potato genomes. *Genetics*, 132, 1141-1160.

Thomas H. (1987). *sid*: A Mendelian locus controlling thylakoid membrane disassembly in senescing leaves of *Festuca pratensis*. *Theor Appl Genet*, 73, 551-555.

Thomas H, C. Evans, H.M. Thomas, M.W. Humphreys, W.G. Morgan, B.D. Hauck, I. S. Donnison (1997). Introgression, tagging and expression of a leaf senescence gene in *Festulolium*. *New Phytologist*, 137, 29-34.

Thomas H, H. J. Ougham, S. Hortensteiner (2001). Recent advances in the cell biology of chlorophyll catabolism. *Advances in Botanical Research*, 35, 1-52

Vicentini F., S. Hortensteiner, M. Schellenberg, H. Thomas and P. Matile (1995). Chlorophyll breakdown in senescent leaves: identification of the biochemical lesion in a stay-green genotype of *Festuca pratensis*. *New Phytologist*, 129, 247-252.

Wang, M.L., M.D. Atkinson, C.N. Chinoy, K.M. Devos & M.D. Gale (1992). Comparative RFLP-based genetic maps of barley chromosome 5 (1H) and rye chromosome 1R. *Theor. Appl. Genet.*, 84, 339-344.

Weng, Y., N.A. Tuleen & G.E. Hart (2000). Extended physical maps and a consensus physical map of the homoeologous group 6 chromosomes of wheat (*Triticum aestivum* L. em Thell). *Theor. Appl. Genet,*. 100, 519-527.

Werner, J.E., T.R. Endo & B.S. Gill (1992) Toward a cytogenetically based physical map of the wheat genome. *Proc. Natl. Acad. Sci. USA*, 89, 11307-11311.

QTL analysis and trait dissection in ryegrasses (*Lolium* spp.)

T. Yamada[1] and J.W. Forster[2]

[1]*National Agricultural Research Center for Hokkaido Region, Sapporo 062-8555, Japan, Email: Toshihiko.Yamada@affrc.go.jp*

[2]*Primary Industries Research Victoria, Plant Biotechnology Centre, La Trobe University, Bundoora, Victoria 3086, Australia and Molecular Plant Breeding Cooperative Research Centre, Australia*

Key points

1. Molecular marker-based genetic analysis permits the dissection of complex phenotypes through resolution of the locations of pleiotropic and interacting genetic factors.
2. Several QTLs for agronomically important characters such as flowering time, winter hardiness and forage quality have been identified in perennial ryegrass by molecular marker-based map analysis.
3. Some QTLs were putatively orthologous to those for equivalent traits in cereals.
4. The identification of co-location between QTLs and functionally-associated genetic markers is critical for the future implementation of marker-assisted selection programs.

Keywords: comparative genomics, flowering time, forage quality, QTL analysis, winter hardiness

Introduction

The majority of traits of interest to pasture grass breeders, such as dry matter yield, forage quality and environmental stress tolerance, show continuous phenotypic variation and are controlled by a variable number of quantitative trait loci (QTL). Substantial advances have been made in the genetic improvement of plant populations through artificial selection of quantitative traits. Most of this selection has been on the basis of observable phenotype, without knowledge of the genetic architecture of the selected characteristics. In major crop species, the development of high-resolution genetic maps has made it possible to identify the chromosomal regions, or in some instances, the individual sequence variants that are responsible for trait variation. There have been relatively few reports to date of QTL analysis for agronomic traits in forage grasses, due to the absence of a sufficiently well developed genetic map. An enhanced molecular marker-based genetic linkage map of perennial ryegrass (*Lolium perenne* L.) has recently been constructed through the activities of the International *Lolium* Genome Initiative (ILGI), using the p150/112 one-way pseudo-testcross reference population (Jones *et al*., 2002a,b). Two genetic mapping populations of perennial ryegrass have been independently developed as successors to the p150/112 population, and have been aligned to the reference map using common markers (Armstead *et al*., 2002, 2004; Faville *et al*., 2005). These genetic maps contain functionally-associated molecular marker information through the inclusion of gene-associated cleaved amplified polymorphic sequences (CAPS) markers, and both restriction fragment length polymorphism (RFLP) and simple sequence repeat (SSR) markers from expressed sequence tags (ESTs), respectively. In addition, high-density molecular marker-based genetic maps have also recently been constructed for other species of *Lolium* and *Festuca* (Alm *et al*., 2003; Inoue *et al*., 2004a; Warnke *et al*., 2004; Saha *et al*., 2005). Genetic markers and maps are consequently available for detailed dissection of complex phenotypes to resolve the locations of pleiotropic and interacting genetic factors. In this chapter, recent results from QTL analysis of the p150/112 reference population and another populations of perennial ryegrass are reviewed, and are related to comparative genomics studies with other Poaceae species.

Flowering time

The timing of flowering during the year is an important adaptive character that impacts on yield and quality in crop and pasture species. Analysis of the molecular basis of several QTLs in *Arabidopsis*, rice, maize and tomato has revealed that in diverse species, orthologous genetic networks can control related complex phenotypes (Paran & Zamir, 2003). For example, QTL cloning studies has revealed that the major flowering-time QTL *ED1* (early day-length insensitive) of the long-day flowering plant *Arabidopsis thaliana* is a novel allele of the blue-light photoreceptor cryptochrome-2 (*CRY2*) (El-Din El-Assal *et al.*, 2001). This allele increased protein stability as a result of a single amino acid substitution that led to early flowering in short days. The genetic control of flowering time control in *Arabidopsis* has been well characterised through the interaction of photoperiod, vernalisation, autonomous and gibberellic acid-dependent pathways (Mouradov *et al.*, 2002; Hayama & Coupland, 2003; Henderson & Dean, 2004; Sung & Amasino, 2004). In rice, a short-day flowering plant, QTL analysis for heading date has been performed using several types of progeny derived from a single cross combination between *Oryza sativa* ssp. *japonica* and *O. sativa* ssp. *indica* cultivars and identified at least 14 QTLs controlling flowering time (Yano *et al.*, 2001). Three heading QTLs (*Hd1, Hd3* and *Hd6*) were mapped at high resolution using near isogenic lines (NILs) and were isolated by a map-based cloning approach (Yano *et al.*, 2000; Kojima *et al.*, 2002; Takahashi *et al.*, 2001). *Hd1* was shown encode a homologue of the *Arabidopsis* photoperiod pathway *CONSTANS (CO)* gene (Yano *et al.*, 2000). *Hd3a* was shown to be related to the *Arabidopsis FLOWERING LOCUS T (FT)* gene, while *Hd6* was shown to encode the α–subunit of casein protein kinase 2 (*CK2α*). Gene interactions in the photoperiod induction pathways of *Arabidopsis* and rice were compared (Izawa *et al.*, 2003).

In cereals, genes that regulate the timing of flowering can be divided three categories based on their interaction with environmental signals: vernalisation response genes (*Vrn*) that regulate flowering using low temperature; photoperiod response genes (*Ppd*) that regulate flowering using day length; and 'earliness' factors that appear to be largely independent of these cues (Laurie, 1997). Conserved genomic locations for genes involved in processes such as vernalisation and photoperiodic induction have been identified between species by comparative genetic studies (Dubcovsky *et al.*, 1998; Laurie *et al.*, 2004). A detailed physical and genetic map of the *Vrn-A^{m}1* (*Vrn1*) region was constructed for the diploid wheat, *Triticum monococcum* L. and found to be colinear with the corresponding region of rice chromosome 3 (Yan *et al.*, 2003). A candidate gene for *Vrn1* was isolated by positional means, and was identified as a relative of the *Arabidopsis apetala-1* gene class (*AP1*). Allelic variation between spring and winter-type growth habit wheats at the *AP1* gene was observed only in the promoter region, suggesting that variation of gene expression was the causal factor for differences between these two varietal groups. A second vernalisation gene, designated *Vrn-A^{m}2*, was assigned to the distal region of chromosome 5A^{m}L within a segment translocated from homoeologous group 4 (Dubcovsky *et al.*, 1998). Yan *et al.* (2004) reported the positional cloning of the *Vrn2* gene, which encodes a dominant repressor of flowering down-regulated by vernalisation. Loss of function at *Vrn2*, whether by natural point mutation or deletion, resulted in lines with spring-type growth habit, which do not require vernalisation to flower.

A single QTL for heading date was observed on linkage group (LG) 4 in the p115/120 reference family of perennial ryegrass (Yamada *et al.*, 2004). However, a number of QTL positions for heading date have been reported from the analysis of single mapping populations in other Poaceae species (Hayes *et al.*, 1993; Laurie *et al.*, 1995; Bezant *et al.*, 1996; Börner *et al.*, 2002). The number of QTLs and their relative importance may vary according to the origin of the genotypes used to construct mapping families. Studies on geographical populations of *Lolium* species covering the climatic range from the Mediterranean region to northern and central Europe revealed a regular cline in flowering responses to temperature and photoperiod (Cooper, 1960). The heterozygous parent of the p150/112 mapping population was derived from a cross between eastern European (Romanian), southern European (north Italian ecotypes) and northern European ('Melle' or 'S23') genotypes, and might be expected to represent a variety of response genes. In rice, the major heading date QTL *Hd6*, which is associated with inhibition of flowering under long day conditions, is located on chromosome 3 (Yamamoto *et al.*, 2000) and encodes an α-subunit of CK2 (Takahashi *et al.*, 2001). Comparative genetic mapping studies between rice and wheat based on the colinearity of four common RFLP markers (Kato *et al.*, 1999) have revealed that the rice *Hd6* locus region on chromosome 3 is syntenic with the *VrnA1* region on chromosome 5AL. RFLP-based mapping of the wheat CK2α gene (*tck2a*) probe detected a genetic locus closely linked (by 1.1 cM) to *VrnA1* (Kato *et al.*, 2002). Two putative CK2α genes (*Lpck2a-1* and *Lpck2a-2*) genes have been isolated from a cDNA library constructed with mRNA isolated from cold-acclimated crown tissues of *Lolium perenne* using sequence information derived from the *tck2a* gene. The *Lpck2a-1* CAPS marker was assigned to LG 4 of the p115/120 reference family near to the location of the QTL for heading date, while the *Lpck2a-2* CAPS marker was assigned to LG 2 (Shinozuka *et al.*, submitted). The location of the *Lpck2a-1* locus supports the inference of conserved synteny between perennial ryegrass LG 4, the Triticeae homoeologous group 5L chromosomes and the corresponding segment of rice chromosome 3 (Yamada *et al.*, 2004).

A mapping population consisting of 184 F_2 genotypes from a cross between a genotype from the perennial ryegrass synthetic variety 'Veyo' and a genotype from the perennial ryegrass ecotype 'Falster' was measured for vernalisation response as days to heading under artificially controlled condition (Jensen *et al.*, 2005). In total, five QTLs were identified on LGs 2, 4, 6 and 7. A CAPS marker derived from the putative orthologue of the *Triticum monococcum VRN1* gene co-located with a major QTL on LG 4 for vernalisation response (Jensen *et al.*, 2005). This data further confirms the presence of flowering time gene orthologues and corresponding QTLs on LG 4.

Heading date QTLs were also identified in one of the second generation reference populations, derived from self-pollination of an F_1 hybrid obtained by crossing individuals from partially inbred lines developed from the two agronomically contrasting cultivars 'Aurora' and 'Perma' (Turner *et al.*, 2001). Genetic mapping of the F_2 (Aurora x Perma) population identified seven linkage groups with a total map length of 628 cM (Armstead *et al.*, 2004), extending the studies of Armstead *et al.* (2002) and consistent with the ILGI reference map (Jones *et al.*, 2002). A major QTL accounting for up to 70% of the variance was identified on LG 7, along with additional small QTLs on LGs 2 and 4 (Armstead *et al.*, 2004). The genomic region associated with the major QTL on LG 7 shows a high degree of conserved synteny with the *Hd3* region of rice chromosome 6.

Two annual ryegrass plants from the cultivar 'Floregon' were crossed with two perennial ryegrass plants from the cultivar 'Manhattan'. From the resultant F_1 populations, two random plants were chosen and crossed to develop a pseudo-F_2 mapping family (ψF_2[MFA-4 x MFB-2]). A total of 235 amplified fragment length polymorphism (AFLP) markers, 81 random amplified polymorphic DNA (RAPD) markers, 16 grass comparative anchor probe RFLPs, 106 SSR markers, 2 isoenzyme loci and 2 morphological characteristics, 8-h flowering and seedling root fluorescence were used to construct a male map 537cM in length and a female map 712 cM in length, each with 7 LGs (Warnke *et al.*, 2004). Two major QTLs influencing photoperiodic control of flowering were identified in locations syntenic with flowering control regions of the wheat and barley genomes (Warnke *et al.*, 2003a,b).

Winter hardiness

Winter hardiness is the outcome of a number of interacting factors that may include vernalisation requirement, photoperiod response, low-temperature tolerance and resistance to snow moulds. An understanding of the genetic basis of these component traits permits more efficient selection based on closely linked molecular marker loci. In the Triticeae cereals, QTL analysis has identified a limited number of conserved genome regions as responsible for the winter hardiness character. The most consistently identified region, on homoeologous group 5 chromosomes, contains QTLs for vernalisation response, low temperature tolerance and photoperiod sensitivity (Pan *et al.*, 1994; Cattivelli *et al.*, 2002). These QTL effects have been described as the effects of single loci. Low-temperature tolerance loci on chromosomes 5A, 5B, and 5D of wheat have been given the locus designation *Fr-A1*, *Fr-B1* and *Fr-D1*, respectively (Sutka & Snape, 1989; Snape *et al.*, 1997; Toth *et al.*, 2003). The vernalisation loci have been assigned a *Vrn* prefix, and orthologues of the *T. monococcum Vrn1* locus map to the homoeologous 5 chromosomes (Cattivelli *et al.*, 2002). The *Vrn* gene nomenclature was standardised and map locations were further refined by Dubcovsky *et al.* (1998). RFLP analysis demonstrated that vernalisation requirement and frost resistance are controlled by two different, but tightly linked loci (*Vrn-A1* and *Fr-A1*) on chromosome 5A of wheat (Galiba *et al.*, 1995; Sutka *et al.*, 1999). Because of the observed large effect on frost resistance, molecular marker-assisted selection for the *Vrn-A1-Fr-A1* chromosomal region has been proposed as a method for improvement of cold hardiness in wheat cultivars (Storlie *et al.*, 1998). Wheat NILs with different vernalisation alleles have been evaluated for cold hardiness in order to assess the viability of this strategy (Koemel *et al.*, 2004). Close genetic linkage between the major genes influencing winter hardiness and genes regulating cold-induced sugar production was also observed in wheat (Galiba *et al.*, 1997). In addition, QTLs controlling traits associated with winter hardiness, such as field survival and crown fructan content, were mapped to the long arm of chromosome 5H in a cross between 'winter' x 'spring' barley varieties (Hayes *et al.*, 1993; Pan *et al.*, 1994). Recently, comprehensive measurements of low temperature tolerance and vernalisation requirement were used for analysis of a new 'winter' x 'spring' barley population, and a QTL for accumulation of proteins encoded by COR (COld Regulated) genes on chromosome 5H (*Cor14b*, *tmc-ap3*) was coincident with a QTL for low temperature tolerance (Francia *et al.*, 2004). C-repeat binding factor (CBF) family genes were also mapped in this region (Francia *et al.*, 2004). In *A. thaliana* the transcription factors encoded by CBF family genes have been shown to be key determinants of low temperature tolerance (Thomashow, 1999; Thomashow *et al.*, 2001).

No significant QTLs for winter survival in the field were identified in the reference map of p150/112 (Yamada *et al.*, 2004). However, a QTL for electrical conductivity corresponding to frost tolerance (Dexter *et al.*, 1930, 1932) was located close to a heading date QTL in a region is likely to show conserved synteny with chromosomal regions associated with both winter hardiness and flowering time variation in cereals, as described above.

The F$_2$ (Aurora x Perma) genetic map population (RASP) was also used to identify QTLs for traits relating winter hardiness, as well as sugar content. Snow mould-resistant varieties accumulate higher levels of fructan and metabolise them at slower rates compared to susceptible varieties (Yoshida *et al.*, 1998). Many of the snow mould fungi, such as *Typhula* spp., *Microdochium nivale* and *Sclerotinia borealis*, can co-infect on single plants, and their interactions may obscure the respective effects on plant survival (Matsumoto & Araki, 1982; Matsumoto *et al.*, 1982). The use of fungicides with a limited spectrum of activity may clarify these specific effects. *Typhula* snow moulds such as *T. ishikariensis* and *T. incarnata* generally occur in the deep snow environment of the western region of Hokkaido, Japan, including Sapporo. In this environment, control of *S. borealis* and *M. nivale* infections with the fungicide iminoctadine-triacetate is an effective method for evaluation of resistance to *Typhula* snow moulds (Takai *et al.*, 2004). Scores of winter survival were measured in the F_2 (Aurora x Perma) population using this control regime, and QTLs for this trait were identified on LGs 2, 4, 6 and 7. Fructan content was also measured by high performance liquid chromatography (HPLC) using crown tissues from plants grown outdoors in December. QTLs for content of high molecular fructan with more than eight degrees of polymerization (DP) were observed on LGs 1, 2 and 4. QTLs for winter survival in LGs 2 and 4 are close to coincident with the QTLs for high molecular weight fructan content.

Two major QTLs for freezing tolerance (*Frf*) and 4 QTL for winter survival have been identified in the closely related pasture grass species meadow fescue (*Festuca pratensis* Huds.) (Rognli *et al.*, 2002). Comparative mapping with heterologous wheat anchor probes indicated that *Frf4_1* on LG4 of *F. pratensis* was orthologous to the frost-tolerance loci *Fr1* and *Fr2* in wheat. The QTLs for winter survival, by contrast, were located on LGs 1, 2, 5, and 6 (Rognli *et al.*, 2002).

Fructosyltransferase genes involved in fructan biosynthesis such as 1-SST, 1-FFT and 6G-FFT were isolated from perennial ryegrass and characterized by heterologous expression in the *Pichia pastoris* system (Hisano *et al.*, in preparation). *Lp1-SST* and *Lp1-FFT* mapped to the upper region of LG 7 in the F_2 (Aurora x Perma) genetic map, but failed to show coincidence with any fructan content QTLs. The *Lp1-SST* gene (Chalmers *et al.*, 2003) was also assigned to the equivalent region of LG 7 as a single nucleotide polymorphism (SNP) locus in the F_1 (NA$_6$ x AU$_6$) second generation reference family (Faville *et al.*, 2005). However, *Lp6G-FFT* mapped to LG 3 close to a QTL for low-molecular weight fructan content (Hisano *et al.*, in preparation). As previously described, evidence from comparative genome studies suggests that the upper part of LG4 in perennial ryegrass may contain a region of conserved synteny with the long arms of the Triticeae homoeologous group 5 chromosome. It is possible that allelic variation in regulatory genes such as those for the CBF transcription factor family may contribute to the QTLs for winter survival and fructan content observed on LG 4.

Herbage quality

Quality is the most important of all agronomic traits for pastures due to the nutritive requirements of grass-fed livestock. The genetic control of nutritive value parameters in pasture species has been reviewed (e.g. Casler, 2001), and genetic variation for specific traits has been established. Digestibility is generally considered to be the most important temperate grass nutritive value trait for either live-weight gain (Wheeler & Corbett, 1989) or dairy production (Smith *et al.*, 1997). Deliberate attempts to improve dry matter digestibility (DMD) in forage crop species have led to rates of genetic gain in the range of 1 - 4.7 % per annum as a proportion of the initial population means (Casler, 2001). Progress in simultaneous improvement of yield and DMD in forage grasses has, however, been variable (Wilkins & Humphreys, 2003).

Forage quality may be directly evaluated by feeding trials using animals, but this approach is costly, laborious and limited for small quantities of herbage from breeding experiments. Indirect methods of assessment include *in vitro* digestibility with rumen liquor (Menke *et al.*, 1979; Tilly and Terry, 1963), enzymatic digestion (De Boever *et al.*, 1986) and chemical analysis of cellular components (van Soest, 1963). The development of near infra-red reflectance spectroscopy (NIRS) analysis for prediction of forage quality has facilitated rapid and non-destructive evaluation of samples from plant breeding programs. NIRS has been used to develop calibrations to predict a wide range of forage quality traits (Marten *et al.*, 1984; Smith and Flinn, 1991) including crude protein (CP) content, estimated *in vivo* dry matter digestibility (IVVDMD), neutral detergent fibre (NDF) content (Smith & Flinn, 1991) and water-soluble carbohydrate (WSC) content (Smith & Kearney, 2000) in perennial ryegrass. NIRS estimates of DMD and related nutritive value traits have been reported in a range of forage systems (e.g. Carpenter & Casler, 1990; Hopkins *et al.*, 1995; Smith *et al.*, 2004).

Lübbestedt *et al.* (1997, 1998) published the first QTL analysis devoted to forage quality in maize. QTLs for cell-wall digestibility and lignification traits in maize were also investigated in two recombinant inbred lines (RIL) progeny by Méchin *et al.* (2001). Cardinal *et al.* (2003) detected 65 QTLs related to fiber and lignin content in maize. The best options for breeding of grasses for improved digestibility was assessed based on a search for genome locations involved in forage quality traits through QTL analysis (Ralph *et al.*, 2004).

Ground herbage samples from genotypes of the p150/112 population were measured for quality traits such as CP, IVVDMD, NDF, estimated metabolisable energy (EstME) and WSC by NIRS analysis (Cogan *et al.*, 2005). A total of 42 QTLs was observed in six different sampling experiments varying by developmental stage (anthesis or vegetative growth), location or year. Coincident QTLs were detected on LGs 3, 5 and 7. The region on LG 3 was associated with variation for all measured traits across various experimental datasets. The region on LG 7 was associated with variation for all traits except CP, and is located in the vicinity of the lignin biosynthesis gene loci *xlpomt1* (caffeic acid-*O*-methyltransferase), *xlpccr1* (cinnamoyl CoA-reductase) and *xlpssrcad2.1* (cinnamyl alcohol dehydrogenase).

WSC provide the most available source of energy for grazing ruminants. In the F_2 (Aurora x Perma) population (RASP), high molecular fructan constituted the major part of the WSC pool was analyzed samples of spring and autumn in tiller bases and leaves with replication of data over years (*ie* collect one replicate each year for several years) (Turner *et al.*, unpublished). Correlation between traits did not always lead to corresponding cluster of QTL and some traits have no reproducible QTL. Tiller base QTL were identified on linkage groups 1 and 5 and leaf QTL on linkage groups 2 and 6, in regions that had previously been identified as important in analyses of single replicates (Humphreys *et al.,* 2003).

Improvements of herbage quality may also be obtained by alteration of the content and ratios of minerals in grasses, to prevent metabolic disorders. Grass tetany (hypomagnesaemia) is caused by low levels of magnesium in the blood of cattle or sheep. Varieties of Italian ryegrass and tall fescue with markedly levels of magnesium have proved to be very effective in maintaining levels of blood magnesium in grazing sheep (Moseley & Baker, 1991) and cattle (Crawford *et al.*, 1998). Milk fever, caused by low blood calcium, produces animal welfare and production problems that could be addressed by reducing potassium content of forage without reducing calcium and magnesium concentrations (Sanchez *et al.*, 1994). Variation in mineral content in grasses may be strongly influenced by genetic factors and is amenable to QTL analysis. Herbage samples of the p150/112 population from four sampling experiment were analyzed for mineral content (aluminum, calcium, cobalt, copper, iron, magnesium, manganese, molybdenum, nickel, phosphorus, potassium, sodium, sulfur and zinc) by inductively-coupled plasma mass spectroscopy (ICP-MS) and a total of 45 QTLs were identified (Cogan *et al.*, in preparation). QTL clusters were observed on LGs 1, 2, 4 and 5. QTLs for the important trait for control of grass tetany, magnesium content were detected on LGs 2 and 5. Field herbage samples from the F_1 (NA$_6$ x AU$_6$) population were also analysed for mineral content by ICP-MS A total of 14 QTLs were identified on the NA$_6$ map, and 9QTLs were identified on the AU$_6$ genetic map. A number of clustered QTL locations showed coincidence between the two different populations.

Morphological traits and other agronomical traits

QTLs were detected for morphological traits such as plant height, tiller size, leaf length, leaf width, fresh weight at harvest, plant type, spikelet number per spike and spike length using the p150/112 genetic map (Yamada *et al.*, 2004). A number of traits were significantly correlated, and coincident QTL locations were identified. For example, coincident QTLs for plant height, tiller size and leaf length were identified on LG 3. The rice *SD1* semi-dwarfing gene, that launched the 'green revolution', encodes a gibberellin biosynthetic enzyme (*GA20ox*), and was assigned to the long arm of rice chromosome 1 (Sasaki *et al.*, 2002). A CAPS marker developed for the perennial ryegrass ortholocus of the *GA20ox* gene was mapped to LG 3 close to the plant height QTL, in a region of conserved synteny with rice chromosome 1 (Kobayashi *et al.*, unpublished). This finding provides further evidence for the utility of the candidate gene-based marker approach.

In Italian ryegrass (*Lolium multiflorum* Lam.), a total of 17 QTLs for six traits related to lodging resistance and heading date were detected by single interval mapping (SIM), while 33 independent QTLs from the male and female parents were detected by composite interval mapping (CIM) (Inoue *et al.*, 2004b). QTLs for plant height were located on LG 1, and for heading date on LGs 4 and 7, potentially in conserved regions with those identified in perennial ryegrass.

Conclusions

This review has demonstrated that although the existing QTL information from in forage grasses is relatively underdeveloped compared to other major crops, the recent establishment of detailed molecular genetic maps is rapidly stimulating QTL analysis and trait dissection. The current genetic maps are largely populated by anonymous genetic markers, with limited diagnostic value. The next generation of molecular genetic markers for forage grasses will be derived from expressed sequences, with an emphasis on functionally-defined genes associated with biochemical and physiological processes that are likely to be correlated with target phenotypic traits (Forster *et al.*, 2004; Faville *et al.*, 2005). Comparative genomics with other Poaceae species such as rice, wheat, barley and maize will support the development of such high value molecular markers through orthologous QTL detection and co-location of candidate genes. Accuracy in phenotypic assessment will be essential for precise QTL detection, and may in future be amenable to automated high-throughput analysis, as for genotyping.

Acknowledgements

The unpublished research from NARCH described in this paper was carried out by Dr. S. Kobayashi, Mrs. H. Hisano and H. Shinozuka. QTL research was carried out in collaboration with Dr. K. F. Smith, Dr. N.O.I. Cogan and Prof. G. S. Spangenberg of Primary Industries Research Victoria, Australia and Prof. M.O. Humphreys, Drs. I.P. Armstead and L. Turner of Institute of Grassland and Environmental Research, UK. This work was supported in part by Grants-in-Aid for Scientific Research (No. 14360160) from the Ministry of Education, Science, Sports and Culture, Japan.

References

Alm, V., C. Fang, C.S. Busso, K.M. Devos, K. Vollan, Z. Grieg & O.A. Rognli (2003). A linkage map of meadow fescue (*Festuca pratensis* Huds.) and comparative mapping with other Poaceae species. Theoretical and Applied Genetics, 108, 25-40.

Armstead, I.P., L.B. Turner, I.P. King, A.J. Cairns & M.O. Humphreys (2002). Comparison and integration of genetic maps generated from F_2 and BC_1-type mapping populations in perennial ryegrass. *Plant Breeding*, 121, 501-507.

Armstead, I.P., L.B.Turner, M. Farrell, L. Skøt, P. Gomez, T. Montoya, I.S. Donnison, I.P. King & M.O. Humphreys (2004). Synteny between a major heading-date QTL in perennial ryegrass (*Lolium perenne* L.) and the *Hd3* heading-date locus in rice. *Theoretical and Applied Genetics*, 108, 822-828.

Bezant, J., D. Laurie, N. Pratchett, J. Chojecki & M. Kearsey (1996). Marker regression mapping of QTL controlling flowering time and plant height in a spring barley (*Hordeum vugare* L.) cross. *Heredity*, 77, 64-73.

Börner, A., G.H. Buck-Sorlin, P.M. Hayes, S. Malyshev & V. Korzun (2002). Molecular mapping of major genes and quantitative trait loci determining flowering time in response to photoperiod in barley. *Plant Breeding*, 121, 129-132.

Cardinal, A.J., M. Lee & K.J. Moore (2003). Genetic mapping and analysis of quantitative trait loci affecting fiber and lignin content in maize. *Theoretical and Applied Genetics*, 106, 866-874.

Carpenter, J.A. & M.D. Casler (1990). Divergent phenotypic selection response in smooth bromegrass for forage yield and nutritive value. *Crop Science*, 30, 17-22.

Casler, M.D. (2001). Breeding forage crops for increased nutritive value. *Advances in Agronomy*, 71, 51-107.

Cattivelli, L., P. Baldi, C. Crosatti, N. Di Fonzo, P. Faccioli, M. Grossi, A.M. Mastrangelo, N. Pecchioni & A.M. Stanca (2002). Chromosome regions and stress-related sequences involved in resistance to abiotic stress in *Triticeae*. *Plant Molecular Bioliology*, 48, 649-665.

Chalmers, J., X. Johnson, A. Lidgett & G.C. Spangenberg (2003). Isolation and characterisation of a sucrose:sucrose 1-fructosyltransferase gene from perennial ryegrass (*Lolium perenne* L.). *Journal of Plant Physiology*, 160, 1385-1391.

Cogan, N.O.I, K.F. Smith, T. Yamada, M.G. Francki, A.C. Vecchies, E.S. Jones, G.C. Spangenberg & J.W. Forster (2005). QTL analysis and comparative genomics of herbage quality traits in perennial ryegrass (*Lolium perenne* L.). *Theoretical and Applied Genetics*, available.on-line.

Cooper, J.P. (1960). Short-day and low-temperature induction in *Lolium*. *Annals Botany*, 24, 232-246.

Crawford, R.J., M.D. Massie, D.A. Sleper & H.F. Mayland (1998). Use of an experimental high-magnesium tall fescue to reduce grass tetany in cattle. *Journal of Production Agriculture*, 11, 491-496.

De Boever, J.L., F.X. Cottyn, F.W. Wainman & J.M. Vanacker (1986). The use of an enzymatic technique to predict digestibility, metabolisable and net energy of compound feedstuffs for ruminants. *Animal Feed Science Technology*, 14, 203-214.

Dexter, S.T., W.E. Tottingham & L.F. Graber (1930). Preliminary results in measuring the hardiness of plants. *Plant Physiology*, 5, 215-223.

Dexter, S.T., W.E. Tottingham & L.F. Graber (1932) Investigations of the hardiness of plants by measurement of electrical conductivity. *Plant Physiology*, 7, 63-78.

Dubcovsky, J., D. Lijavetzky, L. Appendino & G. Tranquilli (1998). Comparative RFLP mapping of *Triticum monococcum* genes controlling vernalisation requirement. *Theoretical and Applied Genetics*, 97, 968-975.

El-Din El-Assal, S, C. Alonso-Blanco, A.J. Peeters, V. Raz & M. Koornneef (2001). A QTL for flowering time in *Arabidopsis* reveals a novel allele of *CRY2*. *Nature Genetics*, 29, 435-440.

Faville, M.J., A.C. Vecchies, M. Schreiber, M.C. Drayton, L.J. Hughes, E.S. Jones, K.M. Guthridge, K.F. Smith, T. Sawbridge, G.C. Spangenberg, G.T. Bryan & J.W. Forster (2005). Functionally associated molecular genetics marker map construction in perennial ryegrass (*Lolium perenne* L.). *Theoretical and Applied Genetics*, available.on-line.

Forster, J.W., E.S. Jones, J. Batley & K.F. Smith (2004). Molecular marker-based genetic analysis of pasture and turf grasses. A. Hopkin *et al.* (eds.) Molecular Breeding of Forage and Turf, 197-238, Kluwer Academic Publishers, the Netherlands.

Francia, E., F. Rizza, L. Cattivelli, A.M. Stanca, G. Galiba, B. Toth, P.M. Hayes, J.S. Skinner & N. Pecchioni (2004). Two loci on chromosome 5H determine low-temperature tolerance in a 'Nure' (winter) x 'Tremois' (spring) barley map. *Theoretical and Applied Genetics*, 108, 670-680.

*Galiba, G., S.A. Quarrie, J. Sutka, A. Morgounov & J.W. Snape (1995). RFLP mapping of the vernalization (*Vrn-A1) and frost resistance (*Fr1) genes on chromosome 5A of wheat.* Theoretical and Applied Genetics, 90, 1174-1179.

Galiba, G., I. Kerepesi, J.W. Snape & J. Sutka (1997). Location of gene regulating cold-induced carbohydrate production on chromosome 5A of wheat. *Theoretical and Applied Genetics*, 95, 265-270.

Hayama, R. & G. Coupland (2003). Shedding light on the circadian clock and photoperiodic control of flowering. *Current Opinion in Plant Biology*, 6, 13-19.

Hayes, P.M., T. Blake, T.H.H. Chen, S. Tragoonrung, F. Chen, A. Pan & B. Liu (1993). Quantitative trait loci on barley (*Hordeum vulgare* L.) chromosome 7 associated with components of winter hardiness. *Genome*, 36, 66-71.

Henderson, I. R. & C. Dean (2004). Control of *Arabidopsis* flowering: the chill before the bloom. *Development*, 131: 3829-3838.

Hopkins, A.A., K.P. Vogel, K.J. Moore, K.D. Johnson & I.T. Carlson (1995). Genotype effects and genotype by environment interactions for traits of elite switchgrass populations. *Crop Science*, 35, 125-132.

Humphreys, M., L. Turner, L. Skøt, L, M. Humphreys, I. King, I. Armstead, I & P. Wilkins (2003). The use of genetic markers in grass breeding. *Czech Journal of Genetics and Plant Breeding*, 39, 112-119. Biodiversity and Genetic Resources as the Bases for Future Breeding. 25th Meeting of the Fodder Crops and Amenity Grasses Section of Eucarpia: Brno, Czech Republic.

Inoue, M., Z. Gao, M. Hirata, M. Fujimori & H. Cai (2004a). Construction of a high-density linkage map of Italian ryegrass (*Lolium multiflorum* Lam.) using restriction fragment length polymorphism, amplified fragment length polymorphism, and telomeric repeat associated sequence markers. *Genome*, 47, 57-65.

Inoue, M., Z. Gao & H. Cai (2004b). QTL analysis of lodging resistance and related traits in Italian ryegrass (*Lolium multiflorum* Lam.). *Theoretical and Applied Genetics*, 109, 1579-1585.

Izawa, T., Y. Takahashi & M. Yano (2003). Comparative biology comes into bloom: genomic and genetic comparison of flowering pathways in rice and *Arabidopsis*. *Current Opinion in Plant Biology*, 6, 113-120.

Jensen, L.B., J.R. Andersen, U. Frei, Y. Xing, C. Taylor, P.B. Holm & T. Lübbestedt (2005). QTL mapping of vernalization response in perennial ryegrass (*Lolium perenne* L.) reveals co-location with an orthologue of wheat *VRN1*. *Theoretical and Applied Genetics*, available. on-line.

Jones, E.S., N.L. Mahoney, M.D. Hayward, I.P. Armstead, J.G. Jones, M.O. Humphreys, I.P. King, T. Kishida, T. Yamada, F. Balfourier, C. Charmet & J.W. Forster (2002a). An enhanced molecular marker-based map of perennial ryegrass (*Lolium perenne* L.) reveals comparative relationships with other Poaceae species. *Genome*, 45, 282-295.

Jones, E.S., M.D. Dupal, J.L. Dumsday, L.J. Hughes & J.W. Forster (2002b). An SSR-based genetic linkage map for perennial ryegrass (*Lolium perenne* L.). *Theoretical and Applied Genetics*, 105, 577-584.

Kato, K., H. Miura & S. Sawada (1999). Comparative mapping of the wheat *Vrn-A1* region with the rice *Hd-6* region. *Genome*, 42, 204-209.

Kato, K., S. Kidou, H. Miura & S. Sawada (2002). Molecular cloning of the wheat *CK2α* gene and detection of its linkage with *Vrn-A1* on chromosome 5A. *Theoretical and Applied Genetics*, 104,1071-1077.

Koemel, J.E. Jr., A.C. Guenzi, J.A. Anderson, E.L. Smith (2004) Cold hardiness of wheta near-isogenic lines differing in vernalization alleles. *Theoretical and Applied Genetics*, 109, 839-846.

Kojima, S., Y. Takahashi, Y. Kobayashi, L. Monna, T. Sasaki, T. Araki & M. Yano (2002). *Hd3a*, a rice orhololog of the *Arabidopsis FT* gene, promotes transition to flowering downstream of Hd1 under short-day condition. *Plant and Cell Physiology*, 43, 1096-1105.

Laurie, D.A., N. Pratchett, J.H. Bezant & J.W. Snape (1995). RFLP mapping of five major genes and eight quantitative trait loci controlling flowering time in a winter x spring barley (*Hordeum vulgare* L.) cross. *Genome*, 38, 575-585.

Laurie, D. A. (1997). Comparative genetics of flowering time. *Plant Molecular Biology*, 35, 167-177.

Laurie, D.A., S. Griffiths, R.P. Dunford, V. Christodoulou, S.A. Taylor, J. Cockram, J. Beales & A. Turner (2004). Comparative genetics approaches to the identification of flowering time genes in temperate cereals. *Field Crops Research* 90, 87-99.

Lübberstedt, T., A.E. Melchinger, D. Klein, H. Degenhardt & C. Paul (1997). QTL mapping in testcrosses of European flint lines of maize: II. Comparison of different testers for forage quality traits. *Crop Science*, 37, 1913-1922.

Lübberstedt, T., A.E. Melchinger, S. Fähr, D. Klein, A. Dally & P. Westhoff (1998). QTL mapping in testcrosses of European flint lines of maize: III. Comparison across populations for forage traits. *Crop Science*, 38, 1278-1289.

Marten, G.C., G.E. Brink, D.R. Buxton, J.L. Halgerson & J.S. Hornstein (1984). Near infra-red reflectance spectroscopy analysis of forage quality in four legume species. *Crop Science*, 24, 1179-1182.

Matsumoto, N. & T. Araki (1982). Field observation of snow mold pathogens of grasses under snow cover in Sapporo, *Research Bulletin of Hokkaido National Agricultural Experiment Station*, 135, 1-10.

Matsumoto, N., T. Sato & T. Araki (1982). Biotype differentiation in the Typhula ishikariensis complex and their allopatry in Hokkaido. Ann. Phytopath Soc. Japan, 48, 275-280.

Méchin, V., O. Argillier, Y. Hebert, E. Guingo, L. Moreau, A. Charcosset & Y. Barrière (2001). Genetic analysis and QTL mapping of cell-wall digestibility and lignification in silage maize. *Crop Science*, 41, 690-697.

Menke, K.H., L. Raab, A. Salewski, H. Steingass, D. Fritz & W. Schneider (1979) The estimation of the digestibility and metabolisable energy content of ruminant feeding stuffs from the gas production when they are incubated with rumen liquor *in vitro*. *Journal of Agricultural Science (Cambridge)*, 93: 217-222.

Moseley, G. & D.H. Baker (1991). The efficacy of a high magnesium grass cultivar in controlling hypomagnesaemia in grazing animals. *Grass and Forage Science*, 46, 375-380.

Mouradov, A., C. F. Cremer & G. Coupland (2002). Control of Flowering Time: Interacting Pathways as a Basis for Diversity. *Plant Cell*, S111-S130 (Suppl.).

Pan, A., P.M. Hayes, F. Chen, T.H.H. Chen, T. Blake, S. Wright, I. Karsai & Z. Bedö (1994). Genetic analysis of the components of winterhardiness in barley (*Hordeum vugare* L.). *Theoretical and Applied Genetics*, 89, 900-910.

Paran, I & D. Zamir (2003). Quantitative traits in plants: beyond the QTL. *TRENDS in Genetics*, 19, 303-306.

Ralph, J., S. Guillaumie, J.H. Grabber, C. Lapierre & Y. Barrière (2004). Genetic and molecular basis of grass cell-wall biosynthesis and degradability. III. Towards a forage grass ideotype. *Comptes Rendus Biologies*, 327, 467-479.

Rognli, O.A., V. Alm, C. Busso, C. Fang, A. Larsen, K. Devos, M. Humphreys, K. Vollan & Z. Grieg (2002). Comparative mapping of quantitative trait loci controlling frost and drought tolerance. In: Abstracts of Plant & Animal Genome X Conference, W167. January 12-16, San Diego,, CA, USA.

Saha, M.C., R. Mian, J.C.Zonitzer, K. Chekhovskiy & A. A. Hopkins (2005). An SSR- and AFLP-based genetic linkage map of tall fescue (*Festuca arundinacea* Schreb.). *Theoretical and Applied Genetics*, available. on-line.

Sanchez, W.K., D.K. Beede & J.A. Cornell (1994). Interactions of sodium potassium, and chloride on lactation, acid-base status, and mineral concentrations. *Journal of Dairy Science*, 77, 1661-1675.

Sasaki, A., M. Ashikari, M. Ueguchi-Tanaka, H. Itoh, A. Nishimura, D. Swapan, K. Ishiyama, T. Sato, M. Kobayashi, G.S. Khush, H. Kitano & M. Matsuoka (2002). A mutant of gibberellin-synthesis gene in rice. *Nature*, 416: 701-702.

Smith, K.F. & P.C. Flinn (1991). Monitoring the performance of a broad-based calibration for measuring the nutritive value of two independent populations of pasture using near infra-red reflectance spectroscopy. *Australian Journal of Experimental Agriculture*, 31, 205-210.

Smith, K.F., K.F.M. Reed & J.Z. Foot (1997). An assessment of the relative importance of specific traits for the genetic improvement of nutritive value in dairy pasture. *Grass and Forage Science*, 52, 167-75.

Smith, K.F. & G.A. Kearney (2000). The distribution of errors associated with genotype and environment during the prediction of the water-soluble carbohydrate concentration of perennial ryegrass cultivars using near infrared reflectance spectroscopy. *Australian Journal of Agricultural Research*, 51, 481-486.

Smith, K.F., R.J. Simpson & R.N. Oram (2004). The effects of site and season on the yield and nutritive value of cultivars and half-sib families of perennial ryegrass (*Lolium perenne* L.) *Australian Journal of Experimental Agriculture*, 44, 763-769.

Snape, J.W., A. Semikhodskii, L. Fish, R.N. Sarma, S.A. Quarrie, G. Galiba & J. Sutka (1997). Mapping frost resistance loci in wheat and comparative mapping with other cereals. *Acta Agron Hungary*, 45, 265-270.

Storlie, E.W., R.E. Allan & M.K. Walker-Simmons (1998). Effect of the *Vrn1-Fr1* interval on cold hardiness levels in near-isogenic wheta lines. *Crop Science*, 38, 483-488.

Sung, S. & R. M. Amasino (2004). Vernalization and epigenetics: how plants remember winter. *Current Opinion in Plant Biology*, 7, 4-10.

Sutka, J. & J.W. Snape (1989). Location of a gene for frost resistance on chromosome 5A of wheat. *Euphytica*, 42, 41-44.

Sutka, J., G. Galiba, A. Vaguifalvi, B.S. Gill & J.W. Snape (1999). Physical mapping of the *Vrn-A1* and *Fr1* genes on chromosome 5A of wheat using deletion lines. *Theoretical and Applied Genetics*, 99, 199-202.

Takai, T., Y. Sanada & T. Yamada (2004). Varietal differences of meadow fescue (*Festuca pratensis* Huds.) in resistance to *Typhula* snow mold. *Grassland Science*, 49, 571-576.

Takahashi, Y., A. Shomura, T. Sasaki & M. Yano (2001). *Hd6*, a rice quantitative trait locus involved in photoperiod sensitivity, encodes the α subunit of protein kinase CK2. *Proceeding of the National Academy of Sciences of the United States of America*, 98, 7922-7927.

Thomashow, M.F. (1999). Plant cold acclimation: Freezing tolerance genes and regulatory mechanisms. *Annual Review of Plant Physiology and Plant Molecular Biology*, 50, 571-599.

Thomashow, M.F., S.J. Gilmour, E.J. Stockinger, K.R. Jaglo-Ottosen & D.G. Zarka (2001). Role of Arabidopsis CBF transcriptional activators in cold acclimation. *Physiologia Plantarum*, 112, 171-175.

Tilly, JMA & R.A. Terry (1963). A two-stage technique for the *in vitro* digestion of forage crops. *Journal of the British Grassland Society*, 18, 104-111.

Toth, B., G. Galiba, E. Fehér , J. Sutka & J. W. Snape (2003). Mapping genes affecting flowering time and frost resistance on chromosome 5B of wheat. *Theoretical and Applied Genetics* 107, 509-514.

Turner, L.B., M.O. Humphryes, A.J. Cairns & C.J. Pollock (2001). Comparison of growth and carbohydrate acculation in seedling of two varieties of *Lolium perenne*. *Journal of Plant Physiology*, 158, 891-897.

van Soest, P.J. (1963). Use of detergents in the analysis of fibrous feeds. *Journal of the Association of Official Agricultural Chemists*, 46, 825-835.

Warnke, S., R.E. Barker, S.-C. Sim, G. Jung & J.W. Forster (2003a). Genetic map development and syntenic relationships an annual x perennial ryegrass mapping population. In: Abstract of Plant & Animal Genome XI Conference, W 206. January 11-15, San Diego, CA, USA.

Warnke, S., R.E. Barker, S.-C. Sim, G. Jung & M.A. R. Mian (2003b). Identification of flowering time QTL's in annual x perennial ryegrass mapping population. In: Abstracts of Molecular Breeding of Forage and Turf, Third International Symposium, p. 122. May 18-22, Dallas, Texas and Ardmore, Oklahoma, USA.

Warnke, S.E., R.E. Barker, G. Jung, S.-C. Sim, M.A. R. Mian, M.C. Saha, L.A. Brilman, M.P. Dupal & J.W. Forster (2004). Genetic linkage mapping of an annual x perennial ryegrass population. *Theoretical and Applied Genetics*, 109, 294-304.

Wheeler, J.L. & J.L.Corbett (1989). Criteria for breeding forages of improved nutritive value: results of a Delphi survey. *Grass and Forage Science*, 44, 77-83.

Wilkins, P.W. & M.O. Humphreys (2003). Progress in breeding perennial forage grasses for temperate agriculture. *Journal of Agricultural Science (Cambridge)*, 140, 129-150.

Yano, M., Y. Katayose, M. Ashikari, U. Yamanouchi, L. Monna, T. Fuse, T. Baba, K. Yamamoto, Y. Umehara, Y. Nagamura & T. Sasaki (2000). *Hd1*, a major photoperiod sensitivity quantitative trait locus in rice, is closely related to the *Arabidopsis* flowering time gene *CONSTANS*. *Plant Cell*, 12, 2473-2484.

Yano, M., S. Kojima, Y. Takahashi, H.X. Lin & T. Sasaki (2001). Genetic control of flowering time in rice, a short-day plant. *Plant Physiology*, 127, 1425-1429.

Yamada, T., E.S. Jones, T. Nomura, H. Hisano, Y. Shimamoto, K.F. Smith, M.D. Hayward & J.W. Forster (2004). QTL analysis of morphological, developmental and winter hardiness-associated traits in perennial ryegrass (*Lolium perenne* L.). *Crop Science*, 44, 925-935.

Yamamoto, T., H. Lin, T. Sasaki & M. Yano (2000). Identification of heading data quantitative locus *Hd6* characterization of its epistatic interaction with *Hd2* in rice using advanced back-cross progeny. *Genetics*, 154. 885-891.

Yan, L, A. Loukoianov, G. Tranquilli, M. Helguera, T. Fahima, J. Dubcovsky (2003) . Positional cloning of the wheat vernalisation gene *VRN1*. *Proceeding of the National Academy of Sciences of the United States of America*, 100, 6263-6268.

Yan, L, A. Loukoianov, A. Blechl, G. Tranquilli, W. Ramakrishna, P. SanMiguel, J.L. Bennetzen, V. Echenique, J. Dubcovsky (2004). The wheat *VRN2* gene is a flowering repressor down-regulated by vernalisation. *Science*, 303, 1640-1644.

Yoshida, M., J. Abe, M. Moriyama, T. Kuwahara (1998). Carbohydrate levels among wheat cultivars varying in freezing tolerance and snow mold resistance during autumn and winter. *Physiologia Plantarum*, 103, 437-444.

Translational genomics for alfalfa varietal improvement

G.D. May

Plant Biology Division, The Samuel Roberts Noble Foundation, 2510 Sam Noble Parkway, Ardmore, Oklahoma 73401, USA, Email: gdmay@noble.org

Key points

1. *Medicago truncatula* is a model legume with available mapping, genome, and RNA, protein and metabolite profiling databases and genetically diverse populations.
2. Genomics resources developed for *M. truncatula* have application in the study and improvement of alfalfa making it an excellent model for this forage legume.

Keywords: *Medicago*, functional genomics, genome, alfalfa

Introduction

With more than 650 genera and 19,000 species, legumes are one of the two most important crop families in the world. Among cultivated plants, legumes are unique in their ability to fix atmospheric nitrogen through a novel symbiotic relationship with bacteria known as Rhizobia. Since they are not limited for nitrogen, legumes have remarkably high levels of protein, a property that is both biologically and agriculturally significant. Nearly 33% of all human nutritional requirement for nitrogen is derived from legumes, and in many developing countries, legumes serve as the single most important source of protein. Legumes synthesize an impressive array of secondary metabolites, including isoflavonoids and triterpene saponins, shown to possess anti-cancer and other health promoting effects. Not surprisingly, legumes play a central role in nearly all crop rotation systems and are universally viewed as essential for secure and sustainable food production.

All major crop legumes are found in the monophyletic subfamily Papilionoideae. Within this subfamily, the tropical legumes include the economically important soybean (*Glycine max*), common bean (Phaseolus spp.), cowpea (*Vigna unguiculata*), and mung bean (*V. radiata*), while temperate legumes include species such as pea (*Pisum sativum*), alfalfa (*Medicago sativa*), lentil (*Lens culinaris*), and chick pea (*Vicia arietinum*). Papilionoid legumes first appeared around 65 million years ago based on fossil records (reviewed in Doyle, 2002), the same time as other important crop families. Because they form a compact monophyletic evolutionary group, comparative genomics among Papilionoid species has huge potential to increase our understanding of this vitally important group of plants. Indeed, a growing body of evidence demonstrating micro- and macrosynteny among Papilionoids suggests that discoveries made in one species can often be extended to other members of the subfamily.

The uniqueness of a plant family is the product of all of its many traits. Some, notably the diagnostic characters that define the family taxonomically, may be truly unique, but most are found, in different combinations, in unrelated groups of plants. At the level of morphology, anatomy, and chemistry, however, characters shared with other families may be analogous, rather than homologous--functionally similar, but derived independently from different ancestors. The molecular basis for analogous characters may involve different genes, either truly non-homologous genes that have arisen independently, or paralogous members of gene families that were recruited independently in different evolutionary lineages to perform similar roles. The underlying theme of much of genomics research, and the basis for the highly successful model system approach, is that there are many features shared among genomes even of very distantly related organisms, permitting generalization.

M. truncatula, also referred to as barrel medic because of the shape of its pods is a forage legume commonly grown in Australia and throughout the Mediterranean. It is closely related to the world's major forage legume, alfalfa, but unlike alfalfa, which is a tetraploid, obligate outcrossing species, *M. truncatula* can be self-pollinated and has a simple diploid genome (with eight pairs of homologous chromosomes). *M. truncatula* has been chosen as a model species for genomic studies in view of its small genome, fast generation time (from seed-to-seed), and high transformation efficiency (Cook, 1999, May & Dixon, 2004). Genes from *M. truncatula* share high sequence identity to their orthologs from alfalfa so it serves as an excellent genetically tractable model for alfalfa. Studies on syntenic relationships (comparisons of genome content and organization between organisms) are establishing links between *M. truncatula*, alfalfa, and pea, as well as *Arabidopsis*.

In 1999, a Center for Medicago Genomics Research was established at the Samuel Roberts Noble Foundation. Scientists at Noble have taken a global approach in studying the genetic and biochemical events associated with the growth, development, and biotic and abiotic interactions of the model legume *M. truncatula*. Approaches taken to dissect the genome of *M. truncatula* and its function include; large-scale EST and genome sequencing, gene expression profiling, the generation of *M. truncatula* transposon-tagged and fast-neutron mutagenized populations and high-throughput protein and metabolite profiling. The resulting multidisciplinary data sets developed in our program are being interfaced to provide scientists with an integrated set of tools to address fundamental questions pertaining to legume biology. Our goal has been to establish a research program that will make significant contributions to the areas of legume molecular biology, biochemistry, and genetics research.

Discussion

There is a strong, cohesive and well-organized *Medicago* research community in the US, Europe, Australia and elsewhere and *Medicago* genomics has advanced rapidly in the past five years. This is due in large part to research at the Samuel Roberts Noble Foundation, and also to projects funded under the NSF Plant Genome Program and in the European Union. These efforts have collectively produced a large number of ESTs, the initiation of a Noble Foundation-funded *Medicago* genome sequencing project, a robust physical map that is well-anchored to the genetic map, two generations of expression microarrays using first spotted cDNAs and then oligonucleotide arrays, programs in protein and metabolite profiling, and the generation of EMS, fast-neutron, and transposon-tagged mutant populations. Following on from this, a group of *Medicago* researchers on both sides of the Atlantic formed the *Medicago* Genome Sequencing Consortium/Initiative and was successful in obtaining funding from the NSF, the EU, BBSRC and INRA/Genoscope to complete the sequencing of the euchromatic portion of the *M. truncatula* genome on a chromosome by chromosome basis using a BAC-based strategy by the end of 2006. The underpinning for these successful proposals was the funding provided by the Noble Foundation to Bruce Roe at the University of Oklahoma. This gave the project a jump-start with approximately 63 Mb of BAC sequence in GenBank at the time that the NSF and international projects were reviewed for funding.

M. truncatula as a reference legume: *Medicago* EST and genome sequencing

The Noble Foundation recently released to NCBI more than 27,000 additional *Medicago* ESTs from our databases. As of January 2005, these sequences push the *M. truncatula* EST total to 216,645 - number eight among all plant species on an EST basis. The Foundation's contribution to this total is 114,913 high-quality EST clones -- approximately 53% of the world's efforts. As we continue EST projects for other species, our *Medicago* EST sequencing efforts will now focus upon characterization of full-length ESTs. The first objective of this proposed activity (a collaboration with C. Town, TIGR) is to generate full-length cDNA sequences for in excess of 20,000 genes (i.e. approximately half of the expected transcriptome) from *M. truncatula*. A total of ~ 10,000 candidate *Medicago* FL-cDNAs have been identified. 2,000 of which have been sequenced to completion. Of the ~ 8,000 remaining candidate cDNAs, 6,000 can be found in the EST collection at Noble. The remainder will be obtained from the construction and sequencing of normalized libraries with a high proportion of full-length sequences. A second objective will be to produce full-length sequence-validated cDNA ORF clones for at least 10,000 of these genes in a Gateway recombination vector system for functional analyses.

The international Medicago genome sequencing project

A whole-genome *M. truncatula* sequence program initially began at the University of Oklahoma. The initial goal of the project was to generate an approximately one-fold whole genome shotgun sequence data of the 500 megabase genome from a plasmid-based genomic library and obtain target shotgun clones for additional primer walking-based sequencing. However, preliminary results from the shotgun approach suggest that the *M. trucatula* genome is highly repetitive. As previously predicted, estimates are that approximately 80% of the genome is highly repetitive and that approximately 80% of the gene-rich regions represent only 20% of the total genome. To reduce the amount of redundant sequence, the sequencing strategy was modified to sequence bacterial artificial chromosome (BAC) clones from *M. truncatula* BAC libraries. More than 800 BACs were identified based on DNA markers or gene content and were sequenced to working draft coverage (four- to five-fold) utilizing a BAC-based shotgun sequencing approach, in the first phase of this project.

The whole-genome shotgun approach resulted in the sequencing of the *M. truncatula* chloroplast genome, since the total genomic DNA preparation not only contained the nuclear genome, but also a significant level of the chloroplast DNA. The DNA sequence of the *M. truncatula* chloroplast genome has now been completed and consists of one contiguous 124,039 base pair circle. Artificially linearizing the sequence at the histidine tRNA prior to the psbA gene allows the *Medicago* chloroplast genomic sequence to be co-linear with the *Arabidopsis*, tobacco, and most other chloroplast genomes. The semi-automated annotation of the *M. truncatula* chloroplast genome using Web-Artemis has been completed, and can be viewed at: http://www.genome.ou.edu/medicago_chloroplast/med_chloro_art.html.

Currently the *Medicago* genome sequencing program involves researchers at the University of Minnesota, The Institute for Genomic Research and the University of Oklahoma in the U.S. and Sanger and Genoscope in the U.K. and Europe, respectively in a chromosome by chromosome approach. As of early 2005, more than 800 BACs are finished and almost 1,400 total BACs are at sequencing phase 1, 2, or 3 with the phase 2 and 3 BAC sequences comprising more than 135 Mbp of non-redundant *M. truncatula* genome sequence. All sequences are available through GenBank and EMBL databases. The anticipated completion date for the *Medicago* genome sequencing project is December 2006.

The other *Medicago* -omics: profiling transcripts, proteins and metabolites

Expression analyses

Two generations of DNA microarray technologies have been established for expression profiling in *Medicago* species. *M. truncatula* genome-wide microarrays are being generated using the *Medicago* Array-Ready Oligonucleotide Set (GS-1700-02) Version 1.0 (Operon). Approximately 16,000, amino-linked, 70-mer oligonucleotides are being printed onto aminosilane-coated "Superamine" slides (Telechem), using Telechem type SMP3 printing pins in Dr. David Galbraith's laboratory at the University of Arizona. Preliminary results in our groups suggest that *M. truncatula* oligonucleotide arrays hybridize well with targets synthesized using *M. sativa* mRNA as a template. These arrays should provide a valuable tool to study complex traits in alfalfa.

The design of an Affymetrix *Medicago* GeneChip array, the composition of which was arrived at after consultations between Affymetrix and the international *Medicago* community has been completed. The array contains probe sets to profile approximately 60,000 gene sequences that were derived by combining all *M. truncatula* EST and annotated genomic sequence data. *M. truncatula* sequences included on the array are; 1) International *Medicago* Genome Annotation Group (IMGAG) high-quality gene prediction from *Medicago* BAC sequences, 2) FGENESH gene predictions from all Phase II and Phase III M. truncatula BACs sequences, and 4) chloroplast ORFs. For tentative consensus sequences (TCs) and singletons that could not be orientated both strands were tiled. In addition, the array includes *M. sativa* sequences that do not have corresponding orthologs in the *M. truncatula* data sets. Also included on the array are *Sinorhizobium meliloti* predicted ORFs from genome and plasmid sequences. It is anticipated that the *Medicago* GeneChip array will be released early summer 2005.

To supplement the expression analyses data generated by microarrays, we have added an "open system" serial analysis of gene expression (SAGE) to our set of transcript profiling tools. Such open system approaches allow for the identification and analysis of genes not previously characterized. With "closed systems" such as microarrays, analysis is limited to only those species previously identified and assigned to an array. SAGE analyses have already been used to study gene expression in plant systems (Matsumura *et al.*, 1999). Among the high-throughput, comprehensive technological methods used to analyze transcript expression levels, array-based hybridization and SAGE are currently the most common approaches.

Molecular mechanisms underlying the initiation and maintenance of embryonic pathways in plants are largely unknown. To gain better insight into these processes, serial analysis of gene expression (SAGE) was used to profile transcript accumulation levels and to identify differentially expressed genes in early stage somatic embryos of *M. truncatula*. A total of more than 131,000 SAGE tags were sequenced and 30,329 unique tags were identified in non-embryogenic, pro-embryo and globular embryo cell cultures. These studies illustrate the power of SAGE technology as a tool for both transcript profiling and gene discovery and its use in examining global changes in plant gene expression patterns. As additional plant genomes such as *M. truncatula* are sequenced and plant-specific SAGE databases become publicly available, the use of SAGE in understanding fundamental changes in gene expression should gain broad appeal in the plant research community.

Protein and metabolite profiling

The protein complement of the genome, the proteome, serves as a biological counterpart to the *Medicago* EST and gene expression analyses. Given that many biological phenomena lack the requirement for *de novo* gene transcription, proteomics studies provide a mechanism to study proteins and their modifications under developmental changes and in response to environmental stimuli.

Two-dimensional polyacrylamide gel electrophoresis (2-D PAGE) has been established as the dominant technique for analysis of complex protein mixtures since its introduction in 1975 (O'Farrell, 1975; Blackstock & Weir, 1999). The technique utilizes isoelectric focusing and polyacrylamide gel electrophoresis for first and second dimension separation, respectively. Currently, 2-D PAGE technology is capable of resolving some 10,000 proteins, with 2,000 proteins being typical experimental results (Klose & Kobalz, 1995). A recent review describes the role of 2-D PAGE in proteomic and genetic studies of plant systems, including its use as a tool to investigate genetic diversity, phylogenetic relationships, mutant characterization, and drought tolerance (Thiellement *et al.*, 1999).

Although 2-D PAGE analysis has been used for the last 20 years in protein profiling, it provides limited information on protein identification. Recent advances in mass spectrometry and the establishment of protein databases have substantially increased the ease and speed with which proteins can be identified (Yates, 1998). The union of these technologies is the foundation for modern proteomic studies. The typical experiment begins with comparative digital imaging of the 2-D gels to detect variations in protein concentration or elution profile. These protein spots are excised, extracted, and identified by using a variety of mass spectrometry techniques.

Basically, two mass spectrometry (MS) techniques are used for protein identification. The first is peptide mass-mapping of proteolytic digested fragments (Wolf *et al.*, 1998; Yates, 1998). The observed mass fragments can be searched against a theoretical list of proteolytic peptide maps predicted by a given database. Increased peptide mass accuracy has increased the success and selectivity of such searches (Jensen *et al.*, 1996). If the database query is unsuccessful, the protein can be sequenced by using tandem mass spectrometry (MS/MS) (Yates, 1998). During the MS/MS experiment, only the peptide mass of interest is isolated or transmitted, thus discriminating against all other components of the mixture with different mass values. After isolation, the peptide is further fragmented by using a unimolecular or bimolecular (collision gas) strategy. Fragments observed in the isolated peptide MS/MS spectrum can then be rationalized to a sequence.

Initial proteome profiling at the Noble Foundation has been performed to generate representative 2-D PAGE protein profiles for stems, leaves, seedpods, roots, flowers, tissues, and suspension cell cultures. Proteins were systematically identified and cataloged by using peptide mass mapping and database searching. An interactive database of the results of these analyses can be found at the following web address: http://www.noble.org/2dpage/search.asp. Analytical and biological variances associated with the 2-D PAGE proteomics approach for *M. truncatula* have been determined and will function as baseline measurements for comparative protein profiling in elicitor-induced *M. truncatula* cell cultures.

Metabolic profiling is the key to understanding how changes at the transcriptional and translational levels affect cellular function. Unlike proteomics, a single analytical technique does not exist that is capable of profiling all the low molecular weight metabolites of the cell. Our approach is to profile metabolites of control and treatment tissues by using an assortment of analytical techniques including: high-performance liquid chromatography (HPLC), capillary electrophoresis (CE), gas chromatography (GC), mass spectrometry (MS), and various combinations of the above techniques such as GC/MS, LC/MS, and CE/MS.

As the program has progressed, the development of methods to extend the profiling range to include metabolite classes such as phenylpropanoids, lignins, terpenoids saponins, soluble sugars, sugar phosphates, complex carbohydrates, amino acids, and lipids has continued. Method development also includes procedures for sequential extraction and parallel analysis. Sequential extraction segregates the metabolome into more manageable classes of chemical compounds with similar physical/chemical properties thereby facilitating the use of parallel analytical profiling techniques. Profiling of elicitor-induced cell cultures and *M. truncatula* natural variants for flavonoids, lignins, other phenylpropanoids and triterpenoids, especially saponins, has been performed as a component of an NSF-funded *Medicago* functional genomics project.

Forward and reverse genetics approaches in *M. truncatula*

Forward and reverse genetic systems for *M. trucatula* are being developed at the Noble Foundation and elsewhere in the *Medicago* research community. Forward genetic systems facilitate efficient identification of genes underlying phenotypic traits of interest, while reverse genetics systems enable the isolation of mutations in genes of known sequence. Kiran Mysore's laboratory at the Noble Foundation, in collaboration with Dr. Pascal Ratet, CNRS, Gif sur Yvette, France, are developing a large-scale, transposon-tagged mutant library of *M. truncatula* using the tobacco retrotransposon Tnt1 (Tadege *et al.*, 2005). Approximately 20,000 tagged *M. truncatula* lines will be generated during the next five years. Transposon-plant genome junctions will be isolated and characterized through DNA sequence analyses. A database of these junction sequences is being created, and these sequences are being mapped to the *M. truncatula* genome for a reverse genetics approach to determine gene function.

Fast-neutron irradiation induces DNA damage and chromosomal deletions. Deletions that occur in known genes can be detected by a shift in the size of PCR amplification products of genes of interest. Dr. Rujin Chen's group at the Noble Foundation is developing a fast-neutron mutagenized population of *M. truncatula*. Of the approximately 10,000 fast-neutron irradiated M1 *M. truncatula* plants generated thus far, two percent display a visible mutant phenotype. It is anticipated that 100,000 M1 *M. truncatula* plants will be screened within the next three years.

Databases and genomics resources

With approximately 200,000 genomics-related visits in 2004, the Foundation's web site continues to benefit the *M. truncatula* and legume research communities by providing access to data and resources developed at Noble (Table 1.). More that 1,200 EST clones from the Foundation's *Medicago* EST collection have been distributed free of charge to researchers in 18 countries. The identity of all requested ESTs are confirmed by 5'-end sequencing. The Foundation is also providing legume researchers access to oligonucleotide microarrays on a cost recovery basis.

A large number of public world wide web-based *Medicago* databases are available (Table 1.) These databases provide the research community with access to the tools and data developed in the *Medicago* DNA sequencing and functional genomics programs.

Table 1 Web addresses for *Medicago* genomics resources

Site	URL
The Center for *Medicago* Genomics Research	http://www.noble.org/medicago/index.html
The Consensus Legume Database	http://www.legumes.org
The Legume Information System	http://www.comparative-legumes.org
The *M. truncatula* Gene Index	http://www.tigr.org/tigr-scripts/tgi/ T_index.cgi?species=Medicago
TIGR *M. truncatula* Genome Resources	http://www.tigr.org/tdb/e2k1/mta1/
M. truncatula Sequencing Resources	http://www.medicago.org/genome/
Medicago Bioinformatics at the University of California – Davis	http://medicago.plantpath.ucdavis.edu/
European Research Programmes on the Model Legume M. truncatula	http://medicago.toulouse.inra.fr/
Medicago Genome Sequencing University of Oklahoma	http://www.genome.ou.edu/medicago.html
Medicago truncatula Functional Genomics and Bioinformatics	http://medicago.vbi.vt.edu/data.html

Conclusion

M. truncatula is a highly developed model legume, with a large research community, that serves as an excellent model for developing new forage varieties. What is still necessary are laser-capture microdissection (LCM) techniques to enable the isolation of specific cells (i.e. specific zones within a nodule or root cap) from complex tissues for subsequent molecular analyses. Tissue preparation and microextraction protocols are being established to allow LCM microsamples to undergo quantitative transcript, protein and metabolite profiling. The application of LCM techniques to established functional genomics technologies has the potential to enhance our understanding of diverse plant cell type-specific biological processes.

The development of bioinformatics tools for the processing, visualization and integration of transcript, protein and metabolite profiles and datasets with the evolving genome sequence is still required. These tools will lead to a correlated view of gene expression and cellular response. The long-term impact will be the integration of transcript, protein, and metabolite data for plant mutants and natural variants, that will advance all aspects of fundamental and applied legume research. This information will be used to develop agronomically important legume species, such as alfalfa that (i) are more resistant to cold, drought and fungal and viral diseases, (ii) will provide higher crop yields while reducing needs for chemical inputs, and (iii) will produce natural chemicals that promote human and animal health and nutrition. Higher yields and lower production costs will enhance the economy of rural agriculture, especially in developing nations, while a reduction in chemical usage will benefit the environment. Value-added traits such as increased levels of nutraceuticals will provide farmers with new crop alternatives and allow them to participate in the high value niche markets.

Acknowledgements

Richard A. Dixon, Maria J. Harrison, Lloyd W. Sumner, Kiran Mysore, Rujin Chen and members of their laboratory teams are to be acknowledged for their efforts. Bruce Roe, Nevin Young and Chris Town are acknowledged for their contributions to U.S. component of the *M. truncatula* genome project. Pedro Mendes for his extensive contributions to the bioinformatics portions of the NSF-funded program "An Integrated Approach to Functional Genomics and Bioinformatics in a Model Legume" (DBI-0109732). This program is supported by the National Science Foundation (DBI-0109732 and DBI-0110206), Forage Genetics International, and the Samuel Roberts Noble Foundation.

References

Blackstock, W.P. & M.P. Weir (1999). Proteomics: quantitative and physical mapping of cellular proteins. *Trends in Biotechnology*, 17, 121-127.

Cook, D.R. (1999). *Medicago truncatula* - a model in the making! *Current Opinion in Plant Biology*, 2, 301-304.

Doyle, J.J., J.L. Doyle, A.H.D. Brown & R.G. Palmer. (2002). Genomes, multiple origins, and lineage recombination in the *Glycine tomentella* (Leguminosae) polyploid complex: histone H3-D gene sequences. *Evolution*, 56, 1388-1402.

May, G.D. & R.A. Dixon (2004). Medicago truncatula. *Current Biology*, 14, 180-181.

Jensen, O.N., A. Podtelejnikov, M. Matthias-Mann (1996). Delayed extraction improves specificity in database searches by matrix-assisted laser desorption/ionization peptide maps. *Rapid Communications in Mass Spectrometry*, 10, 1371-1378.

Klose, J. & U. Kobalz, (1995). Two-dimensional electrophoresis of proteins: An updated protocol and implications for a functional analysis of the genome. *Electrophoresis*, 16, 1034-1059.

Matsumura, H., S. Nirasawa & R. Terauchi, R. (1999). Transcript profiling in rice (Oryza sativa L.) seedlings using serial analysis of gene expression (SAGE). *Plant Journal*, 20, 719-726.

O'Farrell, P.H. (1975). High resolution two-dimensional electrophoresis. *Journal of Biological Chemistry*, 250, 4007-4021.

Tadege, M., Ratet, P. & K.S. Mysore (2005). Insertional mutagenesis: a Swiss Army knife for functional genomics of *Medicago truncatula*. *Trends in Plant Science*, in press.

Thiellement, H., N. Bahrman, C. Damerval, C. Plomion, M. Rossignol, V. Santoni, D. Devienne, & M. Zivy (1999). Proteomics for genetic and physiological studies in plants. *Electrophoresis*, 20, 2013-2026.

Wolf, B.P., L.W. Sumner, S.J. Shields, K. Nielsen, K.A. Gray & D.H. Russell, D.H. (1998). Characterization of proteins utilized in the desulfurization of petroleum products by matrix-assisted laser desorption ionization time-of-flight mass spectrometry. *Analytical Biochemistry*, 260, 117-127.

Yates, J.R. (1998). Mass spectrometry and the age of the proteome. *Journal of Mass Spectrometery*, 33, 1-19.

Application of molecular technologies in forage plant breeding

K.F. Smith[1,3], J.W. Forster[2,3], M.P. Dobrowolski[1,3], N.O.I. Cogan[2,3], N.R. Bannan[1,3], E. van Zijll de Jong[2,3], M. Emmerling[2,3] and G.C. Spangenberg[2,3]

[1]*Primary Industries Research Victoria, Hamilton Centre, Private Bag 105, Hamilton, Victoria 3300, Australia, Email: kevin.f.smith@dpi.vic.gov.au*
[2]*Primary Industries Research Victoria, Plant Biotechnology Centre, La Trobe University, Bundoora, Victoria 3086, Australia*
[3]*Molecular Plant Breeding Cooperative Research Centre, Australia*

Key points

1. A range of molecular breeding technologies have been developed for forage plant species including both transgenic and non-transgenic methodologies.
2. The application of these technologies has the potential to greatly increase the range of genetic variation that is available for incorporation into breeding programs and subsequent delivery to producers in the form of improved germplasm.
3. Further developments in detailing the phenotypic effect of genes and alleles both through research in target species and through inference from results from model species will further refine the delivery of new forage cultivars.

Keywords: plant breeding, molecular markers, forage, grasses, clovers, transgenic

Introduction

The application of molecular breeding technologies in forage plant breeding offers the potential for more accurate development of cultivars with broader adaptation within a shorter generation time. These benefits are already being realised in other plant species, as shown by the development of genetically modified canola (*Brassica napus*) varieties, and the use of molecular markers for trait selection in Australian barley (*Hordeum vulgare*) (Langridge and Barr, 2003) and wheat (*Triticum aestivum*) (Marshall *et al.*, 2001) breeding programs.

In recent years, a concerted research effort has led to the development of tools for the implementation of molecular breeding technologies in forage species. This paper will discuss the application of both molecular marker-based and transgenic technologies in forage plant breeding, using specific examples from our research programs.

Molecular marker technologies in forage plants

Molecular maps have been described for key forage grass species such as perennial ryegrass (*Lolium perenne* L.) (Hayward *et al.*, 1998; Armstead *et al.*, 2002; Jones *et al.*, 2002a, 2002b; Faville *et al.*, 2004), meadow fescue (*Festuca pratensis* Huds.) (Alm *et al.*, 2003), tall fescue (*Festuca arundinacea* Schreb.) (Xu *et al.*, 1995; Saha *et al.*, 2005), as well as pasture legumes such as white clover (*Trifolium repens* L.) (Jones *et al.*, 2003; Barrett *et al.*, 2004), red clover (*Trifolium pratense* L.) (Isobe *et al.*, 2003) and alfalfa/lucerne (*Medicago sativa* L.) (Diwan *et al.*, 2000).

The use of these maps to identify quantitative trait loci (QTL) controlling the expression of key agronomic traits is described elsewhere in these proceedings (Yamada and Forster, 2005). However, it is worth noting that in a relatively short time, QTLs have been identified for a large number of traits and this has led to a large increase in the available information on the genomic location of genes controlling key forage species traits. In perennial ryegrass for instance, QTLs have been identified for resistance to the crown rust pathogen (*Puccinia coronata* f.sp. *lolii*) (Dumsday *et al.*, 2003); various aspects of flowering time and reproductive development (Armstead *et al.*, 2004; Yamada *et al.*, 2004; Jensen *et al.*, 2005); and forage quality traits (Cogan *et al.*, 2005). Some of the identified QTLs have only accounted for relatively small proportions of the phenotypic variation for the relevant quantitative traits. However, in several instances perennial ryegrass QTLs have been identified in regions of conserved synteny with known QTLs or genes for equivalent traits in other Poaceae species. Syntenic relationships have been inferred between a heading date QTL region in perennial ryegrass and the *Hd3* locus in rice (Armstead *et al.*, 2003); a region containing a ryegrass forage digestibility QTL cluster and several lignin biosynthetic genes from wheat (Cogan *et al.*, 2005); and a vernalisation response QTL from perennial ryegrass and the *VRN1* locus from wheat (Jensen *et al.*, 2005). This co-location information helps to assess the likely biological basis and significance of QTL data derived from forage species, and will be crucial for the choice of QTLs for practical implementation in forage programs. More importantly, the co-location of QTLs that are robust across multiple environments with candidate genes involved in physiological processes correlated with target phenotypes will facilitate the identification of gene-associated single nucleotide polymorphisms (SNPs). Development of SNP markers diagnostic for associated with favourable alleles will permit marker assisted selection in outcrossing forage species free of the complexities associated with use of linkage markers (Spangenberg *et al.*, 2005). It is interesting to note that although relatively few studies have been published on QTL locations in forage species, a number of instances of candidate gene-QTL co-location have already been observed. For perennial ryegrass, these include the co-location of the *VRN1* gene with a vernalisation response QTL on LG4 (Jensen *et al.*, 2005); the co-location of a casein protein kinase (*Lpck2a-1*) gene (Shinozuka *et al.*, in preparation) with a flowering time QTL on LG4 identified by Yamada *et al.* (2004); and the co-location of several lignin biosynthetic genes (Heath *et al.*, 1998; Lynch *et al.*, 2002; McInnes *et al.*, 2002) with QTLs for forage digestibility (Cogan *et al.*, 2005). While more extensive research is required to validate positive associations between haplotype and phenotype for putative candidate genes that map to QTL locations, proof-of-concept for this approach is currently being developed for the abiotic stress tolerance gene *Lp*ASRa1 (Spangenberg *et al.*, 2005)

Molecular marker technologies in *Neotyphodium* species

Many of the pasture and turf grasses within the *Lolium-Festuca* complex are hosts to symbiotic fungi of the genus *Neotyphodium* (Christiansen *et al.*, 1993). Although this symbiosis provides positive agronomic benefits to the plant through the mitigation of nutrient and water stress, the fungi also produce a range of alkaloid compounds that are toxic to grazing herbivores, leading to the disorders known as ryegrass staggers and fescue toxicosis. Recently, a number of endophyte strains have been identified that do not produce certain toxins and these are being marketed in both perennial ryegrass (e.g. AR1, AR6, NEA2) and tall fescue (MaxP, ArkPlus) varieties. SSR markers have been developed to detect genetic variability within and between *Neotyphodium* species (van Zijll de Jong *et al.*, 2003) and this technology has been extended to allow *in planta* detection and co-genotyping of *Neotyphodium* strains (van Zijll de Jong *et al.*, 2004). As would be expected for an asexually propagated species, the majority of the variation was present between rather than within species (van Zijll de Jong *et al.*, 2003). However, 'novel' endophyte strains were clearly distinguishable from many of the wild-type variants (Figure 1) (van Zijll de Jong *et al.*, 2005).

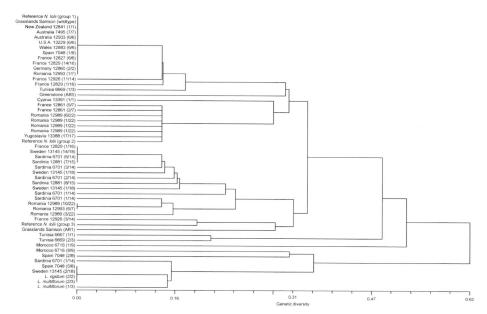

Figure 1 UPGMA phenogram for endophytes in *Lolium* accessions and varieties using measurements of average taxonomic distance based on the genetic diversity of 18 SSR loci. Indicated in brackets following the accession name are the number of genotypes containing this endophyte strain and the total number of genotypes analysed from this accession). Reference *N. lolii* group 1 are isolates from perennial ryegrass varieties Aries, Banks, Ellett, Fitzroy, Vedette and Victorian and an ecotype from North Africa. Reference *N. lolii* group 2 is an isolate from the perennial ryegrass variety Aries. Reference *N. lolii* group 3 is an isolate from an ecotype from Belgium. The low toxin-producing endophytes AR1 (in cultivar Grasslands Samson) and AR5 (in cultivar Greenstone) are included for comparison, along with the resident wild-type endophyte in Grasslands Samson (From van Zijll de Jong *et al.*, 2005).

The ability to distinguish plants infected with novel endophyte strains from those infected with wild-type (and toxic) endophytes is of great benefit during perennial ryegrass breeding. We have recently used SSR markers to assess the endophyte incidence and identity in seed lots of two different breeding lines in a breeding program (Table 1). In this case we were able to discriminate between the seed lots of breeding line 1 based on the endophyte strain they had been inoculated with (AR1 vs. Strain B), and also to demonstrate that the seed lot of breeding line 2 that had been inoculated with Strain B had become contaminated with a strain having an SSR profile similar to the wild-type endophyte found in this population. This inference was supported by toxin profiling of bulked herbage samples from the relevant sub-populations. Molecular marker-based endophyte genotyping consequently shows high applicability for purity analysis during varietal development, and has obvious potential for use as a powerful predictive tool for novel endophyte discovery.

Table 1 The identity and incidence of endophyte strains in perennial ryegrass seed lots classified on the basis of endophyte SSR polymorphism

	Endophyte Inoculated	Number of plants	Number infected	AR1	Strain B	Strain C	Unclassified[a]
Line 1	AR1	95	75	75	0	0	0
Line 1	Strain B	90	82	0	79	0	3
Line 2	Strain B	90	67	0	43	23	0

[a] Unable to discriminate endophyte strain.

Development of transgenic technologies in forage plants

The use of biolistic or *Agrobacterium*-mediated transformation allows for the targeted up- or down-regulation of genes coding for individual enzymes in complex biochemical pathways, such as those involved in fructan metabolism or lignin biosynthesis (Spangenberg *et al.*, 1998; 2001), or induction of plants to produce novel compounds through the transfer of genes from unrelated organisms, such as the introduction of a bacterial gene coding for fructan production into Italian ryegrass (Ye *et al.*, 2001). Genetic modification is particularly useful for elucidation of the role of enzymes in key biosynthetic pathways, and for the modification of traits for which there is no known genetic variation, or that have proven difficult to manipulate through selection and crossing. The number of genes available to plant breeders has rapidly increased with the advent of large-scale gene discovery programs such as those based on expressed sequence tags (ESTs) from relevant target species like perennial ryegrass and white clover (Sawbridge *et al.*, 2003a, 2003b) or the whole genome sequencing of model species such as barrel medic (*Medicago truncatula* L.) (Kulikova *et al.*, 2004), *Lotus japonicus* L. (Stougaard, 2001) and rice (*Oryza sativa* L.) (Goff *et al.*, 2002; Yu *et al.*, 2002). The limitation to adoption of gene technologies in breeding programs is not, therefore, the isolation of the genes themselves, but rather the functional annotation of these genes and the 'proof-of-phenotype' in target species.

This paper highlights the use of transgenic technologies in forage plant breeding using two examples: first, manipulation of forage quality in grasses through the modification of lignin biosynthesis is described, and second, improvement of biotic stress tolerance in white clover through the development of plants immune to infection by alfalfa mosaic virus (AMV) using virus coat protein-mediated resistance.

Genetic modification of forage quality in grasses

To date, most of the functional effects of altering the expression of genes involved in lignin biosynthesis have been described in model plant species such as Arabidopsis thaliana L. or tobacco (Nicotiana tabacum L.), the results of which have been reviewed by Casler (2001). In summary, modification of the expression of genes coding for enzymes early in the lignin synthesis pathway such as phenylalanine ammonialyase (PAL) led to a wide range of negative phenotypes, along with general reductions in lignin concentration (Elkind *et al.*, 1990) and improved digestibility in tobacco plants with down-regulated PAL activity (Sewalt *et al.*, 1997). The results of manipulating the expression of down-stream enzymes in the phenylpropanoid or monolignol pathways have been more promising, with down-regulation of expression of the key enzymes caffeic acid O-methyl transferase (OMT) and cinnamyl alcohol dehydrogenase (CAD) leading to favourable changes in lignin composition and digestibility (Bernard Vailhé et al., 1996; Sewalt et al., 1997). Casler (2001) noted that the phenotypic expression of these changes was similar to that obtained with natural variants selected for altered lignin concentration or digestibility, but that the extreme phenotypes were relatively common compared to the rarity of natural variants. The other aspect of transformation-generated variation is that it will allow the targeted manipulation of combinations of enzymes in complex pathways, to facilitate the development of extreme phenotypes that are outside the range of natural variation.

Recently, perennial ryegrass homologues of 3 key genes in the monolignol biosynthesis pathway have been cloned and characterised: CAD (Lynch et al., 2002), 4-coumarate:CoA-ligase (4CL) (Heath et al., 1998) and cinnamyl CoA reductase (CCR) (McInnes et al., 2002). Transgenic perennial ryegrass plants with sense and anti-sense regulation of these genes are currently being generated, and will provide the opportunity of assessing phenotypic changes in digestibility of perennial ryegrass. The role of down-regulating OMT in altering the digestibility phenotype of transgenic tall fescue plants has already been demonstrated in tall fescue (Chen et al., 2004) with digestibility increased by approximately 10% in some plants, although variation of the effect on increased in digestibility was observed. Further understanding of the cause of this variation will lead to optimisation of transgenic breeding strategies and aid the development of routine phenotypic screening programs for the deployment of transgenic technologies to improve grass digestibility.

Development of AMV resistant white clover

Our most advanced application of molecular breeding of forages using genetic modification is the development of white clover that is immune to alfalfa mosaic virus (AMV) (Emmerling et al., 2004). AMV, white clover mosaic virus (WCMV) and clover yellow vein virus (CYVV) are members of the Bromoviridae, potexvirus group and Potyviridae respectively, and are estimated to cause combined losses to the Australian rural industries of more than $A800 million per year. Infections with these viruses result in reduced foliage yield, reduced nitrogen-fixing capacity and reduced vegetative persistence and can affect the production potential of white clover pastures by up to 30% (Campbell and Moyer, 1984; Dudas et al., 1998; Garrett, 1991; Gibson et al., 1981; Latch and Skipp, 1987; Nikandrow and Chu, 1991).

Even though potential sources of tolerance or resistance to AMV, CYVV or WCMV have been described in Trifolium species and Medicago sativa L. (Barnett and Gibson, 1975; Crill et al., 1971; Gibson et al., 1989; Martin et al., 1997; McLaughlin and Fairbrother, 1993), conventional breeding programs have not been successful. This is mostly due to limitations imposed by virus strain variability, lack of durability of natural resistance and barriers to interspecies sexual and/or somatic hybridisation. Chemical control of insect, fungal or nematode vectors is environmentally unacceptable and economically non-viable for forage legumes.

White clover was transformed with a binary vector carrying a chimeric gene for expression of a cDNA corresponding to AMV RNA4 using Agrobacterium mediated transformation (Ding et al., 2003). After selection, putative transgenic plants were analysed for the presence and copy number of the transgene by Southern hybridisation as well as levels of expression of the transgene by northern and western hybridisation analyses. The plants were clonally propagated and evaluated in a field trial in Hamilton, Victoria. Over a 2-year period, the plants proved to be immune to heavy natural aphid-mediated AMV challenges. White clover plants originating from 2 independent transformation events, H1 and H6, were chosen for the development of elite germplasm due to their high level of expression of the transgene (see Figure 2) and the high titre of the AMV coat protein (AMV-CP; data not shown) as well as their AMV immunity phenotype and agronomical performance during the field trials.

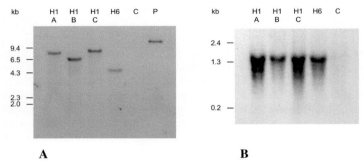

A   **B**

Figure 2 Molecular analysis of T_0 AMV-CP transgenic plants. A) Southern hybridisation analysis with genomic DNA isolated from white clover plants obtained from 4 independent transformation events. C indicates wild type (negative) control, P indicates plasmid (binary vector) control. B) Northern hybridisation analysis with RNA isolated from leaves of the same white clover plants. C indicates wild type (negative) control. Both blots were hybridised with an AMV CP cDNA probe.

The two selected transgenic AMV-resistant white clover lines, H1 and H6, were crossed with the parents of the white clover cultivar "Mink" and subjected to an elite germplasm development strategy designed to bring the transgene to homozygosity while minimising inbred depression (Kalla *et al.*, 2000). More than 8,000 T_2 offspring of these crosses were analysed by real time-PCR (RT-PCR), and a total of 1,300 plants homozygous for the AMV-CP transgene were identified, 888 derived from the H1 event, 412 from the H6 event.

A spaced plant field trial was subsequently established in Hamilton, Victoria, to evaluate the 1,300 transgenic white clover T_2 progeny (see Figure 3). The plants were assessed for virus infection with AMV four and five months after being established in the field. None of the transgenic plants showed any sign of virus infection whereas 28% of the non-transgenic wildtype control plants were infected with AMV.

An initial selection of agronomically superior plants comprised 179 H1-derived and 104 H6-derived elite transgenic clover plants. The selection was based on the basis of plant height, stolon density, leaf length, internode length, flower number, summer growth and survival, and autumn and spring vigour. During the second growth season, a further selection out of the initially selected plants led to the identification of 21 H1-derived and 16 H6-derived elite plants. These plants, resulting from the world's first breeding nursery for white clover, are the Syn_0 parents for the production of agronomically superior transgenic AMV-immune white clover elite cultivars. We are currently working to transfer this AMV immunity into other backgrounds, and to combine AMV immunity with non-transgenic sources of resistance to CYVV.

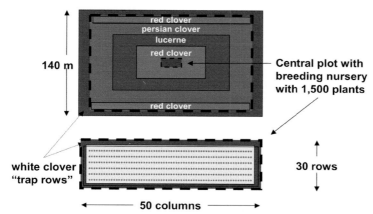

Figure 3 Layout of the spaced plant breeding nursery established in Hamilton, Victoria, to evaluate T_2 of spring of elite transgenic white clover lines homozygous for the AMV-CP transgene (GMAC PR64X2). Checks of non-transgenic control plants (cv. "Mink", total of 200 plants) are uniformly distributed among the 1,300 transgenic T_2 white clover plants.

Conclusions

The application of molecular breeding technologies in forage plant breeding is beginning to deliver on the promises of delivering novel genetic variation (e.g. AMV immune white clover) and more precise understanding of the nature of the genetic variation underlying key phenotypic effects (e.g. genetic variation in *Neotyphodium*, co-location of candidate genes and QTLs for forage quality and flowering time). Recent developments in genomics will greatly increase this genetic knowledge and provide candidate genes that are available for deployment in forage plant breeding. The development of robust phenotypic assays and molecular breeding strategies will ensure that these advances are efficiently captured and utilised to develop improved forage cultivars for the benefit of industry.

Acknowledgements

Parts of the original research described here were supported by the Department of Primary Industries, Victoria, Australia, CSIRO Plant Industry, Dairy Australia, Heritage Seeds Pty. Ltd, Meat and Livestock Australia, the Geoffrey Gardiner Dairy Foundation and the Molecular Plant Breeding CRC.

References

Alm, V., C. Fang, C.S. Busso, K.M. Devos, K. Vollan, Z. Grieg & O.A. Rognli (2003) A linkage map of meadow fescue (*Festuca pratensis* Huds.) and comparative mapping with other Poaceae species. *Theoretical and Applied Genetics,* 108, 25-40.

Armstead, I.P., L.B. Turner, I.P. King, A.J. Cairns & M.O. Humphreys (2002) Comparison and integration of genetic maps generated from F_2 and BC_1 –type mapping populations in perennial ryegrass. *Plant Breeding,* 121, 501-507.

Armstead, I.P., L.B. Turner, M. Farrell, L. Skøt, P. Gomez, T. Montoya, I.S. Donnison, I.P. King & M.O. Humphreys (2004). Synteny between a major heading date QTL in perennial ryegrass (*Lolium perenne* L.) and the *Hd3* heading date locus in rice. *Theoretical and Applied Genetics* 108, 822-828.

Barnett, O.W. & P B. Gibson (1975). Identification and prevalence of white clover viruses and the resistance of *Trifolium species* to these viruses. *Crop Science* 15, 32-37.

Barrett, B., A. Griffiths, M. Schreiber, N. Ellison, C. Mercer, J. Bouton, B. Ong, J. Forster, T. Sawbridge, G. Spangenberg, G. Bryan & D. Woodfield. (2004). A microsatellite map of white clover. *Theoretical and Applied Genetics,* 109, 596-608.

Bernard Vailhé, M.A., J.M. Besle, M.P. Maillot, A. Cornu, C. Halpin & M. Knight (1998). Effect of down-regulation of cinnamyl alcohol dehydrogenase on cell wall composition and on degradability of tobacco stems. *Journal of the Science of Food and Agriculture* 76, 505-514.

Campbell, C.L. & J.W. Moyer (1984). Yield responses of 6 white clover clones to virus infection under field conditions. *Plant Disease* 68, 1033-1035.

Casler, MD. (2001). Breeding forage crops for increased nutritive value. *Advances in Agronomy* 71, 51-107.

Chen, L., C.-K. Auh, P. Dowling, J. Bell, D. Lehmann & Z.-Y. Wang (2004). Transgenic down-regulation of caffeic acid *O*-methyltransferase (COMT) led to improved digestibility in tall fescue (*Festuca arundinacea*). *Functional Plant Biology* 31, 235-245.

Cogan, N.O.I., K.F. Smith, T. Yamada, M.G. Francki, A.C. Vecchies, E.S. Jones, G.C. Spangenberg, & J.W. Forster, (2005). QTL analysis and comparative genomics of herbage quality traits in perennial ryegrass (*Lolium perenne* L.). *Theoretical and Applied Genetics* 110, 64-380.

Crill P., E.W. Hanson & D.J. Hagedorn (1971). Resistance and tolerance to alfalfa mosaic virus in alfalfa. *Phytopathology* 61, 371-379.

Ding, Y.L., G. Aldao-Humble, E. Ludlow, M. Drayton, Y.H. Lin, J. Nagel, M. Dupal, G. Zhao, C. Pallaghy, R. Kalla, M. Emmerling & G. Spangenberg (2003). Efficient plant regeneration and *Agrobacterium*-mediated transformation in *Medicago* and *Trifolium* species. *Plant Science* 165, 1419-1427.

Diwan, N., J.H. Bouton, G. Kochert & P.B. Cregan (2000). Mapping of simple sequence repeat (SSR) DNA markers in diploid and tetraploid alfalfa. .....

Dudas B., D.R. Woodfield, P.M. Tong, M.F. Nicholls, G.R. Cousins, R. Burgess, D.W.R. White, D.L. Beck, T.J. Lough & R.L.S. Forster (1998). Estimating the agronomic impact of white clover mosaic virus on white clover performance in the North Island of New Zealand. *New Zealand Journal of Agricultural Research* 41, 171-178.

Dumsday, J.L., K.F. Smith, J.W. Forster & E.S. Jones (2003). SSR-based genetic linkage analysis of resistance to crown rust (*Puccinia coronata* f. sp. *Lolii*) in perennial ryegrass (*Lolium perenne*). *Plant Pathology* 52, 628-637.

Elkind, Y., R. Edwards, M. Mavandad, S.A. Hedrick, O. Riback, R.A. Dixon & C.J. Lamb (1990). Abnormal plant development and down-regulation of phenylpropanoid biosynthesis in transgenic tobacco containing a heterologous phenylalanine ammonia-lyase gene. *Proceedings of the National Academy of Science, USA* 87, 9057-9061.

Emmerling, M., K.F. Smith, P. Chu, R. Kalla & G. Spangenberg, (2004) Field evaluation of transgenic white clover with AMV immunity and development of elite transgenic germplasm. Molecular breeding of forage crops. Hopkins, A.; Wang, Z.-Y.; Mian, R.; Sledge, M.; Barker, R.E. (Eds.). Proceedings of the 3rd International Symposium, Molecular Breeding of Forage and Turf, Dallas, Texas, and Ardmore, Oklahoma, USA, May 18-22, 2003.

Faville, M.J., A.C. Vecchies, M. Schrieber, M C. Drayton, L.J. Hughes, E.S. Jones, K.M. Guthridge, K.F. Smith, T. Sawbridge, G.C. Spangenberg, G.T. Bryan & J.W. Forster (2004). Functionally associated molecular genetic marker map construction in perennial ryegrass (*Lolium perenne* L.). *Theoretical and Applied Genetics* 110, 12-32.

Garrett R.G. (1991). Impact of viruses on pasture legume productivity. In: Proceedings of Department of Agriculture, Victoria white clover conference. Pp 50-57.

Gibson, P.B., O.W. Barnett, H.D. Skipper & M.R. McLaughlin (1981). Effects of 3 viruses on growth of white clover. *Plant Disease* 65, 50-51.

Gibson, P.B., O.W. Barnett, M.R. Pedersen, M.R. McLaughlin, W.E. Knight, J.D. Miller, W.A. Cope & S.A. Tolin (1989). Registration of southern regional virus resistant white clover germplasm. *Crop Science* 29, 241-242.

Goff, S.A., D. Ricke, D, T.H. Lan, G. Presting, R. Wang, M. Dunn. *et al.* (2002) A draft sequence of the rice genome (*Oryza sativa* L. ssp. *japonica*). Science, 296, 92-100.

Hayward, M.D., J.W. Forster, J.G. Jones, O. Dolstra, C. Evans, N.J. McAdam, K.G. Hossain, M. Stammers, J.A.K. Will, M.O. Humphreys & G.M. Evans (1998). Genetic analysis of *Lolium.* I. Identification of linkage groups and the establishment of a genetic map. *Plant Breeding,* 117, 451-455.

Heath, R., H. Huxley, B. Stone & G. Spangenberg (1998). cDNA cloning and differential expression of three caffeic acid *O*-methyltransferase homologues from perennial ryegrass (*Lolium perenne* L.). *Journal of Plant Physiology,* 153, 649-657.

Isobe, S., I. Klimenko, S. Ivashuta, M. Gau & N.N. Kozlov (2003). First RFLP linkage map of red clover (*Trifolium pratense* L.) based on cDNA probes and its transferability to other red clover germplasm. *Theoretical and Applied Genetics,* 108, 105-112.

Jensen, L.B., J.R. Andersen, U. Frei, Y. Xing, C. Taylor, P.B. Holm & T. Lübberstedt (2005). QTL mapping of vernalization response in perennial ryegrass (*Lolium perenne* L.) reveals comparative relationships with other Poaceae species. *Theoretical and Applied Genetics,* available on-line.

Jones, E.S., N.L. Mahoney, M.D. Hayward, I.P. Armstead, J.G. Jones, M.O. Humphreys, I.P. King, T. Kishida, T. Yamada, F. Balfourier, G. Charmet & J.W. Forster (2002a). An enhanced molecular marker-based map of perennial ryegrass (*Lolium perenne* L.) reveals comparative relationships with other Poaceae species. *Genome,* 45, 282-295.

Jones, E.S., M.D. Dupal, J.L. Dumsday, L.J. Hughes & J.W. Forster (2002b). An SSR-based genetic linkage map for perennial ryegrass (*Lolium perenne* L.). *Theoretical and Applied Genetics*, 105, 577-584.

Jones, E., L. Hughes, M. Drayton, M. Abberton, T. Michaelson-Yeates, C. Bowen & J. Forster (2003). An SSR and AFLP molecular marker-based genetic map of white clover (*Trifolium repens* L.). *Plant Science* 165, 531-539.

Kalla, R., P. Chu & G. Spangenberg (2000) Molecular breeding of forage legumes for virus resistance. In: Spangenberg G (ed.). Molecular breeding of forage crops, Kluwer Academic Publishers, Dordrecht/Boston/London, p. 219-237.

Kulikova, O., R. Geurts, M, Lamine, D.J. Kim, D.R. Cook, J. Leunissen, h. de Jong, B.A. Roe & T. Bisseling (2004) Satellite repeats in the functional centromere and pericentromeric heterochromatin of *Medicago truncatula*. *Chromosoma*, 113, 276-283.

Langridge, P. & A.R. Barr (2003). Better barley faster: the role of marker assisted selection – Preface. *Australian Journal of Agricultural Research,* 54, i-iv.

Latch, G.C.M. & R.A. Skipp (1987). Diseases. In: White clover. Baker, M.J. & Williams, W.M. (eds.), CAB international, UK, pp. 421-446.

Lynch, D., A. Lidgett, R. McInnes, H. Huxley, E. Jones, N. Mahoney & G. Spangenberg (2002.) Isolation and characterisation of three cinnamyl alcohol dehydrogenase homologue cDNAs from perennial ryegrass (*Lolium perenne* L.). *Journal of Plant Physiology* 159, 653-660.

Marshall, D.R., P. Langridge & R. Appels (2001). Wheat breeding in the new century – Preface. *Australian Journal of Agricultural Research,* 52, i-iv.

McInnes R., A. Lidgett, D. Lynch, H. Huxley, E. Jones, N. Mahoney & G. Spangenberg (2002). Isolation and characterisation of a cinnamoyl-CoA reductase gene from perennial ryegrass (*Lolium perenne* L.). *Journal of Plant Physiology* 159, 415-422.

McLaughlin, M.R. & T.E. Fairbrother (1993) Selecting subclover for resistance to clover yellow vein virus. *Phytopathology* 83: 1421.

Martin, P.H., B.E. Coulman & J.F. Peterson (1997) Genetics to resistance to alfalfa mosaic virus in red clover. *Canadian J. Plant Sci.* 77: 601-605.

Nikandrow. A. & P.W.G. Chu (1991) Pests and diseases. The NSW experience. In: Proceedings White Clover Conference. Department of Agriculture Victoria, pp. 64-67.

Saha, M.C., R. Mian, J.C. Zwonitzer, K. Chekhovsky & A.A. Hopkins (2005). An SSR- and AFLP-based genetic linkage map of tall fescue (*Festuca arundinacea* Schreb.). *Theoretical and Applied Genetics*, available on-line

Sawbridge, T., E.-K. Ong, C. Binnion, M. Emmerling, R. McInnes, K. Meath, N. Nguyen, K. Nunan, M. O'Neill, F. O'Toole, C. Rhodes, J. Simmonds, P. Tian, K. Wearne, T. Webster, A. Winkworth & G. Spangenberg (2003a) Generation and analysis of expressed sequence tags in perennial ryegrass (*Lolium perenne* L.). *Plant Science*, 165, 1089-1100.

Sawbridge, T., E.-K. Ong, , C. Binnion, M. Emmerling, K. Meath, K. Nunan, M. O'Neill, F. O'Toole, J. Simmonds, K. Wearne, A. Winkworth & G. Spangenberg (2003b) Generation and analysis of expressed sequence tags in white clover (*Trifolium repens* L.). *Plant Science*, 165, 1077-1087.

Sewalt, V.J.H., W. Ni, H.G. Jung & R.A. Dixon (1997). Lignin impact on fiber degradation: Increased enzymatic digestibility of genetically engineered tobacco (*Nicotiana tabacum*) stems reduced in lignin content. *Journal of Agricultural and Food Chemistry* 45, 1977-1983.

Spangenberg, G., Z.-Y. Wang & I. Potrykus (1998). Biotechnology in forage and turf grass improvement. *In* R. Frankel *et al.* (eds.) Monographs on Theoretical and Applied Genetics. Volume 23, 10 chapters, Springer Verlag, Heidelberg.

Spangenberg, G., R. Kalla, A. Lidgett, T. Sawbridge, E.K. Ong & U. John (2001). Breeding forage plants in the genome era. 1-39. *In* G. Spangenberg. (ed.) Molecular breeding of forage and turf. Kluwer Academic Publishers, Dordrecht, Netherlands.

Stougaard, J. (2001) Genetics and genomics of root symbiosis. *Current Opinion in Plant Biology* 4, 328-335.

van Zijll de Jong, E., K.M. Guthridge, G.C. Spangenberg & J.W. Forster (2003). Development and characterization of EST-derived simple sequence repeat (SSR) markers for pasture grass endophytes. *Genome* 46, 277-290.

van Zijll de Jong, E., N.R. Bannan, J. Batley, K.M. Guthridge, G.C. Spangenberg, K.F. Smith & J.W. Forster (2004). Genetic diversity in the perennial ryegrass fungal endophyte *Neotyphodium lolii*. pp. 155-164. *In* A. Hopkins *et al.* (ed.) Molecular breeding of forage and turf. Kluwer Academic Publishers, Dordrecht, Netherlands.

van Zijll de Jong, E., K.F Smith, G.C. Spangenberg & J.W. Forster (2004) Molecular genetic marker-based analysis of the forage grass host – *Neotyphodium* endophyte interaction. *In* C.P. West *et al*. (ed.) *Neotyphodium* in Cool Season Grasses: Current Research and Applications, in press.

Xu, W.W., D.A. Sleper & S. Chao (1995). Genome mapping of polyploid tall fescue (*Festuca arundinacea* Schreb.) with RFLP markers. *Theoretical and Applied Genetics* 91, 947-955.

Yamada, T., E.S. Jones, N.O.I. Cogan, A C. Vecchies, T. Nomura, H. Hisano, Y. Shimamoto, K.F. Smith, M.D. Hayward & J.W. Forster (2004). QTL analysis of morphological, developmental, and winter hardiness-associated traits in perennial ryegrass. *Crop Science* 44, 925-935.

Ye, X.D., X.L. Wu, H. Zhao, M. Frehner, J. Nösberger, I. Potrykus & G. Spangenberg (2001). Altered fructan accumulation in transgenic *Lolium multiflorum* plants expressing a *Bacillus subtilis sacB* gene. *Plant Cell Reports* 20, 205-212.

Yu J., S. Hu, J. Wang, G.K.-S. Wong, S. Li, B. Liu *et al*. (2002) A draft sequence of the rice genome sequence (*Oryza sativa* L. ssp. *indica*). *Science*, 296, 79-91.

A computational pipeline for the development of comparative anchor tagged sequence (CATS) markers

L. Schauser[1], J. Fredslund[1], L. Heegaard Madsen[2], N. Sandal[2] and J. Stougaard[2]
[1]Bioinformatics Research Center, University of Aarhus, Ny Munkegade, Bldg. 540 8000 Aarhus Denmark
[2]Laboratory of Gene expression, Department of Molecular Biology, Aarhus University, Gustav Wieds Vej 10, DK-8000 Aarhus C, Denmark, Email: schauser@daimi.au.dk

Key points:

1. Molecular markers that allow the transfer of map information from one species to another are vital in comparative genetics.
2. To identify potential anchor marker sequences more efficiently, we have established a bioinformatic pipeline that combines multi-species EST- and genome- sequence data.
3. Taking advantage of information from a few related species, comparative EST sequence analysis identifies evolutionary conserved sequences in less well-characterised species in the same family.
4. Alignment of evolutionary conserved EST sequences with corresponding genomic sequences defines sets of PCR primer sites flanking introns.
5. Markers identified by this procedure will be readily transferable to other species since they are selected on the basis of their common evolutionary origin.
6. We exemplify our procedure on legumes and grasses, where model plant studies and the genome- and EST-sequence data available have a potential impact on breeding crop species.

Keywords: bioinformatics, expressed sequence tags, molecular markers, comparative anchor tagged sequences, polymorphism ascertainment

Introduction

Precise comparison of plant gene maps requires common anchor loci as landmarks for the alignment of conserved chromosomal segments. With the completion of the genomic sequences of Arabidopsis and rice, and large collections of ESTs at hand, comparative genome mapping carries the promise for rapid increase in knowledge about the large and repetitive genomes of many crop species. A common observation is conservation of linkage organization of homologous genes in species from diverse plant lineages. Comparative genome mapping allows the transfer of knowledge from one species to another related species. This information transfer can go two ways: (i) from well characterized model species with detailed genetic maps and / or complete genome sequence information to large genome crop species which are the target of breeding programs, and (ii) from the crop where quantitative trait loci (QTL) have been mapped, to a relevant model species where the gene content of this region is known. Map comparisons are hampered by the fact that genomes are not static in their arrangement, but often undergo chromosomal rearrangements, such as inversions, translocations, duplications, deletions and cycles of polyploidization followed by diploidization. Plants, given their sexual promiscuity and potential for vegetative reproduction are particularly prone to genome rearrangements (Bennetzen, 2000). For example, whole genome duplications have occurred at several occasions during the evolution of modern plant species (Paterson et al., 2004). In the diploid phase, members of a duplicated gene pair are retained or deleted at random in the two duplicated regions, obscuring the common past. This process results in problems with congruency between two genomes that are separated by a polyploidization-diplodization cycle. Hence, in order to succeed, any attempt of comparative genome mapping must carefully choose the species of comparison.

A central step in genome comparisons is the identification of sets of sequences that can readily be identified in the genomes of the species to be compared, and serve as "anchors" of their respective genetic maps. Commonly used markers, such as microsatellite or AFLP markers, can give high resolution genetic maps, but they are of little comparative value because they are not conserved across several species. Anchor sequences should be chosen such that they maximize the potential to serve as markers in several species, and also maximize congruency between the genetic maps of the organisms. Previous comparative maps haves relied on hybridisation of homologous probes and their scoring as RFLP markers (Fulton *et al.*, 2002, Draye *et al.*, 2001). Hybridization markers are time consuming, labour intensive and often involve the handling of radioactivity. Furthermore, it is not easy to generate specific hybridization markers, as they often cross-hybridize to other genomic regions. PCR based markers are much more efficient, as they are amenable to high throughput automation and, if well designed, of high specificity. Towards this goal we employ a strategy based on the identification of single copy number evolutionary conserved sequences within transcriptomes of representative species of the lineage under study. These sequences are used as PCR primer annealing sites for the amplification of intervening intronic sequences that are subject to subsequent polymorphism discovery. This approach ensures that unique, gene rich regions of the genome are the primary target of this effort.

Our bioinformatic approach is based on differences in the evolutionary rate of DNA changes in a genome. During evolution, many functional sequences, such as coding regions and regulatory elements, are under strong purifying selection. In contrast, intron sequences are less constrained and will display a higher degree of mutational variation between any two ecotypes / varieties. Although the evolutionary constraints on the exact sequence of the intron are relaxed, the position and approximate length of the intron is usually conserved, even over long evolutionary distances (Roy *et al.*, 2003). An automated primer-finding algorithm proposes primers pairs in regions of high conservation for the PCR amplification of intron sequences that have a high probability of capturing polymorphisms between varieties of any species within the clade under study. Subsequent sequencing of intron-spanning PCR products in mapping parents will reveal the presence of any polymorphism that can be used as a molecular marker. Here we present an automated pipeline for the generation of CATS and apply it to two plant lineages of major interest to agriculture: grasses and legumes.

Methods

The algorithm designed to identify conserved anchor tagged sequences (CATS, Lyon *et al.*, 1997) is best illustrated as a succession of three comparative filters and a primer-finding step. *1) Identifying expressed evolutionary conserved sequences (ECS) from different plant species.* Regions of strong homology between collections of ESTs from different species were identified. *2) Counting copy numbers in the Arabidopsis / rice proteome.* In order to avoid gene families and to get a score for the information content of an ECS, we counted the number of highly homologous sequences in the Arabidopsis /rice proteome. *3) ECS-genome alignment.* The presence and length of introns in reference genomes in scored at this step. Introns are highly conserved features, even among distant species. *4) Primer design* Multiple alignments of ECSs with indication of intron position are generated and primers are designed using this alignment as a guide. The order of application of the comparative filters does not influence the results and should hence be organized in a way that minimizes the computational cost.

Sequences: The EST clusters used for this analysis were retrieved from the Institute of Genome Research (TIGR). We downloaded the gene indices (clustered EST collections,

Quackenbush *et al.*, 2000, Pertea *et al.*, 2003) for legumes (*Lotus japonicus, Medicago truncatula* and *Glycine max*) and selected grasses (*Hordeum vulgare*, and *Sorghum bicolor*).

The Arabidopsis and rice (*Oryza sativa*) genome, proteome and coding sequences were downloaded from the TIGR FTP site. The *Lotus japonicus* and *Medicago truncatula* genomic sequences were retrieved using NCBIs ENTREZ.

The Blast package (Altschul *et al.,* 1997) was obtained from the NCBI. For comparison of nucleotide sequences, we used the megablast program with a wordsize of 20 and cutoff e-value of 2e-40. For DNA-protein comparisons, we used the blastx program (e-value 10e10-6). A series of Python scripts were generated to parse the Blast outputs and assemble sequence collections of ECS. Multiple alignments were generated by ClustalW (Chenna *et al.*, 2003), and automated primer design was achieved through application of the PriFi program (Fredslund *et al.,* manuscript in preparation).

DNA extraction, PCR amplification and sequencing of amplicons were performed using standard laboratory protocols.

Results

We modified the CATS algorithm (Lyons *et al.*, 1997) to reduce the potential pitfalls induced by gene families and paralogous copies. The filtering of sequences is divided into four operational steps and is best illustrated as a pipeline adding consecutive comparative selection criteria (Figure 1). The filters can be applied in any order without affecting the result.

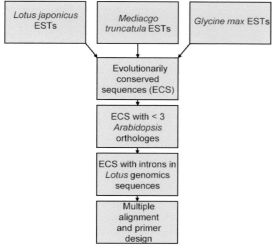

Figure 1 Pipeline of the marker candidate algorithm exemplified in legumes. In the first step, EST collections of selected species are compared. Evolutionary conserved sequences are passed on to the next step. Here the number of sequences with homology to the *Arabidopsis* reference proteome is estimated. Sequences with one or two homologues in the *Arabidopsis* proteome are considered because Arabidopsis has undergone a recent whole genome duplication. ECSs passing this criterion are compared to *Lotus* and *Medicago* genomic sequences and ranked according to overall length of the ECS and optimal length of introns. The ECSs are multiply aligned and primers are designed using this alignment as input.

Selecting species for the comparative approach in legumes and grasses

Our aim is to exploit colinearity between genomes of species with dense genetic maps and crops with important agronomic traits. In plants this colinearity erodes rather fast with phylogenetic distance. It is therefore crucial to choose the resources that allow maximal information transfer between species. Parameters that we considered include the amount of EST information and their phylogenetic relationship. For legumes, the resources originate from *Lotus japonicus*, *Medicago truncatula* and *Glycine max*. Their phylogenetic relationship is depicted in figure 2a. For the grasses we chose the species *Hordeum vulgare*, *Oryza sativa* and *Sorghum bicolor* (figure 2b). The CATS primers developed by our pipeline should amplify PCR products in all species within these clades. They may also be relevant to species outside these clades, such as peanut (*Arachis*) for legumes and banana (*Musa)* for grasses (not shown).

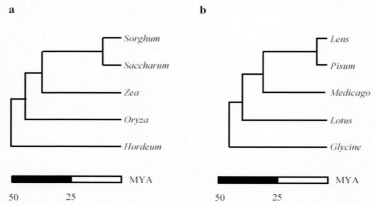

Figure 2 Phylogenetic relationship of the species in a) the grasses and b) the legumes. Species with sequence information used in this study together with selected other species are shown.

Selecting genes and primer design

Filter 1. Identifying expressed ECS from different representative species of the same lineage. We have here used the ready clustered EST collections downloadable from TIGR (gene indices) as input (Table 1), but any assembled collection of ESTs could serve as entry points. Stringent comparisons between different gene indices (using megablast with wordsize 20 and a cutoff e-value of 2e-40) revealed those sequences that display a high degree of conservation. ECSs represent exons that have been under strong purifying selection during evolution i.e. they display a higher-than-average conservation between species.

Table 1 TIGR Gene indices for the species used in this study

	Hordeum Vulgare	Sorghum bicolor	Medicago truncatula	Lotus japonicus	Glycine max
Number of GIs	50,453	39,148	36,976	28,460	63,676

Filter 2. Counting copy numbers in the proteome of a reference species (Arabidopsis for legumes, rice for grasses). In order to get a score for the information content of an ECS, we counted the number of highly homologous sequences encoded by the ECS in the reference species proteome. Repeated sequences are not useful for mapping purposes, since polymorphisms might reflect paralogous origin rather than allelic variation. Furthermore, allelic variation at a candidate marker locus can be partially or completely masked by the presence of paralogues, reducing the information content of this marker. Several rounds of genome duplication and gene family amplification have occurred prior to the split between Leguminosae (Rosid I) and Brassicacae (Rosid II), and also prior to the diversification of the grasses (Paterson *et al.*, 2004). The Arabidopsis genome has been subject to at least one round of duplication since the Rosid divide. The diploid legume species *Lotus japonicus* and *Medicago truncatula* do not seem to have undergone a similar duplication (Bowers *et al.*, 2003). Therefore counts of Arabidopsis genes can be taken as an overestimate of the legume count and we allow for two homologues in our pipeline. Some grasses (*Zea mays* and others) have undergone rounds of whole genome duplications, and care must be taken when assessing co-linearity of the genetic maps.

Filter 3. ECS-genome alignment. In order to maximize chances of detecting polymorphisms at later steps, introns interrupting a given ECS are scored by aligning the ECSs with corresponding genomic sequences. For grasses, we assess the presence of introns by comparing rice ECS sequences with the rice genome. For legumes we make use of the *Lotus japonicus* and *Medicago truncatula* genome sequences, but this could easily be extended to the Arabidopsis genomic sequence. This would still be informative, since the presence, location and approximate length of introns are highly conserved features, even among distantly related species (Roy *et al.*, 2003). We also score the length of introns. This quantity is of interest for two reasons: (i) short introns are less likely to be polymorphic than longer ones, and thus longer introns are of interest and (ii) the final PCR reaction using degenerate primers is limited to the maximum amplicon size of ~3 kb using standard polymerases.

Filter 4. Primer design. Multiple alignments of ECSs with indication of intron position are generated and forward and reverse primers are designed using this alignment as a guide. A number of criteria are scored which have to do with the number of species in the alignment the melting temperature and GC content of the proposed primer and the length of the intron(s) which separates two primers. A conservation score reflects the degree of similarity between the most evolutionarily divergent species in the alignment. Finally a score is given for the distance from primer site to the exon-intron junction. This score is introduced as a means of selecting primers that allow the identification of the PCR product as being derived from a homologous locus in a subsequent sequencing step. A combined score for each primer pair allows their comparison and ranking within and between candidate regions.

Legumes

The collections of gene indices (preclustered EST collections) for Legumes were downloaded from TIGR (Table 2). In order to estimate the number of homologues in Arabidopsis, these sequences were compared to the proteome of Arabidopsis. Since Arabidopsis has undergone a recent whole genome duplication, we considered sequences with both one and two hits in the Arabidopsis proteome (Table 2). Most gene indices have several (>2) Arabidopsis homologues. Next, we tested for the presence of introns in the corresponding genomic regions of *Lotus* (122 Mbp of genomic sequence) and *Medicago* (143 Mbp of genomic sequence). If no corresponding genomic sequence was identified, we ignored the gene index. The fractions of gene indices which have genomic regions sequenced is slightly higher in *Lotus* than in *Medicago*, indicating that the *Lotus* genome project covers more genes than the *Medicago* genome project (18% vs.13%)

Table 2 Information content of the collections of gene indices. The species name is followed by the Release version (in parenthesis), and the number of gene indices (clustered ESTs and singleton ESTs) are indicated for each species. These sequences are binned according to the count of homologous genes in Arabidopsis. The numbers in parenthesis indicate the size of the subset with an intron in the respective genomic sequence: *Lotus* GIs were compared to *Lotus* genomic sequences, whereas *Medicago* GIs were compared to *Medicago* genomic sequences.

	Lotus japonicus (v. 3.0)	*Medicago truncatula* (v. 7.0)	*Glycine max* (v. 12.0)
Number of gene indices	28,460	36,976	63,676
One Arabidopsis homolologe	2,282 (394)	3,088 (397)	4,281
Two Arabidopsis homologes	1,606 (306)	2,151 (265)	3,349

Next, we compared the gene indices with intron information to the other EST collections. It is striking, that if a sequence is found in both *Lotus* and *Medicago*, it has a very high probability of being present in Glycine as well. There was about 25% redundancy between the sequences identified through *Lotus* and *Medicago* genomic information (introns), reflecting the unfinished state of the two genomes. These collections of three sequences were the basis for a ClustalW multiple sequence alignment followed by our CATS primer finding algorithm.

Table 3 Comparative CATS identification. Sequences with introns and one or two Arabidopsis homologues were successively compared to the EST collections of the other legumes, generating sets of three sequences. These sets of sequences were the basis for ClustalW multiple sequence alignment and an automated CATS primer finding algorithm.

Query sequences:	Compared to:		Number of CATS primer pairs identified
Lotus japonicus	*Medicago* only	*Medicago* & *Glycine*	
GIs with one Arabidopsis homologue and intron information	288	269	48
GIs with two Arabidopsis homologues and intron information	186	166	22
Medicago truncatula	*Lotus* only	*Lotus* and *Glycine*	
GIs with one Arabidopsis homologue and intron information	220	207 (57)	22
GIs with two Arabidopsis homologues and intron information	128	118 (27)	18

Grasses

When the pipeline was applied to the grasses by simply exchanging all the relevant data files, we were able to identify 1335 CATS primer pairs. The selected gene indices originated from *Hordeum vulgare* (Release 9.0) and *Sorghum bicolor* (Release 8.0). These were compared to the rice (*Oryza sativa*) genome and annotated CDS.

Testing CATS primers

We tested the potential of our pipeline to generate CATS markers by randomly choosing 36 of the legume CATS primer sets and attempting to develop them as markers in the legumes *Phaseolus vulgare* (common bean) and *Arachis* (Peanut spp.). 70 % of these primer sets amplified the correct product in the relatively closely related *P. vulgare*, whereas this figure dropped to 62 % in the outgroup species *Arachis*. Of these, up to 90% were polymorphic, depending on the mapping parents used.

Discussion

In this presentation we exploit evolutionary conserved sequences for developing molecular markers useful as anchors when comparing genetic maps of different species. Our goal is to use these sequences as conserved and unique sites for primer annealing. A pair of such sites can then be used for amplifying intervening intronic sequences that subsequently can be scanned for polymorphisms distinguishing breeding varieties or ecotypes. We have shown that our automated bioinformatic algorithm is a versatile tool allowing the quick generation of marker candidates useful for map construction projects in legumes and grasses and by extension, to any phylogenetic clade with appropriate comparative sequence information.

Since only unique sequences are useful as markers, we are interested in the number of paralogous sequences in the genome. An approximation to this number is obtained by counting homologous sequences in the proteome of a reference species. Strictly speaking, we are not able to discern between orthologous and paralogous origin of homologous sequences. However, for those sequences with only one homologue in the Arabidopsis / rice genome, we can reasonably assume orthology. Although this criterion maximises congruency when comparing maps, it by no means guarantees it. Both clades studied here have a common ancestor that at some point has undergone a whole-genome duplication (Paterson *et al.*, 2004), potentially obstructing colinearity through differential gene loss. The degree of microsynteny depends on the timing of the duplication event relative to the most recent ancestor of the clade. In any case, our selection filters out genes that are prone to duplication and hence are members of gene families.

A main application of this algorithm is the transfer of genome information between model species and closely related large genome crops. In plant breeding programs traits of economic importance are screened out of large populations. Breeders introduce variation through crosses between varieties and in some cases also wild relatives but there is rarely a simple method for following the segregation of the trait or allele of interest. Instead of screening for the traits per se, which can be difficult to score due to environmental conditions, late onset or small contributions to the phenotype, breeders often use linked markers as indicators of inheritance. For this purpose, molecular markers are best suited, since they can be co-dominant, cheap and readily scored. Dense genetic maps spanning all linkage groups are of invaluable help for breeding purposes. If dense genetic maps are not available for the species at hand, comparison with other species can help in the development of new markers and qualified guesses at candidate genes in the region under investigation. It is therefore generally advisable to initially map markers that can serve as anchors, connecting genetic maps of as many species as possible. The phylogenetic distances that limit such an approach should be considered. Most macrosyntenic information is lost between evolutionary diverse lineages. This information loss is dramatically enhanced when a whole genome duplication event has occurred in one of the lineages. For example, no macrosynteny is recognizable between *Medicago truncatula* and Arabidopsis (Zhu *et al.*, 2003). On the other hand, microsynteny is generally much better conserved (Zhu *et al.*, 2003, see also Krusell *et al.*, 2002). Within a given clade, such as legumes or grasses, both micro- and macrosynteny are well-conserved (Choi *et al.*, 2004, Bennetzen, 2000, Draye *et al.*, 2001).

As for any marker, the success of applying our pipeline depends on the variation between any two varieties used for mapping. We have shown that the pipline produces valid marker candidates when applied to outgroups of the clade under consideration, as in the *Arachis* example. For related legumes the pipeline should be of value as a tool to bridge the genetic maps of model and crop legumes. The density of CATS is not very high in legumes. This could be changed by lowering the requirements to consider pairwise comparisons, instead of comparisons between three species. Another improvement could be gained when looking for introns. Here, we have here only exploited the incomplete genome sequence information of *Lotus japonicus* and *Medicago truncatula*. Given the observed conservation of intron positions it should even be possible to use the Arabidopsis genome as a reference, a strategy which has recently been employed by Choi *et al.* (2004). Thus we should be able to generate more than a thousand CATS marker candidates for any legume cross in the near future. The comparative approach described here is broadly applicable to all EST resources collected from species with appropriate phylogenetic distance and reference genome information. The phylogenetic distance and the amount of sequence information for the species chosen will determine success. For grasses, we found 1335 CATS primer pairs, illustrating this point. When developed as markers and mapped in several species, these could add considerable density to existing comparative mapping databases such as Gramene (Ware *et al.*, 2002).

Conclusion

Our automated bioinformatic pipeline for the generation of CATS is an efficient approach to generating anchor markers, allowing rapid information transfer between traits of interest to the breeders and the dense genetic maps of model plants.

References

Altschul SF, Madden TL, Schaffer AA, Zhang J, Zhang Z, Miller W, Lipman DJ. (1997) Gapped BLAST and PSI-BLAST: a new generation of protein database search programs. Nucleic Acids Res. 25:3389-402.
Bennetzen JL. (2000) Comparative sequence analysis of plant nuclear genomes: microcolinearity and its many exceptions. Plant Cell. 12:1021-9.
Bowers JE, Chapman BA, Rong J, Paterson AH. (2003) Unravelling angiosperm genome evolution by phylogenetic analysis of chromosomal duplication events. Nature. 422(6930):433-8.
Chenna R, Sugawara H, Koike T, Lopez R, Gibson TJ, Higgins DG, Thompson JD. (2003) Multiple sequence alignment with the Clustal series of programs. Nucleic Acids Res. 31:3497-500.
Choi HK, Mun JH, Kim DJ, Zhu H, Baek JM, Mudge J, Roe B, Ellis N, Doyle J, Kiss GB, Young ND, Cook DR. (2004) Estimating genome conservation between crop and model legume species. Proc Natl Acad Sci U S A. 101:15289-94.
Draye X, Lin YR, Qian XY, Bowers JE, Burow GB, Morrell PL, Peterson DG, Presting GG, Ren SX, Wing RA,
Fedorov A, Roy S, Fedorova L, Gilbert W. (2003) Mystery of intron gain. Genome Res. 13:2236-41

Fulton TM, Van der Hoeven R, Eannetta NT, Tanksley SD. (2002) Identification, analysis, and utilization of conserved ortholog set markers for comparative genomics in higher plants. Plant Cell 14:1457-67.

Krusell L, Madsen LH, Sato S, Aubert G, Genua A, Szczyglowski K, Duc G, Kaneko T, Tabata S, de Bruijn F, Pajuelo E, Sandal N, Stougaard J. (2002) Shoot control of root development and nodulation is mediated by a receptor-like kinase. Nature 420:422-6.

Lyons, L. A. T. F., Laughlin, N. G. Copeland, N. A. Jenkins, J. E. Womack, S. J. O'Brien (1997) Comparative anchor tagged sequences (CATS) for integrative mapping of mammalian genomes. Nature Genetics 15, 47 - 56.

Paterson AH. (2001) Toward integration of comparative genetic, physical, diversity, and cytomolecular maps for grasses and grains, using the sorghum genome as a foundation.
Plant Physiol. 125:1325-41.

Paterson, AH., J. E. Bowers and B. A. Chapman (2004) Ancient polyploidization predating divergence of the cereals, and its consequences for comparative genomics. PNAS 101, 9903-9908

Pertea G, Huang X, Liang F, Antonescu V, Sultana R, Karamycheva S, Lee Y, White J, Cheung F, Parvizi B, Tsai J, Quackenbush J. (2003) TIGR Gene Indices clustering tools (TGICL): a software system for fast clustering of large EST datasets. Bioinformatics. 19:651-2.

Quackenbush J, Cho J, Lee D, Liang F, Holt I, Karamycheva S, Parvizi B, Pertea G, Sultana R, White J. (2001) The TIGR Gene Indices: analysis of gene transcript sequences in highly sampled eukaryotic species. Nucleic Acids Res. 29:159-64.

Zhu H, Kim DJ, Baek JM, Choi HK, Ellis LC, Kuester H, McCombie WR, Peng HM, Cook DR. (2003) Syntenic relationships between *Medicago truncatula* and Arabidopsis reveal extensive divergence of genome organization. Plant Physiol. 131:1018-26.

Ware DH, Jaiswal P, Ni J, Yap IV, Pan X, Clark KY, Teytelman L, Schmidt SC, Zhao W, Chang K, Cartinhour S, Stein LD, McCouch SR. (2002) Gramene, a tool for grass genomics. Plant Physiol. 130:1606-13.

Future directions in the molecular breeding of forage and turf

G.C. Spangenberg[1,3], J.W. Forster[1,3], D. Edwards[1,3], U. John[1,3], A. Mouradov[1,3], M. Emmerling[1,3], J. Batley[1,3], S. Felitti[1,3], N.O.I. Cogan[1,3], K.F. Smith[2,3] and M.P. Dobrowolski[2,3]

[1]*Primary Industries Research Victoria, Plant Biotechnology Centre, La Trobe University, Bundoora, Victoria 3086, Australia, Email: german.spangenberg@dpi.vic.gov.au*
[2]*Primary Industries Research Victoria, Hamilton Centre, Mount Napier Road, Hamilton, Victoria 3300, Australia*
[3]*Molecular Plant Breeding Cooperative Research Centre, Australia*

Key points

1. Molecular breeding of forage and turf plants and their endosymbionts has entered the post-genomic era with a large amount of structural genomics information and genomic resources available for key forage and turf species and relevant model systems.
2. A primary future challenge is the conversion of this information into useful functional knowledge for the development of molecular breeding technologies and products that address a range of high impact outcome scenarios in forage and turf.
3. High-throughput approaches for spatial and temporal analysis, from genome to phenome, and the respective data integration in a systems biology context will be critical for the establishment of stringent gene-function correlations.
4. Translational genomics will permit results obtained using model systems to have major impact on the understanding of the molecular basis of plant processes and direct application to the molecular breeding of forage and turf plants.
5. These developments will be enhanced through applications of transgenesis and functionally-associated genetic markers in forage and turf molecular breeding building on genomic and post-genomic discoveries in these target species.

Keywords: systems biology, functional and translational genomics, bioinformatics, transgenesis, candidate gene based marker systems

Introduction

Major advances in genome technologies have revolutionised biology. High-throughput gene discovery through expressed sequence tag (EST) sequencing and genome shotgun sequencing provides a powerful tool for the analysis of biological systems at different hierarchical levels, from organism to ecosystem (or, from genomes to biomes). As well as providing access to sequence information from the transcribed component of genomes, EST-derived resources also provide the basis for the analysis of gene expression, genetic diversity and molecular evolution. Large-scale EST discovery has been performed for key forage and turf grasses and legumes allowing the generation of unigene cDNA microarrays for transcriptome analysis and the development of functionally-associated molecular marker systems. The analysis of genomic and post-genomic data and the integration of information from the related fields of transcriptomics, proteomics, metabolomics and phenomics has been facilitated by developments in bioinformatics. This has enabled the identification of genes and gene products, and the elucidation of functional relationships between genotype and phenotype, thereby allowing a system-wide analysis from genome to phenome. The effective application of systems biology approaches to forage and turf plants will close gaps in our understanding of the underlying genetics, physiology and biochemistry of many complex plant pathways and developmental processes. This is likely to advance progress in many applications of transgenesis in forage and turf plant improvement, building on gene technology as a powerful tool for the generation of the required functional genomics knowledge. It will further require improved transformation methodologies for the development of 'market-ready' transformation events of likely 'pyramided' gene technologies following a sensible choice of high impact targets. Representative examples of these developments for perennial ryegrass (*Lolium perenne* L.), white clover (*Trifolium repens* L.), and grass fungal endophytes of the genera *Epichloë* and *Neotyphodium* are outlined in this chapter. Analogous developments in the forage model system, barrel medic (*Medicago truncatula* Gaertn.), are described elsewhere in this volume.

Systems biology and translational genomics for forage and turf molecular breeding

Genomics

High-throughput gene discovery by EST sequencing, initiated in 1991 (Adams *et al.*, 1991), led to the development of significant genomic resources for a range of species, including the chief temperate forage and turf grasses such as perennial ryegrass (Sawbridge *et al.*, 2003a), forage legumes such as white clover (Sawbridge *et al.*, 2003b), and forage and turf plant endosymbionts such as *Epichloë* and *Neotyphodium* grass fungal endophytes (Felitti *et al.*, 2004; Spangenberg *et al.*, 2005). It has also established bioinformatics requirements for large and searchable sequence databases including tools for data analyses, sequence annotation and mining of complex interacting datasets. Although EST sequencing is still the standard procedure for gene discovery in many crops, a reduction in the cost of DNA sequencing has led to a move towards whole-genome sequencing and sequencing of gene-rich genomic regions.

Plant genomics was revolutionised by the release of the complete *Arabidopsis thaliana* genome sequence by the *Arabidopsis* Genome Initiative in 2000, four years ahead of schedule (The Arabidopsis Genome Initiative, 2000). Two years later, the completion of the rice (*Oryza sativa* L ssp. *japonica* Nipponbare) genome sequence by public consortia was announced. This research was complemented by rice sequencing work undertaken by the agribusinesses Monsanto (Barry, 2001) and Syngenta (Goff *et al.*, 2002), and a separate research project at the Beijing Genomics Institute that sequenced the rice subspecies *indica* (Yu *et al.*, 2002). Particularly relevant for translational genomics in forage legumes have been the ongoing efforts to completely sequence of gene-rich genomic regions of the model legume *Medicago truncatula* (Roe & Kupfer, 2004) and the development of other integrated *Medicago* functional genomics resources for improvement of alfalfa (May, 2004; this volume). Due to the anticipated similarities at the genomic level between the model plant species arabidopsis, barrel medic and rice and important forage crops such as white clover, alfalfa and perennial ryegrass, completion of these genome sequences is expected to have significant future impact on forage and turf molecular breeding. In this context, bioinformatics developments will play a fundamental role in enabling translational genomics for forage and turf plants, by translating and allowing exploitation of genomic information from the model to the agriculturally relevant species. Examples of such developments in gene discovery and associated establishment of bioinformatics tools and databases for representative forage and turf species are outlined below.

A resource of 44,524 perennial ryegrass ESTs was generated from single pass DNA sequencing of randomly selected clones from 29 cDNA libraries that represent a range of plant organs and developmental stages (Sawbridge *et al.*, 2003a). EST redundancy was resolved through assembly with CAP3, identifying a total of 12,170 ryegrass unigenes within this dataset. Similarly, a resource of 42,017 white clover ESTs from 16 cDNA libraries and corresponding to 15,989 unigenes was established (Sawbridge *et al.*, 2003b). Each of these sequences has been annotated by comparison to GenBank and SwissProt public sequence databases and automated intermediate Gene Ontology (GO) annotation has also been determined. All sequences and annotation are maintained within ASTRA format MySQL databases, with web-based access for text searching, BLAST sequence comparison and GO hierarchical tree browsing. Each ryegrass sequence was mapped onto an EnsEMBL genome viewer for comparison with the complete genome sequence of rice and expressed sequences from related species. Similarly, each white clover sequence was mapped onto an EnsEMBL genome viewer for comparison with the complete *A. thaliana* genome sequence and expressed sequences from related legumes. Analogous developments in forage and turf symbio-genomics, e.g. large-scale gene discovery in grass endophytes (Spangenberg *et al.*, 2001; 2005) have led to the production of 13,964 ESTs collectively from the grass endophytes *Neotyphodium coenophialum, N. lolii* and *Epichloë festucae* thus identifying at total of 7,585 unigenes (Felitti *et al.*, 2004; Spangenberg *et al.*, 2005).

Complementing these genomic resources, large insert genomic libraries of white clover (50,302 BAC clones with 101 kb average insert size, corresponding to 6.3 genome equivalents and 99% genome coverage), perennial ryegrass (50,304 BAC clones with 113 kb average insert size, corresponding to 3.4 genome equivalents and 97% genome coverage) and the ryegrass fungal endophyte *N. lolii* (6,000 BAC clones with 120 kb average insert size and 15-fold genome coverage) have been established for physical mapping, map-based cloning, novel gene and promoter discovery.

The availability of large sequence datasets permits mining for biological features, such as single nucleotide polymorphism (SNP) (Barker *et al.*, 2003; Batley *et al.*, 2003; Somers *et al.*, 2003) and simple sequence repeat (SSR) (Robinson *et al.*, 2004) molecular markers, that can be then applied to molecular breeding research such as genetic trait mapping and the development of functionally-associated genetic markers (Morgante *et al.*, 2003; Andersen & Lübberstedt, 2003). The representative examples of EST resources for forage and turf molecular breeding described above, led to the identification of 3,214 ryegrass, 5,407 white clover, and 1,047 endophyte SSR molecular markers using SSRPrimer and to the design of associated specific PCR amplification primers. In addition, AutoSNP permitted the identification of: 1,817 candidate SNPs and 1,706 insertion-deletion (InDel) molecular markers across 1,409 white clover loci; 2,716 candidate SNPs and 345 InDels in 493 perennial ryegrass loci; and 1,636 candidate SNPs and 326 InDels across 300 grass endophyte loci. The integration of different analysis tools and the resulting data within central resources permits researchers to identify novel candidate genes and associated molecular genetic markers that can be applied to functional genomics and germplasm enhancement in these forage and turf plants and their endosymbionts.

The availability of complete-genome sequences enables further mining for novel promoter sequences (Qui, 2003; Ettwiller *et al.*, 2003) and other regulatory features such as micro-RNAs (Rhoades *et al.*, 2003; Nelson *et al.*, 2003). This tertiary level annotation provides links to both the phenotype and the complex regulatory mechanisms that govern development and response to the environment (Edwards & Batley, 2004).

Transcriptomics

Knowledge of the expression pattern of genes provides a valuable insight into gene function and role in determining the observed heritable phenotype. The application of microarrays and sequence-based methods to transcriptome analysis and expression profiling has added an extra dimension to current genomic data and has founded several statistics-based disciplines within bioinformatics. Owing to their extended linear dynamics, sequence-based methods have the potential to determine more accurately quantitative levels of gene expression. Furthermore, they do not require prior sequence information and so have the advantage of being able to identify novel genes or to assess gene expression in uncharacterized plant species. With the increasing scale of EST sequencing efforts, it is becoming possible to mine these datasets to estimate expression information (Rafalski *et al.*, 1998), although this remains more a by-product of EST discovery than a true transcriptome analysis tool.

While the predominant methods for sequence-based expression analysis are serial analysis of gene expression (SAGE) (Velculescu *et al.*, 1995) and massively parallel signature sequencing (MPSS) (Brenner *et al.*, 2000a, 2000b), hybridisation-based microarrays have become the transcriptomic tool of choice. High–density cDNA and oligonucleotide microarrays represent powerful tools for transcriptome analysis to gain an understanding of gene expression patterns for thousands of genes. The rapid implementation of microarrays has been followed by a growth in the bioinformatics of microarray data analysis (Moreau *et al.*, 2003; Goodman, 2002). Microarray technology continues to expand: cDNA arrays are being produced for gene expression analysis in many plant species, and complete oligonucleotide-based unigene arrays have been developed for major crop plants such as rice, wheat and barley.

Internationally coordinated efforts in transcriptome analyses and sharing of microarray resources will benefit the advancement of our understanding of gene function in forage and turf species. In this context, high-density cDNA microarrays representing approximately 15,000 unique genes for each of perennial ryegrass and white clover (Sawbridge *et al.*, 2003a; b) and unigene microarrays allowing the interrogation of over 5,000 *Neotyphodium* and *Epichloe* genes (Nchip™ microarray and EndoChip™ microarray) (Felitti *et al.*, 2004; Spangenberg *et al.*, 2005) have been developed. These microarrays have been applied in hybridisations with labelled total RNA isolated from a variety of genotypes, plant organs, developmental stages, and growth conditions. The collated results enabled validation of functions predicted through comparative sequence annotation and suggested roles for novel genes lacking comparative sequence annotation. This data allowed expression analysis of genes associated with selected metabolic pathways and developmental processes, to dissect these at the transcriptome level, and to identify novel genes co-regulated with template genes known to be involved in these processes. Furthermore, these microarrays have enabled applications for gene and promoter discovery when used in concert with BAC libraries established for each of these target species. An *International Transcriptome Initiative for Forage and Turf* (ITIFT) to facilitate international efforts in microarray-based transcriptome analyses for key forage and turf plants and their endosymbionts is proposed here. Support of ITIFT will be provided through contributed access to and transcriptional profiling with unigene microarrays for key forage and turf species, namely perennial ryegrass, white clover and grass endophytes (*Neotyphodium*/*Epichloe*), as well as access to a range of platforms for microarray spotting, hybridisation and scanning operationally integrated through a Scierra laboratory workflow system. An ITIFT database with a web-based front-end portal for secure access by the research community to appropriate data and information will be developed and maintained.

Proteomics

The term 'proteomics' was coined in the mid 1990s (Wilkins *et al.*, 1996, 1997) on the back of the success of 'genomics' and has since come to incorporate many aspects of protein biochemistry. Knowledge of the structure and function of every protein would revolutionise the field of proteomics. Proteomics has significant prospects for advancing our understanding in plant biotechnology owing to its direct relationship with gene and transcript data. Proteomes also have a strong influence on the measured phenotype of the plant, either directly through protein content or function or indirectly through the relationship of a protein with the metabolome. The potential for bioinformatics to structure and integrate –omic data, therefore, relies on an ability to model both the proteome and its interactions (Edwards & Batley, 2004). With the first reports on comprehensive proteome analysis in model plant systems now becoming available (Canovas et al., 2004), there is still substantial scope for analogous developments in forage and turf plants and significant challenges ahead in the translation of complete genome DNA sequence data into protein structures and predicted functions.

Metabolomics

Like proteomics, metabolomics was derived from the field of biochemistry and involves the analysis (usually high-throughput or broad scale) of small-molecule metabolites. The foundations of metabolomics are descriptions of biological pathways, and current metabolomic databases such as Kyoto Encyclopedia of Genes and Genomes (Kanehis et al., 2000, 2002) are frequently based on well-characterized pathways. Metabolomics might be considered to be the key to integrated systems biology because it frequently provides direct measure of the desired phenotype (Fiehn, 2002), with relevance to both quantitative and qualitative traits. Moreover, metabolomes can be correlated with genetics through proteomes, transcriptomes and genomes and complement the more traditional quantitative trait locus (QTL) approach to molecular breeding. One of the challenges for bioinformatics will be the structuring and integration of these diverse types of data for the emerging field of systems biology (Fernie, 2003; Sumner et al., 2003; Weckworth, 2003). Emerging applications of metabolome analysis in forage and turf are currently focused primarily on *M. truncatula* ecotypes and elicited cell cultures (Sumner et al., 2003; May, 2004), as well as on metabolic fingerprinting and metabolic profiling of the mutualistic interaction between perennial ryegrass and its fungal endophyte within the context of a spatial and temporal systems biology approach (Spangenberg et al., 2005).

Bioinformatics, phenomics and data integration from genome to phenome

Bioinformatics arose from the need to structure and to interrogate the ever-increasing quantity and types of biological data that is generated by the developing '–omic' technologies. Genomics has spawned a plethora of related –omics terms that frequently relate to established fields of research. Of these terms, 'phenomics', the high-throughput analysis of phenotypes, has probably the biggest application in molecular breeding. Phenomics platforms using large scale detailed phenotypic data collection strategies based on series of distinct growth stages that describe the entire plant life cycle have been developed in recent years by agribusinesses for *A. thaliana* (e.g. Paradigm Genetics; K. R. Davies, pers. comm.) and rice (e.g. CropDesign; W. Broekaert, pers. comm.). As these technologies continue to grow, so does the need for interrogation across the various types of data and scientific disciplines. Precise data integration requires the formal annotation of data with relational terms, and this is an essential driver behind applications of bioinformatics in the development of systems biology (Edwards & Batley, 2004). Recent bioinformatics developments for systems biology in forage and turf plants have primarily focused on tools for data integration and target selection in *M. truncatula* (Wang & Zhang, 2004), perennial ryegrass and white clover (Love et al., 2005).

Transgenesis for forage and turf molecular breeding

Transgenesis offers the opportunity to generate unique genetic variation for molecular breeding of forage and turf, specifically when pre-existing variability is either absent or shows very low levels of heritability (Spangenberg *et al*., 2001). Transgenic forage and turf plants with simple 'engineered' traits have been generated and characterised over the last decade (Hartman *et al*., 1994; Austin-Phillips and Ziegelhoffer, 2001; Kalla *et al*., 2001; Wang *et al*., 2001; Xu *et al*., 2001; Chen *et al*., 2003; Chen *et al*., 2004; Luo *et al*., 2004; Petrovska *et al*., 2004), with few selected transformation events having already reached the stage of field-evaluation and transgenic germplasm development (Harriman *et al*., 2003; Emmerling *et al*., 2004; Smith *et al*., 2005).

In the near future, applications of transgenesis to forage and turf plant improvement are likely to remain primarily focused on the development of transformation events with unique genetic variation, on studies on the molecular genetic dissection of key developmental processes (e.g. leaf senescence, transition to flowering, pollen development), metabolic pathways (e.g. lignin biosynthesis, fructan metabolism, fatty acid biosynthesis, proanthocyanidin biosynthesis), and on the assessment of candidate genes for enhancing tolerance to biotic and abiotic stresses.

An emphasis on key target traits for the application of transgenesis to forage and turf improvement will be retained for forage quality, disease and pest resistance, nutrient acquisition efficiency, tolerance to abiotic stresses and the manipulation of growth and development. Corresponding experimental approaches and representative examples in temperate forage and turf grasses and forage legumes have been reviewed in detail over recent years (Spangenberg *et al*., 1998; 2001; 2005; Wang *et al*., 2001; Gruber *et al*., 2001; Humphreys & Abberton, 2004; Dixon, 2004) and are consequently not considered in any further detail here.

Increasingly, transgenesis applications in forage and turf improvement will address outcome scenarios associated with animal health and welfare, such as transgenic white clover and alfalfa with 'bloat safety' through foliar accumulation of proanthocyanidins, transgenic alfalfa with foliar expression of foot and mouth disease antigens, as well as environmental and human health, such as transgenic ryegrasses with down-regulation of main pollen allergens, transgenic forage grasses and legumes with modified fatty acid biosynthesis for animal products with enhanced levels of 'healthier fats', transgenic turf grasses for bioremediation applications, transgenic alfalfa with salt tolerance, and transgenic white clover with enhanced phosphorus acquisition efficiency.

As described above, significant advances in gene discovery for key forage and turf plants and their endosymbionts (Sawbridge *et al*., 2003a; 2003b; Felitti *et al*., 2004), and opportunities arising from translational genomics derived from EST discovery and sequencing of gene-rich regions of relevant model systems such as *M. truncatula* (May, 2004; Roe & Kupfer, 2004) and whole genome sequencing of rice (Goff *et al*., 2002; Yu *et al.,* 2002) have established an overwhelming amount of information concerning the physical structure of thousands of specific genes. This has led to the requirement of efficient experimental approaches for high-throughput functional analysis of the large number of genes for which limited or no functional information is available.

While robust methodologies for genetic transformation of key forage and turf grasses and forage legumes have been established over the last decade (for reviews refer to McKersie &

Brown, 1997; Spangenberg *et al.*, 1997; 1998; 2001; Forster & Spangenberg, 1999; Wang *et al.*, 2001), these methods are generally low-throughput in nature and often lead to multiple gene insertions. They are thus largely inadequate for use in forward and reverse genetic approaches to determine plant gene function. Consequently, efficient methods and tools for high-throughput gene silencing (e.g. high-throughput construction hairpin RNA vectors, virus-induced gene silencing, biolistic delivery of siRNAs), for large-scale, fast and low-cost transient gene function assays as well as the development of T-DNA insertional mutant collections and T-DNA activation tagging lines are required for key forage and turf plants and their endosymbionts, such as white clover, perennial ryegrass and *Neotyphodium* grass endophytes.

The effective application of systems biology approaches to forage and turf plants will close gaps in our understanding of the underlying genetics, physiology and biochemistry of many complex plant pathways and developmental processes. This is likely to advance progress in many applications of transgenesis in forage and turf plant improvement, building on gene technology as a powerful tool for the generation of the required functional genomics knowledge. Furthermore, requisite improved transformation methodologies for the development of 'market-ready' (i.e. single transgene inserts; free of selectable markers; deploying primarily host gene sequences) transformation events of prospective 'pyramided' gene technologies (that is, deployment of multiple transgenes in a single transformation event) following a sensible choice of high impact targets, and permitting the stable transfer of large inserts (e.g. BAC clones for QTL transfer) will be developed and applied in molecular breeding of forage and turf plants in the not too distant future.

Accompanying these developments, biosafety research including modelling of transgenic pastoral production systems and the development of transgene tracking and tracing assays will be required to meet needs of regulatory frameworks currently based on extreme interpretations of the precautionary principle.

Functionally-associated genetic markers for forage and turf molecular breeding

As previously described (Forster *et al.*, 2004), the majority of research to date on molecular marker development and validation in outcrossing forage species has been based on anonymous genetic markers, such as genomic DNA-derived SSRs and amplified fragment length polymorphisms (AFLPs) (Jones *et al.*, 2002a,b; Jones *et al.*, 2003). The paradigm for marker-assisted selection (MAS) that was established in autogamous plant species such as tomato, rice and wheat involves the use of such markers to construct linkage maps, genetic trait dissection through QTL analysis, and selection of linked markers in selection schemes such as donor-recipient recurrent selection. The obligate outbreeding nature of many pasture grasses and legumes clearly presents major limitations to the ready implementation of the inbreeding paradigm.

The use of markers that are in linkage of varying strength, rather than directly associated with the gene of interest, is a problem for both inbreeding and outbreeding species. For this reason, closely linked markers, ideally flanking the target region, are preferred. However, given fixation of the target region in an inbred background, to generate a homogeneous variety, the problem of potential reversal of linkage between favourable gene variant and selected marker allele is eliminated. In the context of a genetically heterogeneous synthetic population, complete fixation of a target genomic region will be difficult and slow to achieve, and consequently, the probability of recombination to decouple the favourable marker-trait allele combination will be high, especially in the absence of closely linked flanking markers. This logic implies that diagnostic genetic markers are of even higher potential value for outbreeding than inbreeding crops.

A further problem for the implementation of inbreeding paradigm MAS for pasture species is the large number of parental genotypes that are generally used in the polycross design for synthetic development. The number of foundation individuals varies between restricted base varieties (4-6 parents) and non-restricted base varieties (6-100+ parents) (Guthridge *et al.*, 2001; Forster *et al.*, 2001). Even for restricted base varieties, the process of tagging each gene variant in the parental genotypes with linked markers would imply multiple cycles of genetic trait-dissection in pair cross-derived mapping families. This is in contrast to the situation for inbreeders, in which the conduct of the trait-dissection process in a sib-ship derived from crossing of the future donor and recipient lines provides all relevant information for subsequent recurrent selection. One way to overcome this multiplicity of marker-trait allele associations would be to pre-introgress the desired combination into each of the selected parents. However, this implies a prior round of MAS, and does not adequately address the logistical complexity problem for molecular breeding of outcrossing forages.

The most obvious solution to these problems is to develop candidate gene-based markers that show a functional association with the target trait region (Andersen & Lübberstedt, 2003). Based on the population biology of perennial ryegrass (outbreeding with relatively large effective population sizes, at least for ecotypic populations), linkage disequilibrium (LD) is expected to extend over relatively short molecular distances, although variations of recombination frequency in different genotypes and different genomic regions may complicate this scenario (Forster *et al.*, 2004). In this instance, it should be possible to identify diagnostic variants for the selection of individual parental genotypes on the basis of superior allele content, given selection of the correct candidate gene. This will allow more efficient use of germplasm collections for parental selection. In addition, such 'perfect' markers will allow highly effective progeny selection.

The large-scale gene sequence collections generated by both incremental and EST discovery in perennial ryegrass and white clover provide the resource for functionally-associated marker development, with c. 15,000 unigenes currently defined for each species (Sawbridge *et al.*, 2003a; b). Selected genes have already been mapped as gene-associated RFLP and SSR loci (Barrett *et al.*, 2004; Faville *et al.*, 2005). RFLP markers are not readily implemented in molecular breeding, and SSRs are only present in a sub-set (generally less than 10%) of target genes. However, genic SNP markers can in principle be developed for any gene, and show the benefits of locus-specificity, high data fidelity and high-throughput analysis.

Our gene-associated SNP discovery process is based on a four-part strategy. The 'ultra fast-track' approach is based on validation of SNPs detected *in silico* in EST contigs. The perennial ryegrass and white clover EST collections are derived from multiple heterogeneous individuals from cultivars Grasslands Nui (Sawbridge *et al.*, 2003a) and Grasslands Huia (Sawbridge *et al.*, 2003b), respectively. SNPs may be distinguished from sequencing errors based on co-segregation data generated by applications such as AutoSNP (Barker *et al.*, 2003), and are validated by transfer and segregation analysis in pair cross-derived mapping families such as $F_1(NA_6 \times AU_6)$ (Faville *et al.*, 2005). Current data from perennial ryegrass suggests that c. 15% of *in silico* SNPs detect polymorphic loci in the mapping family. However, the majority of targeted genes may not been accessible to this approach. The other three approaches are based on *in vitro* discovery and are differentiated by the number of SNPs obtained: 'fast-track' involves short ESTs, providing single SNP loci for structured map enhancement; 'medium-track' involves full-length cDNAs, providing several SNP loci and partial haplotypic data; 'slow-track' is based on full-length genes with intron-exon structure, providing multiple SNP loci and determination of complete haplotype structures.

The experimental method for SNP discovery is based on cloning and sequencing of gene-specific amplicons from the heterozygous parents of two-way pseudo-testcross mapping families. Although this method is costly and time-consuming compared to direct sequencing of PCR products (Zhu *et al.*, 2003; Rickert *et al.*, 2003), it has several important advantages. Ambiguities in sequence traces due to heterozygous InDels may be readily distinguished, and haplotype structures within amplicons are directly determined (Zhang & Hewitt, 2003). In addition, paralogous sequence amplified by conserved primers are discriminated, and may be used to distinguish between members of multigene families. The putative SNPs are then validated in the progeny set, and cross-validated in other sib-ships and diverse germplasm.

'Proof-of-concept' for this process was obtained with the perennial ryegrass *Lp*ASRa2 gene, which shows 89% nucleotide identity with the rice *OsAsr1* gene. The *Asr* gene family encodes a group of proteins that are transcriptionally-induced by ABA treatment and water stress, and during fruit ripening. Osmotic and saline stress leads to up-regulation of the rice gene (Vaidyanathan *et al.*, 1999), and the maize *Zm-Asr1* gene co-locates with QTLs for traits responsive to mild water stress (Jeanneau *et al.*, 2002). The *Lp*ASRa2 gene consequently provides an excellent candidate for the assessment of correlation between genic sequence polymorphism and phenotypic variation. The full-length cDNA (890 bp) was tiled with 4 amplicons, covering 716 bp of the 5'-UTR, CDS and 3'UTR, and including a single 100 bp intron. A total of 9 SNPs were detected within and between the parents of the F_1(NA$_6$ x AU$_6$) mapping family. Of these, 7 SNP loci showing segregation in the progeny were assigned to coincident locations on linkage group (LG) 4 of the NA$_6$ parental genetic map, directly adjacent to the corresponding RFLP locus (Figure 1A). Partial haplotypic data for *Lp*ASRa2 reveals the maximum variant number of four, three of which are closely related, while the fourth is more divergent, defining two putative haplogroups (Olsen *et al.*, 2004). Although the majority of the exon-located changes define synonymous amino acid changes (Figure 1B), two SNPs defined amino acid substitutions, one of which (glutamate to glutamine at coordinate 136) produces a radical change, and may be functionally significant. Alternatively, the characterised SNPs may be in LD with functionally-significant changes in the transcriptional control regions (Paran & Zamir, 2003), given haplotype stability over gene-length distances.

Preliminary analysis of haplotype variation has been performed using a diversity panel including representatives of all *Lolium* taxa, including the three outbreeding cultivated species and several wild inbreeding species, with 8 validated *Lp*ASRa2 SNPs. Because of the uncertainty over *cis-trans* relationships between polymorphic SNP loci in heterozygous genotypes, a minimum-maximum range of haplotype numbers must be defined. However, inferences based on the prevalence of certain haplotype combinations in the homozygous state, as proposed by Clark (1990), suggest that the ryegrasses, like maize, may possess limited numbers of diverse haplotypes (Rafalski & Morgante, 2004).

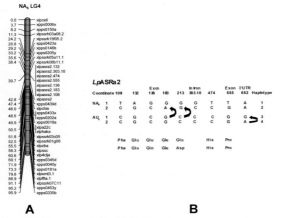

Figure 1 A. Genetic linkage map of LG4 of the NA₆ parental map, showing the SNP loci (indicated as xlpasra2.coordinate number) in close linkage with the corresponding RFLP locus (xlpasra2). B. *Lp*ASRa2 haplotype structures within and between the NA₆ and AU₆ parental genotypes. Arrows show putative mutational changes between members of the second haplogroup (haplotypes 2, 3 and 4), and predicted translation products of exon-located SNP loci are indicated.

By the start of 2005, over 150 genes had been introduced into the *in vitro* SNP discovery, of which 98 were sequenced and aligned, with a total of 1542 putative SNPs across 78 genes. SnuPe-validated SNPs were detected for 64 genes. Over a total of 83 kb of re-sequenced DNA, a relatively high SNP frequency (for four haplotypes) of 1/54 bp was observed, with an anticipated higher incidence in intron compared to exon sequences. For full-length herbage quality genes (the lignin biosynthetic genes *Lp*CCR1 and *Lp*CAD2 and the fructan metabolism genes *Lp*FT1 and *Lp*1-SST), a large number of SNPs have been identified (up to 265). Polymorphic SNPs have been validated at positions up to 6 kb distant in these genes, providing the basis for LD studies over gene-length distances.

Success in the detection of functionally-associated SNP variation must be followed by demonstration of causal correlation between genotypic diversity and phenotypic variation. An important pre-requisite for such analysis is the observation of candidate gene-QTL co-location. For perennial ryegrass, such coincidences have been observed for the orthologue of wheat *VRN1* with a QTL for vernalisation response on LG4 (Jensen *et al.*, 2005), the orthologue of the rice semi-dwarf gene *SD1* with a QTL for plant height on LG3 (T. Yamada, pers. comm.), and the lignin biosynthetic genes *Lp*OMT1, *Lp*CCR1 and *Lp*CAD2 with QTLs for herbage digestibility on LG7 (Cogan *et al.*, 2005). Experiments are currently being performed to evaluate the association between haplotype variation in genes such as *Lp*ASRa2 and *Lp*CCR1 and various aspects of phenotypic variation, such as gene transcription levels, metabolite synthesis and field phenotypes such as drought tolerance and herbage digestibility, respectively.

Apart from practical applications, candidate gene-based marker development provides an auxiliary source of information for functional genomics analysis of gene function. Given success in the validation of functionally-associated markers, an optimised strategy for forage biotechnology may be devised, in which the transgene-based technologies described in previous sections are deployed in highly adapted germplasm derived by marker-assisted breeding.

Positional cloning of genes from forage and turf species

The development of high-resolution genetic maps and functionally-associated markers, along with BAC libraries with high genome coverage, provides the basis for highly targeted physical mapping and eventual positional cloning of forage plant genes. The criteria for adoption of such an approach will be identification of genes controlling specific (initially qualitative) variation for important agronomic traits, for which minimal biochemical data to infer gene function is available. The *Festuca pratensis* 'stay-green' gene *sid* belongs to this category, and has been the subject of physical mapping studies (Donnison *et al.*, 2003). The highest priority targets in our studies are the *S* and *Z* genes that control gametophytic self-incompatibility (SI) in grasses (Cornish *et al.*, 1979). Isolation and construction of allele-specific markers for *S* and *Z* will be crucial for the effective implementation of molecular breeding, particularly in the context of restricted-base varietal development. The SI genes have been mapped on LGs 1 and 2 of perennial ryegrass in locations predicted through conserved synteny with rye and *Phalaris coerulescens* (Thorogood *et al.*, 2002). Fine mapping data from *P. coerulescens* (Bian *et al.*, 2004), including the location of a thioredoxin gene in close linkage to *S* (Baumann *et al.*, 2000), suggests a strategy for BAC contig construction in the equivalent *L . perenne* region. Similar data is now available for the Z gene-region of cereal rye (*Secale cereale* L.) (Hackauf & Wehling, 2005). Other genomic regions will be targeted as more data on gene-trait variation associations becomes available.

Conclusion

Forage and turf plants molecular breeding has now clearly entered the post-genomic era, reliant on high-throughput approaches for spatial and temporal analysis from genome to phenome, the respective data integration in a systems biology context for the establishment of stringent gene-function correlations. A primary challenge ahead is now the conversion of the vast amount of structural genomics information available into useful functional information. Model plant systems such as *Arabidopsis*, rice and *M. truncatula* will play a major role in the elucidation of gene function in higher plants. Translational genomics will allow for results obtained using these model systems to have major impact on understanding the molecular basis of plant processes and have direct application to the molecular breeding of forage and turf plants. These developments will be enhanced through applications of transgenesis and functionally-associated genetic markers in forage and turf molecular breeding building on genomic and post-genomic discoveries in these target species.

Acknowledgements

The support by the Victorian Department of Primary Industries, the Molecular Plant Breeding Cooperative Research Centre, Dairy Australia Ltd., Meat & Livestock Australia Ltd., the Gardiner Dairy Foundation, the Victorian Microarray Technology Consortium, the Victorian Bioinformatics Consortium, the Victorian Centre for Plant Functional Genomics, the Australian Centre for Plant Functional Genomics and the Argentine Research Council is gratefully acknowledged.

References

Adams, M.D., J.M. Kelley, J.D. Gocayne, M. Dubnick, M.H. Polymeropoulos, H. Xiao, C.R. Merril, A. Wu, B. Olde, R.F. Moreno *et al.* (1991) Complementary DNA sequencing: expressed sequence tags and human genome project. *Science*, 252, 1651-1656.
Andersen, J.R. & T. Lübberstedt (2003) Functional markers in plants. *Trends in Plant Science*, 8, 554-560.

Austin-Phillips, S. & T. Ziegelhoffer (2001) The production of value-added proteins in transgenic alfalfa. In: G. Spangenberg, (ed.). Molecular Breeding of Forage Crops. Kluwer Academic Press, 285-301.

Barker, G., J. Batley, H. O'Sullivan, K.J. Edwards & D. Edwards (2003) Redundancy-based detection of sequence polymorphisms in expressed sequence tag data using AutoSNP. *Bioinformatics*, 19, 412-422.

Barrett, B., A. Griffiths, M. Schreiber, N. Ellison, C. Mercer, J. Bouton, B. Ong, J. Forster, T. Sawbridge, G. Spangenberg, G. Bryan & D. Woodfield, D. (2004) A microsatellite map of white clover (*Trifolium repens* L.). *Theoretical and Applied Genetics*, 109, 596-608.

Barry, G.F. (2001) The use of the Monsanto draft rice genome sequence in research. *Plant Physiol.*, 125, 1164-1165.

Batley, J., G. Barker, H. O'Sullivan, K.J. Edwards & D. Edwards (2003) Mining for single nucleotide polymorphisms and insertions/deletions in maize expressed sequence tag data. *Plant Physiol.*, 132, 84-91.

Baumann, U., J. Juttner, B. Xueyu & P. Langridge (2000) Self incompatbility in the grasses. *Annals of Botany*, 85 (Supplement A), 203-209.

Bian, X.Y., A. Friedrich, J.R. Bai, U. Baumann, D.L. Hayman, S.J. Barker & P. Langridge (2004) High-resolution mapping of the *S* and *Z* loci in *Phalaris coerulescens*. *Genome*, 47, 918-930.

Brenner, S., M. Johnson, J. Bridgham, G. Golda, D.H. Lloyd, D. Johnson, S. Luo, S. McCurdy, M. Foy, M. Ewan, R. Roth, D. George, S. Eletr, G. Albrecht, E. Vermaas, S.R. Williams, K. Moon, T. Burcham, M. Pallas, R.B. DuBridge, J. Kirchner, K. Fearon, J. Mao & K. Corcoran (2000a) Gene expression analysis by massively parallel signal sequencing (MPSS) on microbead arrays. *Nature Biotechnology*, 18, 630-634.

Brenner, S., S.R. Williams, E.H. Vermaas, T. Storck, K. Moon, C. McCollum, J.I. Mao, S. Luo, J.J. Kirchner, S. Eletr, R.B. DuBridge, T. Burcham & G. Albrecht (2000b) *In vitro* cloning of complex mixtures of DNA on microbeads: physical separation of differentially expressed cDNAs. *Proc. Natl. Acad. Sci.*, 97, 1665-1670.

Canovas, F.M., E. Dumas-Gaudot, G. Recorbet, J. Jorrin, H.P. Mock & M. Rossignol (2004) Plant proteome analysis. *Proteomics*, 4, 285-298.

Chen, L., C.K. Auh, P. Dowling, J. Bell, F. Chen, A. Hopkins, R.A. Dixon & Z.Y. Wang (2003) Improved forage digestibility of tall fescue (*Festuca arundinacea*) by transgenic down-regulation of cinnamyl alcohol dehydrogenase. *Plant Biotechnology Journal*, 1, 437-449.

Chen, L., C.K. Auh, P. Dowling, J. Bell, D. Lehman & Z.Y. Wang (2004) Transgenic down-regulation of caffeic acid *O*-methyltransferase (COMT) led to improved digestibility in tall fescue (*Festuca arundinacea*). *Functional Plant Biology*, 31, 235-245.

Clark, A.G. (1990) Inference of haplotypes from PCR-amplified samples of diploid populations. *Molecular Biology and Evolution*, 7, 111-122.

Cogan, N.O.I., K.F Smith, T. Yamada, M.G. Francki, A.C. Vecchies, E.S. Jones, G.C Spangenberg & J.W Forster (2005) QTL analysis and comparative genomics of herbage quality traits in perennial ryegrass (*Lolium perenne* L.). *Theoretical and Applied Genetics*, 110, 364-380.

Cornish, M.A., M.D. Hayward & M.J. Lawrence (1979) Self-incompatibility in ryegrass. I. Genetic control in diploid *Lolium perenne* L. *Heredity*, 43, 95-106.

Dixon, R.A.(2004) Molecular improvement of forages – from genomics to GMOs. In: A. Hopkins, Z.-Y. Wang, M. Sledge, R.E. Barker (eds.). Molecular Breeding of Forage and Turf. Kluwer Academic Press, 1-19.

Donnison, I., D. O'Sullivan, A. Thomas, H. Thomas, K. Edwards, H.-M. Thomas, I. King (2003) Plant and Animal Genome XI, San Diego, California, W218.

Edwards, D. & J. Batley (2004) Plant bioinformatics: from genome to phenome. *Trends in Biotechnology*, 22, 232-237.

Emmerling, M., P. Chu, K. Smith, R. Kalla & G. Spangenberg (2004) Field evaluation of transgenic white clover with AMV Immunity and development of elite transgenic germplasm. In: A. Hopkins, Z.-Y. Wang, M. Sledge, R.E. Barker (eds.). Molecular Breeding of Forage and Turf. Kluwer *Academic Press, 359-366.*

Ettwiller, L.M., J. Rung & E. Birney (2003) Discovering novel cis-regulatory motifs using functional networks. *Genome Res.*, 13, 883-895.

Faville, M., A.C. Vecchies, M. Schreiber, M.C. Drayton, L.J. Hughes, E.S. Jones, K.M. Guthridge, K.F. Smith, T. Sawbridge, G.C. Spangenberg, G.T. Bryan & J.W. Forster (2005) Candidate gene-based molecular marker map construction in perennial ryegrass (*Lolium perenne* L.). *Theoretical and Applied Genetics*, 110, 12-32.

Felitti, S., K. Shields, M. Ramsperger, P. Tian, T. Webster, B. Ong, T. Sawbridge & G. Spangenberg (2004) Gene discovery and microarray-based transcriptome analysis in grass endophytes. In: A. Hopkins, Z.-Y. Wang, M. Sledge, R.E. Barker (eds.). Molecular Breeding of Forage and Turf. Kluwer Academic Press, 145-153.

Fernie, A.R. (2003) Metabolome characterisation in plant systems analysis. *Funct. Plant Biol.*, 30, 111-120.

Fiehn, O. (2002) Metabolomics – the link between genotypes and phenotypes. *Plant Mol. Biol.*, 48, 155-171.

Forster, J.W. & G. Spangenberg (1999) Forage and turf grass biotechnology: principles, methods and prospects. In: J.K. Setlow (ed.) Genetic Engineering: Principles and Methods, Vol 21, Kluwer Academic Publishers, 191-237.

Forster, J.W., E.S. Jones, R. Kölliker, M.C. Drayton, J. Dumsday, M.P. Dupal, K.M. Guthridge, N.L. Mahoney, E. van Zijll de Jong & K.F. Smith, (2001) Development and implementation of molecular markers for forage crop improvement. In: G. Spangenberg, (ed.). Molecular Breeding of Forage Crops. Kluwer Academic Press, 101-133.

Forster, J.W., E.S. Jones, J. Batley & K.F. Smith (2004) Molecular marker-based genetic analysis of pasture and turf grasses. In: A. Hopkins, Z.-Y. Wang, M. Sledge, R.E. Barker (eds.). Molecular Breeding of Forage and Turf. Kluwer Academic Press, 197-239.

Goodman, N. (2002) Biological data becomes computer literate: new advances in bioinformatics. *Curr. Opin. Biotechn.*, 13, 68-71.

Gruber, M.Y., H. Ray & L. Blahut-Beatty (2001) Genetic manipulation of condensed tannin synthesis in forage crops. In: G. Spangenberg, (ed.). Molecular Breeding of Forage Crops. Kluwer Academic Press, 189-201.

Guthridge, K.M., M.D. Dupal, R. Kölliker, E.S. Jones, K.F. Smith & J.W. Forster (2001) AFLP analysis of genetic diversity within and between populations of perennial ryegrass (*Lolium perenne* L.). *Euphytica*, 122, 191-201.

Goff, S.A., D. Ricke, T.H. Lan, G. Presting, R. Wang, M. Dunn, J. Glazebrook, A. Sessions, P. Oeller, H. Varma, D. Hadley, D. Hutchinson, C. Martin, F. Katagiri, B.M. Lange, T. Moughamer, Y. Xia, P. Budworth, J. Zhong, T. Miguel, U. Paszkowski, S. Zhang, M. Colbert, W.L. Sun, L. Chen, B. Cooper, S. Park, T.C. Wood, L. Mao, P. Quail, R. Wing, R. Dean, Y. Yu, A. Zharkikh, R. Shen, S. Sahasrabudhe, A.Thomas, R. Cannings, A. Gutin, D. Pruss, J. Reid, S. Tavtigian, J. Mitchell, G. Eldredge, T. Scholl, R.M. Miller, S. Bhatnagar, N. Adey, T. Rubano, N. Tusneem, R. Robinson, J. Feldhaus, T. Macalma, A. Oliphant & S. Briggs (2002) A draft sequence of the rice genome (*Oryza sativa* L. ssp *japonica*). *Science* 296, 92-100.

Hackauf, B. & P. Wehling (2005) Approaching the self-incompatibility locus Z in rye (*Secale cereale* L.) via comparative genetics. *Theoretical and Applied Genetics*, available on-line.

Harriman, R.W., E. Nelson & L. Lee (2003) Enhance turfgrass performance with biotechnology. In: I.K. Vasil (ed.) Plant Biotechnology 2002 and Beyond, Proceedings of the 10th IAPTC7B Congress, Kluwer Academic Publishers, 503-506.

Hartman, C.L., L. Lee, P.R. Day & E.T. Nilgun (1994) Herbicide resistant turfgrass (*Agrostis palustris* Huds.) by biolistic transformation. *Bio/Technology*, 12, 919-923.

Humphreys, M.O. & M.T. Abberton (2004) Molecular breeding for animal, human and environmental welfare. In: A. Hopkins, Z.-Y. Wang, M. Sledge, R.E. Barker (eds.). Molecular Breeding of Forage and Turf. Kluwer Academic Press, 165-180.

Jeanneau, M., D. Gerentes, X. Foueillassar, M. Zivy, J. Vidal, A. Toppan, P. Perez (2002) Improvement of drought tolerance in maize: towards the functional validation of the *Zm-Asr1* gene and increase of water use efficiency by over-expressing C4-PEPC. *Biochimie*, 84, 1127-1135.

Jensen, L.B., J.R. Andersen, U. Frei, Y. Xing, C. Taylor, P.B. Holm, T. Lübberstedt (2005) QTL mapping of vernalization response in perennial ryegrass (*Lolium perenne* L.) reveals co-location with an orthologue of wheat *VRN1*. *Theoretical and Applied Genetics*, 110, 527-536.

Jones, E.S., N.L. Mahoney, M.D. Hayward, I.P. Armstead, J.G. Jones, M.O. Humphreys, I.P. King, T. Kishida, T. Yamada, F. Balfourier, C. Charmet & J.W. Forster (2002) An enhanced molecular marker-based map of perennial ryegrass (*Lolium perenne* L.) reveals comparative relationships with other Poaceae species. *Genome*, 45, 282-295.

Jones, E.S., M.D. Dupal, J.L. Dumsday, L.J. Hughes & J.W. Forster (2002) An SSR-based genetic linkage map for perennial ryegrass (*Lolium perenne* L.). *Theoretical and Applied Genetics*, 105, 577-584.

Jones, E.S., L.J. Hughes, M.C. Drayton, M.T. Abberton, T.P.T. Michaelson-Yeates & J.W. Forster (2003) An SSR and AFLP molecular marker-based genetic map of white clover (*Trifolium repens* L.). *Plant Science*, 165, 531-539.

Kalla, R., P. Chu & G. Spangenberg (2001) Molecular breeding of forage legumes for virus resistance. In: G. Spangenberg (ed.) Molecular Breeding of Forage Crops, Kluwer Academic Publishers, 219-237.

Kanehisa, M. & S. Goto (2000) KEGG: Kyoto encyclopedia of genes and genomes. *Nucleic Acids Res.*, 28, 27-30.

Kanehisa, M., S. Goto, S. Kawashima and A. Nakaya (2002) The KEGG databases at GenomeNet. *Nucleic Acids Res.*, 30, 42-46.

May, G.D. (2004) From model to crops: integrated *Medicago* genomics for alfalfa improvement. In: A. Hopkins, Z.-Y. Wang, M. Sledge, R.E. Barker (eds.). Molecular Breeding of Forage and Turf. Kluwer Academic Press, 325-332.

Moreau, Y., S. Aerts, B. De Moore, B. De Strooper & M. Dabrowski (2003) Comparison and meta-analysis of microarray data: from the bench to the computer desk. *Trends Genet.*, 19, 570-577.

Morgante, M. & F. Salamini (2003) From plant genomics to breeding practice. *Curr. Opin. Biotechnol.*, 14, 214-219.

Love, C.L., T.A. Erwin, E.G. Logan, G. Barker, N. James, S. May, G. Spangenberg & D. Edwards (2005) ASTRA: integrated bioinformatics tools for ryegrass and white clover genomics. Plant and Animal Genome XIII, San Diego, California, W098.

Luo, H., Q. Hu, K. Nelson, C. Longo & A.P. Kausch (2004) Controlling transgene escape in genetically modified grasses. In: A. Hopkins, Z.-Y. Wang, M. Sledge, R.E. Barker (eds.). Molecular Breeding of Forage and Turf. Kluwer Academic Press, 245-254.

McKersie B.D. & D.C.W. Brown (1997) Biotechnology and the improvement of forage legumes. In: Biotechnology in Agricultural Series No. 17. CAB International, Wallingford, 444.

Nelson, P., M. Kiriakidou, A. Sharma, E. Maniataki & Z. Mourelatos (2003) The microRNA world: small is mighty. *Trends Biochem. Sci.*, 28, 534-540.

Olsen, K.M., S.S. Halldorsdottir, J.R. Stinchcombe, C. Weinig, J. Schmitt & M.D. Purugganan, (2004) Linkage disequilibrium mapping of *Arabidopsis CRY2* flowering time alleles. *Genetics*, 167, 1361-1369.

Paran, I. & D. Zamir (2003) Quantitative traits in plants: beyond the QTL. *Trends in Genetics*, 19, 303-306.

Petrovska, P., X. Wu, R. Donato, Z.Y. Wang, E.K. Ong, E. Jones, J. Forster, M. Emmerling, A. Sidoli, R. O'Hehir & G. Spangenberg (2004) Transgenic ryegrasses (*Lolium* spp.) with down-regulation of main pollen allergens. *Molecular Breeding*, 14, 489-501.

Qui, P. (2003) Recent advances in computational promoter analysis in understanding the transcriptional regulatory network. *Biochem. Biophys. Res. Commun.*, 309, 495-501.

Rafalski, J.A., M. Hanafey, G.H. Miao, A. Ching, J.M. Lee, M. Dolan & S. Tingey (1998) New experimental and computational approaches to the analysis of gene expression. *Acta Biochim. Pol.*, 45, 929-934.

Rafalski, A. & M. Morgante, (2004) Corn and humans: recombination and linkage disequilibrium in two genomes of similar size. *Trends in Genetics*, 20, 103-111.

Rhoades, M.W., B.J. Reinhart, L.P. Lim, C.B. Burge CB, B. Bartel & D.P. Bartel (2003) Prediction of plant microRNA targets. *Cell*, 110, 513-520.

Rickert, A.M., J.H. Kim, S. Meyer, A. Nagel, A. Ballvora, P.J. Oefner & C. Gebhardt, (2003) First-generation SNP/InDel markers tagging loci for pathogen resistance in the potato genome. *Plant Biotech. J.*, 1, 399-410.

Robinson, A.J., Love, C.G., Batley, J., Barker, G., Edwards, D. (2004) Simple sequence repeat marker loci discovery using SSR primer. *Bioinformatics* 10.1093/bioinformatics/bth104 (www.bioinformatics.oupjournals.org).

Roe, B.A. & D.M. Kupfer (2004) Sequencing gene rich regions of *Medicago truncatula*, a model legume. In: A. Hopkins, Z.-Y. Wang, M. Sledge, R.E. Barker (eds.). Molecular Breeding of Forage and Turf. Kluwer Academic Press, 333-344.

Sawbridge, T., E.-K. Ong, C. Binnion, M. Emmerling, R. McInnes, K. Meath, N. Nguyen, K. Nunan, M. O'Neill, F. O'Toole, C. Rhodes, J. Simmonds, P. Tian, K. Wearne, T. Webster, A. Winkworth & G. Spangenberg, (2003a) Generation and analysis of expressed sequence tags in perennial ryegrass (*Lolium perenne* L.). *Plant Science*, 165, 1089-1100.

Sawbridge, T., E.-K. Ong, , C. Binnion, M. Emmerling, K. Meath, K. Nunan, M. O'Neill, F. O'Toole, J. Simmonds, K. Wearne, A. Winkworth & G. Spangenberg, (2003b) Generation and analysis of expressed sequence tags in white clover (*Trifolium repens* L.). *Plant Science*, 165, 1077-1087.

Smith, K.F., J.W. Forster, M.P. Dobrowolski, N.O.I. Cogan, N.R. Bannan, E. van Zijll de Jong, M. Emmerling & G.C. Spangenberg (2005) Application of molecular technologies in forage plant breeding. In: Molecular Breeding of Forage and Turf, Aberystwyth.

Somers, D.J., R. Kirkpatrick, M. Moniwa & A. Walsh (2003) Mining single-nucleotide polymorphisms from hexaploid wheat ESTs. *Genome*, 46, 431-437.

Spangenberg, G., Z.Y. Wang, R. Heath, V. Kaul & R. Garrett (1997) Biotechnology in pasture plant improvement: methods and prospects. In: Proceedings XVIII International Grassland Congress, Winnipeg and Saskatoon, Vol 3.

Spangenberg, G., Z.Y. Wang & I. Potrykus (1998) Biotechnology in forage and turf grass improvement. In: R. Frankel (ed.) Monographs on Theoretical and Applied Genetics, Vol 23, Springer Verlag, Heidelberg, pp192.

Spangenberg, G., R. Kalla, A. Lidgett, T. Sawbridge, E.K. Ong & U. John (2001) Breeding forage plants in the genome era. In: G. Spangenberg (ed.) Molecular Breeding of forage crops, Kluwer Academic Publishers, 1-39.

Spangenberg, G., M. Emmerling, U. John, R. Kalla, A. Lidgett, E.K. Ong, T. Sawbridge & T. Webster (2003) Transgenesis and genomics in molecular breeding of temperate pasture grasses and legumes. In: I.K. Vasil (ed.) Plant Biotechnology 2002 and Beyond, Proceedings of the 10th IAPTC7B Congress, Kluwer Academic Publishers, 497-502.

Spangenberg, G., S. Felitti, K. Shields, M. Ramsperger, P. Tian, E.K. Ong, D. Singh, E. Logan & D. Edwards (2005) Gene discovery and microarray-based transcriptome analysis of the grass-endophyte association. In: C.A. Roberts, C.P.West & D.E. Spiers (eds.). *Neotyphodium* in Cool-Season Grasses. Blackwell Publishing, 103-121.

Sumner, L.W., P. Mendes & R.A Dixon (2003) Plant metabolomics: large-scale phytochemistry in the functional genomics era. *Phytochemistry*, 62, 817-836.

The *Arabidopsis* Genome Initiative (2000) Analysis of the genome of *Arabidopsis thaliana*. *Nature*, 408, 796-815.

Thorogood, D., W.J. Kaiser, J.G. Jones & I. Armstead (2002) Self-incompatibility in ryegrass. 12. Genotyping and mapping the S and Z loci of *Lolium perenne* L. *Heredity*, 88, 385-390.

Vaidyanathan, R., S. Kuruvilla & G. Thomas (1999) Characterisation and expression pattern of an abscisic acid and osmotic stress responsive gene from rice. *Plant Science*, 140, 25-36.

Velculescu, V.E., L. Zhang, B. Vogelstein & K.W. Kinzler (1995) Serial analysis of gene expression. *Science*, 270, 484-487.

Wang, L. & Y. Zhang (2004) Data integration and target selection for *Medicago* genomics. In: A. Hopkins, Z.-Y. Wang, M. Sledge, R.E. Barker (eds.). Molecular Breeding of Forage and Turf. Kluwer Academic Press, 275-288.

Wang, Z.Y., X.D. Ye, J. Nagel, I. Potrykus & G. Spangenberg (2001) Expression of a sulphur-rich sunflower albumin gene in transgenic tall fescue (*Festuca arundinacea* Schreb.) plants. *Plant Cell Rep.*, 20, 213-219.

Weckworth, W. (2003) Metabolomics in systems biology. *Annu. Rev. Plant Physiol.*, 54, 660-689.

Wilkins, M.R., C. Pasquali, R.D. Appel, K. Ou, O. Golaz, J.C. Sanchez, J.X. Yan, A.A. Gooley, G. Hughes, I. Humphrey-Smith, K.L. Williams & D.F. Hochstrasser (1996) From proteins to proteomes: large scale protein identification by two-dimensional electrophoresis and amino acid analysis. *Nature Biotechnology*, 14, 61-65.

Wilkins, M.R., K.L. Williams, R.D. Appel and D.F. Hochstrasser (eds.) (1997) Proteome Research: New Frontiers in Functional Genomics, Springer Verlag

Xu, J., J. Schubert & F. Altpeter (2001) Dissection of RNA mediated virus resistance in fertile transgenic perennial ryegrass (*Lolium perenne* L.). *Plant Journal*, 26, 265-274.

Yu, J. S. Hu, J. Wang, G. KS Wong, S. Li, B. Liu, Y. Deng, L. Dai, Y. Zhou, X. Zhang, M. Cao, J. Liu, J. Sun, J. Tang, Y. Chen, X. Huang, W. Lin, C. Ye, W. Tong, L. Cong, J. Liu, Q. Qi, J. Liu, L. Li, T. Li, X. Wang, H. Lu, T. Wu, M. Zhu, P. Ni, H. Han, W. Dong, X. Ren, X. Feng, P. Cui, X. Li, H. Wang, X. Xu, W. Zhai, Z. Xu, J. Zhang, S. He, J. Zhang, J. Xu, K. Zhang, X. Zheng, J. Dong, W. Zeng, L. Tao, J. Ye, J. Tan, X. Ren, X. Chen, J. He, D. Liu, W. Tian, C. Tian, H. Xia, Q. Bao, G. Li, H. Gao, T. Cao, J. Wang, W. Zhao, P. Li, W. Chen, X. Wang, Y. Zhang, J. Hu, J. Wang, S. Liu, J. Yang, G. Zhang, Y. Xiong, Z. Li, L. Mao, C. Zhou, Z. Zhu, R. Chen, B. Hao, W. Zheng, S. Chen, W. Guo, G. Li, S. Liu, M. Tao, J. Wang, L. Zhu, L. Yuan & H. Yang (2002) A draft sequence of the rice genome (*Oryza sativa* L. ssp *indica*). *Science*, 296, 79-92.

Zhang, D.-X. & G.M. Hewitt (2003) Nuclear DNA analyses in genetic studies of populations: practice, problems and prospects. *Molecular Ecology*, 12, 563-584.

Zhu, Y.L., Q.J. Song, D.L. Hyten, C.P. Van Tassell, L.K. Matukumalli, D.R. Grimm, S.M. Hyatt, E.W. Fickus, N.D. Young & P.B. Cregan, (2003) Single-nucleotide polymorphisms in soybean. *Genetics*, 163, 1123-1134.

Application of molecular markers in genetic resources management of perennial ryegrass

R. van Treuren

Centre for Genetic Resources, the Netherlands, Wageningen University and Research Centre, P.O. Box 16, 6700 AA Wageningen, the Netherlands, Email: robbert.vantreuren@wur.nl

Key points

1. Molecular markers are effective tools to support traditional approaches in plant genetic resources management.
2. Genetic diversity assessed for perennial ryegrass by AFLP analysis revealed differentiation of populations occurring in traditional Dutch grasslands from commercial varieties, but not from populations occurring in Dutch nature reserves.
3. No specific conservation measures were recommended to maintain the genetic diversity of perennial ryegrass occurring in traditional Dutch grasslands.
4. Pollination rates estimated by microsatellite analysis for a rejuvenated population of a perennial ryegrass genebank accession were very well described by an inverse quadratic function of inter-plant distance between potential mating pairs, while recorded flowering characteristics contributed only to a minor extent.
5. Compared to variation in pollination rates, the genetic integrity of the rejuvenated perennial ryegrass accession was found to be more threatened by contamination, which indicated the need for improved regeneration protocols to prevent gene flow from other germplasm.

Keywords: AFLPs, conservation, microsatellite, perennial ryegrass, seed multiplication

Introduction

Loss of genetic diversity has been widely recognised as a major threat for the maintenance and adaptive potential of species. Therefore, *ex situ* as well as *in situ* strategies have been developed for many plant species to conserve the extant genetic diversity. *Ex situ* conservation of genetic resources includes the storage of species samples in genebanks, which is intended to represent the genetic diversity of a crop as much as possible. It has been estimated that world-wide more than 1300 collections are stored in genebanks, collectively containing more than six million accessions (FAO, 1996). To manage genetic resources effectively the ability to identify genetic variation is indispensable. Characterisation of diversity has long been based on morphological traits mainly. However, morphological variation is often found to be restricted and genotype expression may be affected by environmental conditions, thereby constraining the analysis of genetic variation. Therefore, molecular marker techniques are of increasing interest to genebanks as complementary tools to traditional approaches in the management of plant genetic resources (Bretting & Widrlechner, 1995). Acquisition, maintenance, characterisation and utilisation are considered the four main categories of activities in plant genetic resources management, which may all benefit from the application of molecular genetic markers (Brown & Kresovich, 1996).

During the last few decades a wide variety of techniques to analyse genetic variation have emerged, including Amplified Fragment Length Polymorphisms (AFLPs) and microsatellites (Whitkus *et al.*, 1994; Karp *et al.*, 1996; Parker *et al.*, 1998). AFLPs are DNA fragments obtained from endonuclease restriction of DNA and amplification by the Polymerase Chain Reaction (PCR). The AFLP technique simultaneously generates fragments from multiple genomic sites that are separated by gel-electrophoresis and are generally scored as dominant markers. Because of the robust and highly informative fingerprinting profiles that are generally obtained, AFLPs can be applied in studies involving genetic identity, parentage, identification of clones and cultivars and gene mapping studies (Vos *et al*., 1995). Microsatellites are DNA fragments consisting of tandem repeat units of very short nucleotide motif. Polymerase slippage during DNA replication, or slipped strand mispairing, is considered to be the main cause of variation in the number of repeat units, resulting in length polymorphisms that can be detected by gel-electrophoresis following amplification of microsatellites by PCR. Due to their robustness, high level of polymorphism and codominant inheritance, microsatellites are informative markers in population genetic and gene mapping studies (Morgante & Olivieri, 1993; Jarne & Lagoda, 1996). In this paper the application of AFLPs and microsatellites in plant genetic resources management is illustrated by two studies in perennial ryegrass (*Lolium perenne* L.).

Perennial ryegrass is a major fodder crop and the main species in cultivated grasslands in temperate climate zones (Holmes, 1980). Apart from being used for forage, perennial ryegrass is also important as a turf species in sports fields, recreational areas and lawns. The primary centre of perennial ryegrass diversity is located in the European-Siberian region of diversity (Zeven & de Wet, 1982). Optimisation of fodder production has strongly reduced the biodiversity present in temperate grasslands during the last few decades. Germplasm collections were therefore developed by genebanks in order to safeguard genetic resources for present and future utilisation. It has been estimated that about 7348 perennial ryegrass accessions are collectively maintained in European genebanks (Sackville Hamilton, 1999).

Perennial ryegrass is one of the most important species of Dutch grasslands. However, during recent decades the original diversity of many grasslands has been replaced through the widespread use of more uniform commercial cultivars, adapted to the application of high doses of nitrogenous fertiliser. Therefore, grasslands that have not been resown and that have been treated with no or only limited amounts of nitrogen fertiliser have become scarce in the Netherlands. The Centre for Genetic Resources, the Netherlands (CGN) managed to identify about 50 such grasslands in 1998 by an inventory of Dutch farms still in agricultural use. Because it was unknown whether unique diversity still existed among these traditional grasslands, a diversity study using AFLPs was undertaken for perennial ryegrass to support conservation policies.

Genetic resources conserved as seed populations need periodic rejuvenation by means of seed multiplication because of decreased germinability of seeds by ageing and reduced seed supplies due to distribution. During seed multiplication, genebanks aim to avoid loss of genetic integrity of accessions as much as possible. The potential for genetic change particularly applies to species that are characterised by highly heterogeneous populations, including the outbreeding, wind-pollinated, perennial ryegrass (Forster *et al*., 2001; Guthridge *et al*., 2001). For forage species conserved in European genebanks it has been estimated that about 20% of the accessions are in desperate need of rejuvenation (Marum *et al*., 1998). Both from a genetic and economic viewpoint, optimisation of the regeneration protocols is therefore indispensable, but the necessary empirical data are largely absent. Loss of genetic integrity may result from a variety of reasons, including contamination, variation in pollination rates and differential seed production among plants. Microsatellites were used to analyse mating patterns in a regenerated perennial ryegrass accession and to investigate how well observed pollination rates could be predicted from data on the spatial and temporal distribution of pollen release.

Materials and methods

Assessment of diversity in traditional Dutch grasslands

From the 50 traditional grasslands identified by CGN in 1998, 16 were selected for the present study based on variation in geographic location and variation in soil type. In order to enable comparison of the diversity observed within the traditional grasslands with the diversity present in varieties, eight cultivars were selected that have played a major role in Dutch grassland cultivation during the last 50 years. These cultivars were 'Perma', 'Lamora', 'Barenza', 'Pelo', 'Vigor', 'Barmaco', 'Tresor' and 'Semperweide'. In addition, six nature reserves were selected to enable comparison with the genetic diversity of grasslands that are already conserved *in situ*. These grasslands were also managed extensively and displayed a similar geographic range as the traditional grasslands. From each grassland and cultivar, 36 plants per population were investigated by AFLP analysis.

DNA extractions were carried out using about 100 mg of freeze dried tissue material taken from young fresh leaves and largely followed the procedures described by Fulton *et al.* (1995). AFLP procedures basically followed the protocol of Vos *et al.* (1995) using about 300 ng of total genomic DNA and restriction enzymes *Eco*RI and *Mse*I. PCR reactions were carried out on a PE 9600 thermo cycler. Amplified fragments using ^{33}P were separated on 6% denaturing polyacrylamide gels. All samples were screened for the primer combinations E32/M51 and E32/M54. Details about experimental procedures are presented by Treuren *et al.* (in press).

AFLP fragments on the autoradiograms were scored manually for presence or absence in the range of approximately 50-500 bp. Band frequencies of AFLP markers were calculated in each of the populations. Genetic relationships among populations were investigated by an UPGMA cluster analysis (e.g. Nei, 1987) using the software package TFPGA (Miller, 1997). To determine to what extent the AFLP data were able to distinguish between populations, an assignment test was performed using the procedures of Paetkau *et al.* (1995). These analyses were carried out by a custom-designed computer programme.

Analysis of mating patterns during regeneration

Perennial ryegrass accession BA 11894 (Institute of Grassland & Environmental Research, Aberystwyth, UK) was regenerated in an experimental field at the Grassland Research Station Rožnov-Zubří (Czech Republic). The 49 plants used as parents for the rejuvenation were arranged in a 7x7 matrix with 50 cm distance between plants. During the flowering season, the daily number of inflorescences with open flowers was recorded for each of the 49 plants. From these data, the flowering period of each plant could be determined and estimates of pollen production could be derived. For the regenerated accession, a paternity exclusion analysis was carried out by microsatellite analysis of the 49 parental plants and 46 offspring from each of 12 different progenies.

DNA extraction procedures followed those described for the AFLP study, with the exception that a combination of the protocol of Fulton *et al.* (1995) and the DNeasy 96 Plant Kit was used. The parental plants were screened for 38 microsatellites, of which 25 appeared to be useful. Details for the majority of the microsatellites used are presented by Kubik *et al.* (1999; 2001) and Jones *et al.* (2001). Microsatellites were amplified on Peltier thermo cyclers (PTC-200) and PCR products were separated by capillary electrophoresis using an ABI Prism 3700

DNA Analyzer. Microsatellite alleles were sized using the program GeneScan and the resulting output was analysed with the software package Genotyper, version 3.5 NT (Perkin Elmer). All offspring samples were analysed for an initial set of 12 informative microsatellites. Samples were not investigated any further in case of identified paternity, otherwise samples were screened for additional microsatellites. In case of remaining ambiguities, offspring were screened by AFLP analysis, together with their maternal plant and the remaining potential pollen donors.

To relate observed pollination rates to data on the spatial and temporal distribution of pollen release, first several functions were considered that may describe the effect of inter-plant distance on pollen deposition rates between pairs of plants, including functions that assume an exponential reduction in pollen deposition rates with increasing inter-plant distance. From these functions the relative paternal contribution of plants to other plants was estimated, assuming an effect of distance only. Estimates were compared to observed data to evaluate the goodness of fit of the examined functions. The recorded numbers of inflorescences per plant were then used to estimate the contribution of each plant to the total pollen pool and to estimate the paternal contribution of each plant to the progenies of other plants on each day of the flowering season. Subsequently, paternal contributions of plants to progenies were calculated for the entire flowering season in order to estimate relative contributions of plants to individual total progenies. Estimates of relative paternal contributions of plants to progenies based on inter-plant distance and flowering data were then combined and compared to observed pollination rates. Self-fertilisation was disregarded in all estimation procedures. Microsoft Excel and custom-designed computer programs were used for data analyses.

Results

Assessment of diversity in traditional Dutch grasslands

Within the total sample, 101 variable bands (88.6%) were scored. The average fraction of polymorphic loci was lower for the cultivars (0.582) than for the traditional grasslands (0.682) and the grasslands from nature reserves (0.703), indicating lower levels of variability within the cultivars. Within the group of traditional grasslands 95 polymorphic bands were observed, of which 94.7% and 93.7% were also found polymorphic in the group of cultivars and the group of nature reserves, respectively. Polymorphisms absent from the latter two groups involved low-frequency alleles (<10%) in the group of traditional grasslands in all cases, indicating that the common alleles are present in all three groups. Differential fixation of alleles between populations was observed for none of the investigated markers. However, band frequencies of markers could differ substantially between populations, particularly for the reference cultivars.

Assignment tests using the observed band frequencies showed low fractions of genotypes correctly assigned to their population of origin for both the traditional grasslands (range: 0.0% to 27.8%) and the nature reserves (range: 6.3% to 50.5%), indicating low population differentiation among the investigated grasslands (Table 1). Better results were found for the reference cultivars (range: 61.1% to 94.4%), indicating a clearer identity among this material. When separate analyses were performed in each of the three groups, some of the cultivars even showed 100% correctly assigned genotypes.

Table 1 Percentage of genotypes assigned correctly to the population of origin based on the assessed AFLP variation. Analyses were carried out for the entire set of populations (total sample), and separately (own group) for the traditional grasslands (A), the grasslands from nature reserves (B) and the reference cultivars (C).

Population	Total sample	Own group	Population	Total sample	Own group
Traditional grasslands			*Nature reserves*		
Grijpskerk	13.9	19.4	Oldenzaal	8.6	28.6
Milheeze	2.9	11.4	Schin op Geul	38.7	58.1
Zundert	0.0	0.0	Culemborg	6.3	37.5
Warder	11.4	17.1	Achtmaal	30.4	56.5
SintJansklooster	8.3	13.9	Zegveld	22.9	31.4
Driewegen	5.7	17.1	Oostereind	50.5	67.9
Idsegahuizum	16.7	19.4			
Halle	11.1	19.4	*C. Reference cultivars*		
Losser	8.6	11.4	Perma	83.3	100.0
Eckelrade	8.6	11.4	Semperweide	72.2	97.2
Britswerd	20.6	20.6	Lamora	83.3	88.9
Castricum	19.4	30.6	Barenza	61.1	75.0
Wageningen	27.8	30.6	Pelo	63.9	75.0
Thesinge	8.3	25.0	Vigor	63.9	91.2
Mijdrecht	13.9	19.4	Barmaco	94.4	100.0
Oudewater	17.6	23.5	Tresor	72.2	86.1

Genetic relationships between populations were represented by a dendrogram based on Nei's unbiased estimate of standard genetic distance (Figure 1). Genetic distances among the grassland populations appeared small, all clustering together within a single group. As could be expected based on the results of the assignment tests, larger genetic distances were found among the reference cultivars and between the reference cultivars and the grassland populations. Genetic relationships between populations were also investigated by Principal Component Analysis (results not shown), which were in line with the results of the UPGMA cluster analysis. PCA plots supported the observed distinction between the reference cultivars and the grassland populations and the finding that the traditional grasslands and the grasslands from nature reserves basically cover the same range of genetic diversity.

Analysis of mating patterns during regeneration

Flowering was observed for all 49 plants of the regeneration plot and none of the plants showed non-overlapping flowering periods with other plants. Daily registration of the number of inflorescences with open flowers was performed over a 50-day period. Microsatellite data of the parental plants indicated tetraploidy instead of diploidy in six cases based on the observation of three or four alleles for the majority of microsatellites investigated. Tetraploidy of the six plants was supported by morphological examination of the plants and confirmed by flow cytometry.

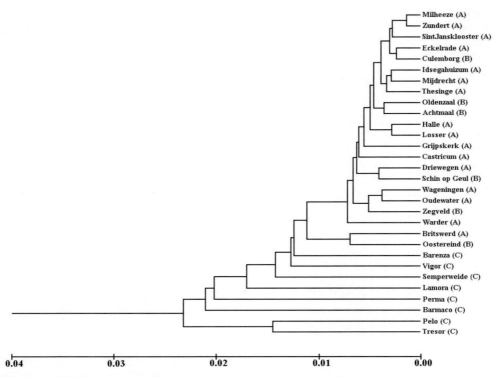

Milheeze (A)
Zundert (A)
SintJansklooster (A)
Eckelrade (A)
Culemborg (B)
Idsegahuizum (A)
Mijdrecht (A)
Thesinge (A)
Oldenzaal (B)
Achtmaal (B)
Halle (A)
Losser (A)
Grijpskerk (A)
Castricum (A)
Driewegen (A)
Schin op Geul (B)
Wageningen (A)
Oudewater (A)
Zegveld (B)
Warder (A)
Britswerd (A)
Oostereind (B)
Barenza (C)
Vigor (C)
Semperweide (C)
Lamora (C)
Perma (C)
Barmaco (C)
Pelo (C)
Tresor (C)

0.04   0.03   0.02   0.01   0.00

Figure 1 UPGMA cluster analysis of perennial ryegrass populations based on Nei's unbiased estimate of standard genetic distance derived from AFLP data. Traditional grasslands are denoted by (A), grasslands from nature reserves by (B) and reference cultivars by (C).

Out of the 551 offspring analysed, 451 actual pollen donors (81.9%) were identified based on the microsatellite data. Remaining ambiguities could be resolved with additional AFLP analysis, except in four cases (Table 2). For nine offspring, microsatellites alleles were observed at multiple loci that were unknown to the parental population. These included five cases of matching alleles with the maternal plant that were considered pollen contaminants and four cases of mismatches with the mother that were regarded seed contaminants. Despite the fact that selfing rates are generally assumed to be low in perennial ryegrass because of a self-incompatibility system, 19 cases of self-fertilisation were observed. Self-fertilisation was most pronounced for the tetraploid plant 21 (Table 2). Based on the molecular data, no cross-pollination between plants of different ploidy level was observed in the study plot. Consequently the two groups of different ploidy level were treated as two reproductively isolated populations in further analysis of the data.

Out of the total number of identified paternal plants, 61.9% were located within 1 m of the maternal plant, indicating that inter-plant distance had a large effect on pollination probabilities between plants (Table 2). High pollination rates observed within progenies involved neighbouring plants in nearly all cases, as is shown for progeny 22 in Figure 2. In contrast to the impact of inter-plant distance on pollination rates, no clear effect was observed of pollen production and extent of temporal overlap in flowering.

Table 2 Paternity data obtained from microsatellite and AFLP analysis after the regeneration of perennial ryegrass accession BA 11894 using 49 parental plants. The 12 analysed progenies are denoted by the code of the maternal plant (9, 11, …, 46), plants 21 and 46 having a tetraploid genome. Paternity data are classified by self-fertilisation, inter-plant distance in cm between the parental pair, absence of matching with any of the 49 parental plants and remaining ambiguous paternity.

Progeny	Selfing	Inter-plant distance				No match	Ambiguity	Total
		1-100	101-200	201-300	301-400			
9	0	27	11	6	0	0	1	45
11	0	32	12	1	-	1	0	46
19	1	24	16	4	-	1	0	46
22	0	38	7	1	-	0	0	46
30	3	15	16	7	2	2	1	46
33	0	28	13	4	0	1	0	46
35	0	31	9	5	0	0	1	46
39	0	32	9	4	0	0	1	46
41	0	23	13	7	2	1	0	46
43	0	19	24	1	-	2	0	46
21 (4n)	15	-	30	-	-	1	0	46
46 (4n)	0	45	1	-	-	0	0	46
Total	19	314	161	40	4	9	4	551

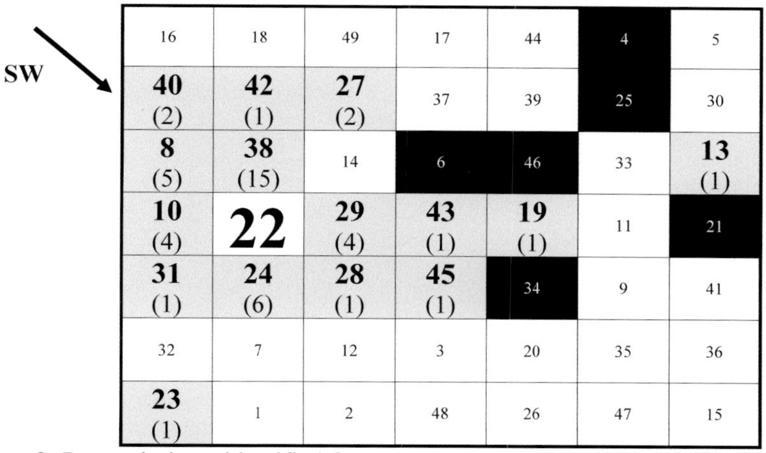

Figure 2 Paternal plants identified for progeny 22 by microsatellite and AFLP analysis after the regeneration of perennial ryegrass accession BA 11894. The regeneration plot consisted of 49 plants arranged in a 7x7 matrix with 50 cm inter-plant distance. Paternal plants are indicated by the grey shade of the cells within the matrix with the number of observed matings given between brackets. Six plants of the regeneration population appeared to be tetraploid and are indicated by black shading. The prevailing wind direction during the flowering period (SWW) is denoted by an arrow.

To analyse the effects of inter-plant distance and flowering characteristics in closer detail, pollination data of different progenies were combined and classified into observed distance categories. Because inter-plant distances within the plot showed a discrete distribution, pollination rates as a function of inter-plant distance were presented in a cumulative manner. Subsequently, observed cumulative pollination rates were compared to expected values based on the progeny samples sizes and various functions relating pollen deposition rates to inter-plant distance. These analyses showed that expected values based solely on an inverse quadratic function of inter-plant distance already fitted the observed pollination rates very well (Figure 3). Using the sum of squared differences between observed and expected values of all 430 potential mating pairs as a measure of goodness of fit, the inverse quadratic function displayed a 51.0% better fit than a random mating model, a 28.4% better fit than an inverse distance function and a 9.5 % better fit than an inverse third power distance function. The fit of the inverse quadratic function improved with only 0.77% when the flowering data were included in the calculation of expected values (Figure 3), indicating that spatial proximity was the main cause of variation in pollination rates within the study plot.

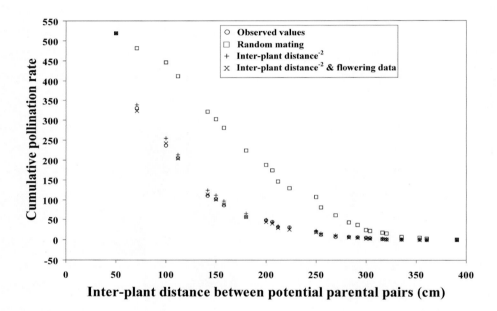

Figure 3 Cumulative number of pollinations in relation to inter-plant distance between pairs of plants presented for the observed data, estimates based on random mating and estimates based on an inverse quadratic function of inter-plant distance. For the latter function, estimates were derived both excluding and including the recorded flowering data.

Discussion

Assessment of diversity in traditional Dutch grasslands

High levels of variation were found within the grassland populations, which were in line with the results from other studies in perennial ryegrass (e.g. Roldán-Ruiz *et al.*, 2000; Creswell *et* al., 2001; Guthridge et al., 2001). Compared to the grassland populations, lower levels of diversity were observed within the reference cultivars, which were mainly the result of the absence of rare alleles occurring in grasslands. These results suggested that diversity has been sampled thoroughly from grasslands at the onset of modern plant breeding.

Despite the fact that the traditional grasslands, reference cultivars and grasslands from nature reserves shared the major part of the AFLP variation, band frequencies of AFLP markers could vary considerably between populations. These results were in accordance with the results of an AFLP study of perennial ryegrass populations from Portugal that did not reveal diagnostic bands but showed that the total AFLP profile of a genotype was a reliable indicator of the population of origin (Cresswell et al., 2001). In the present study this was observed to some extent only for the reference cultivars. These results indicated distinction between the reference cultivars and the grassland populations and low levels of population differentiation within the latter group. This was supported by cluster analyses of the investigated material, grouping the traditional grasslands and grasslands from nature reserves together and separating these populations from the reference cultivars. Genetic relationships among the investigated material were in agreement with a morphological study carried out on the same genotypes from the traditional grasslands and reference cultivars (Treuren et al., in press).

The low levels of population differentiation among the grasslands suggested that in the past the perennial ryegrass populations have not gone through severe bottlenecks, nor have experienced restricted gene flow. Based on the distinction between the grasslands and the reference cultivars, the grassland populations do not seem to have experienced substantial introgression from cultivated material (see also Loos, 1994). The finding of large differences in marker frequencies for subsets of the markers may indicate linkage to genes under selection. Differences in frequencies for these markers observed between some traditional grasslands were accompanied by strong differences in agronomic characters, notably heading date (Treuren et al., in press). It has been suggested that alleles that are common only in certain populations may indicate adaptive significance and, therefore, that conservation should focus on such populations (Marshal & Brown, 1975; Allard, 1992). This could apply to some of the traditional grasslands. However, the fact that the traditional grasslands and grasslands from nature reserves covered basically the same range of variation did not support specific conservation measures for traditional grasslands. Proper in situ conservation of the nature reserves is considered sufficient to maintain the genetic diversity of perennial ryegrass occurring in traditional grasslands. A similar conclusion was reached from an accompanying study in white clover carried out for the same traditional grasslands and nature reserves (Treuren et al., in press). Whether this conclusion also will apply to species that usually display higher levels of population differentiation, such as less outcrossing species, is still an open question. Moreover, in the present study only genetic diversity within species was considered. However, conservation issues may also include other aspects of biodiversity, such as species diversity.

Analysis of mating patterns during regeneration

A large effect of inter-plant distance on pollination rates was observed for the study plot, which was in accordance with the generally observed leptokurtotic distribution of pollen dispersal in perennial ryegrass (Giddings et al., 1997a; 1997b; Cunliffe et al., 2004). Pollination rates were best described by an inverse quadratic function of inter-plant distance, which corresponds to a pollen cloud rapidly diluting with increasing distance in two dimensions. Flowering characteristics appeared to have only a minor effect on pollination rates. However, the magnitude of this effect strongly depends on the variation in flowering characteristics among plants and, therefore, may be higher for populations displaying larger variance values. Curve fitting was performed on cumulative data obtained from the combined results of different progenies. Analysis of individual progenies suggested that incompatibility for specific parental combinations was at least one additional factor affecting pollination rates. Other influencing factors may include disturbance of natural pollination rates by daily visit of the plot to record flowering data. Other studies in perennial ryegrass have suggested an effect of wind parameters on pollen dispersal (Giddings *et al.*, 1997a; 1997b), but no clear relationship was found in the present study.

Differential pollination rates between plants may cause unequal contributions of plants to the next generation. Skewed contributions by plants will reduce the effective population size and, therefore, increase the inbreeding coefficient (Falconer, 1981). This will compromise the objective of genebanks to maintain the genetic integrity of accessions as much as possible. However, if the spatial arrangement of plants is the main factor influencing pollination rates, paternal contributions and the variance therein between plants can be estimated from the inverse quadratic function of inter-plant distance. Modifications of the regeneration protocol that minimise unequal contributions by plants will reduce loss of genetic integrity. At least two modifications could be envisaged. First, arranging the plants into a more linear plot design will reduce the variance in paternal contributions. For example, a single row of 49 plants will reduce the variance by 72% compared to a 7x7 matrix, while the variance can be reduced to zero using a single circle of 49 plants. However, it can be questioned whether the observed distance function will hold for such plot designs. Moreover, apart from practical disadvantages, a linear or circular plot design will make the pollination of plants more dependent on its direct neighbours, while overall pollination rates will be strongly affected by prevailing wind directions. Second, unequal paternal contributions could be compensated by harvesting differential numbers of seeds from individual plants. Seed numbers required per plant can easily be determined by a simple computer program, minimising the variance in the sum of maternal and paternal contributions between plants. Differential seed harvesting may be relevant to genebanks that already practise 'balanced bulking' to correct for differences in seed production between plants (e.g. Johnson *et al.*, 2002; 2004). Implementation of this modification could be supported by validation of the general application of the distance function.

A total of nine contaminants were detected among the 551 offspring analysed (1.63%), which at first sight may seem only a minor contamination rate. However, based on this contamination level, the probability of selecting at least one contaminant as parent for a next regeneration with 49 plants is 57%. Thus, even 'minor' contamination levels may cause persistence of foreign germplasm in accessions. This will even be more severe in case the original germplasm has a selective disadvantage compared to the contaminants. In particular, the six tetraploid plants detected among the parental plants were worrisome because investigation of the original seed lot of the accession by flow cytometry did not indicate that they originated from the accession. Additional data sources showed that contamination must have occurred in the preparatory phase of the regeneration, somewhere in between seed sowing and establishment of the study plot. Although no genetic data from other populations were available for confirmation, pollen contaminations were likely the result of pollen flow from other regeneration plots, while seed contaminants were probably due to post-harvest seed handling. Thus, the present study indicated that contamination may occur at various stages of the regeneration process, and can be considered more threatening to the genetic integrity of perennial ryegrass accessions than variation in pollination rates between plants.

Therefore, apart from considering modifications to reduce the variance in contributions by plants, regeneration procedures need to be carefully re-examined and extended with safety precautions to avoid mix-up of different germplasm.

Acknowledgements

The project on genetic diversity in traditional Dutch grasslands was carried out within the framework of the research program Agrobiodiversity, financed by the Netherlands Ministry of Agriculture, Nature and Food Quality. The study of mating patterns was carried out within the framework of a research project aiming at the optimisation of regeneration protocols for perennial European forage species (ICONFORS), which was funded by the European Union (QLRT-1999-30621).

References

Allard, R.W. (1992). Predictive methods for germplasm identification. In: H.T. Stalker & J.P. Murphy (eds.) Plant Breeding in the 1990s. CAB International, Wallingford, UK, 119-146.

Bretting, P.K. & M.P. Widrlechner (1995). Genetic markers and plant genetic resource management. *Plant Breeding Reviews*, 31, 11-86.

Brown, S.M. & S. Kresovich (1996). Molecular characterization for plant genetic resources conservation. In: H. Paterson (ed.) Genome Mapping of Plants. Academic Press, San Diego, 85-93.

Cresswell, A., N.R. Sackville Hamilton, A.K. Roy & B.M.F. Viegas (2001). Use of amplified fragment length polymorphism markers to assess genetic diversity of *Lolium* species from Portugal. *Molecular Ecology*, 10, 229-241.

Cunliffe, K.V., A.C. Vecchies, E.S. Jones, G.A. Kearney, J.W. Forster, G.C. Spangenberg & K.F. Smith (2004). Assessment of gene flow using tetraploid genotypes of perennial ryegrass (*Lolium perenne* L.). Australian Journal of Agricultural Research, 55, 389-396.

FAO (1996). FAO State of the World's Plant Genetic Resources for Food and Agriculture. Food and Agriculture Organisation of the United Nations, Rome.

Falconer, D.S. (1981). Introduction to Quantitative Genetics. Longman, London.

Forster, J.W., E.S. Jones, R. Kölliker, *et al.* (2001). DNA profiling in outbreeding forage species. In: R. Henry (ed.) Plant Genotyping – the DNA Fingerprinting of Plants. CABI Press, 299-320.

Fulton, T.M., J. Chunwongse, S.D. Tanksley (1995). Microprep protocol for extraction of DNA from tomato and other herbaceous plants. *Plant Molecular Biology Reporter*, 13, 207-209.

Giddings, G.D., N.R. Sackville Hamilton & M.D. Hayward (1997a). The release of genetically modified grasses. 1: Pollen dispersal to traps in *Lolium perenne*. *Theoretical and Applied Genetics*, 94, 1000-1006.

Giddings, G.D., N.R. Sackville Hamilton & M.D. Hayward (1997b). The release of genetically modified grasses. 2: The influence of wind direction on pollen dispersal. *Theoretical and Applied Genetics*, 94, 1007-1014.

Guthridge, K.M., M.P. Dupal, R. Kölliker, E.S. Jones, K.F. Smith & J.W. Forster (2001). AFLP analysis of genetic diversity within and between populations of perennial ryegrass (*Lolium perenne* L.). *Euphytica*, 122, 191-201.

Holmes, W. (1980). Grass, its Production and Utilization. Blackwell scientific publications, Oxford, UK.

Jarne, P. & P.J.L. Lagoda (1996). Microsatellites, from molecules to populations and back. *Trends in Ecology and Evolution*, 11, 424-429.

Johnson, R.C., V.L. Bradley & M.A. Evans (2002). Effective population size during grass germplasm seed regeneration. *Crop Science*, 42, 286-290.

Johnson, R.C., V.L. Bradley & M.A. Evans (2004). Inflorescence sampling improves effective population size of grasses. *Crop Science*, 44, 1450-1455.

Jones, E.S., M.P. Dupal, R. Kölliker, M.C. Drayton & J.W. Forster (2001). Development and characterisation of simple sequence repeat (SSR) markers for perennial ryegrass (*Lolium perenne* L.). *Theoretical and Applied Genetics*, 102, 405-415.

Karp, A., O. Seberg & M. Buiatti (1996). Molecular techniques in the assessment of botanical diversity. *Annals of Botany*, 78, 143-149.

Kubik, C., W.A. Meyer & B.S. Gaut (1999). Assessing the abundance and polymorphism of simple sequence repeats in perennial ryegrass. *Crop Science*, 39, 1136-1141.

Kubik, C., M. Sawkins, W.A. Meyer & B.S. Gaut (2001). Genetic diversity in seven perennial ryegrass (*lolium perenne* L.) cultivars based on SSR markers. *Crop Science*, 41, 1565-1572.

Loos, B.P. (1994). Morphological variation in Dutch perennial ryegrass (*Lolium perenne* L.) populations, in relation to environmental factors. *Euphytica*, 74, 97-107.

Marshall, D.R. & A.H.D. Brown (1975). Optimum sampling strategies in genetic conservation. In: O.H. Frankel & J.G. Hawkes (eds.) Genetic Resources for Today and Tomorrow. Cambridge University Press, Cambridge, UK.

Marum, P., I.D. Thomas & M. Veteläinen (1998). Summary of germplasm holdings. In: L. Maggioni, P. Marum, N.R. Sackville Hamilton, I. Thomas, T. Gass & E. Lipman (eds.) Report of a Working Group on Forages (Sixth Meeting, 6-8 March 1997, Beitostølen, Norway). European Co-operative Programme for Crop Genetic Resources Networks (ECP/GR). International Plant Genetic Resources Institute, Rome, Italy, 184-190.

Miller, M.P. (1997). Tools for population genetic analyses (TFPGA) 1.3: A Windows program for the analysis of allozym and molecular population genetic data. Computer software distributed by author.

Morgante, M. & A.M. Olivieri (1993). PCR-amplified microsatellites as markers in plant genetics. *The Plant Journal*, 3, 175-182.

Paetkau D, W. Calvert, I. Sterling, & C. Strobeck (1995). Microsatellite analysis of population structure in Canadian polar bears. *Molecular Ecology*, 4, 347-354.

Parker, P.G., A.A. Snow, M.D. Schug, G.C. Booton & P.A. Fuerst (1998). What molecules can tell us about populations: choosing and using a molecular marker. *Ecology*, 79, 361-382.

Roldán-Ruiz, I., J. Dendauw, E. van Bockstaele, A. Depicker & M. de Loose (2000). AFLP markers reveal high polymorphic rates in ryegrass (*Lolium* spp.). *Molecular Breeding*, 6, 125-134.

Sackville Hamilton, N.R. (1999). European *Lolium* and *Trifolium repens* Databases. In: L. Maggioni, P. Marum, N.R. Sackville Hamilton, M. Hulden & E. Lipman (eds.) Report of a Working Group on Forages (Seventh Meeting, 18-20 November 1999, Elvas, Portugal). European Co-operative Programme for Crop Genetic Resources Networks (ECP/GR). International Plant Genetic Resources Institute, Rome, Italy, 42-46.

Treuren, R. van, N. Bas, P.J. Goossens, H. Jansen & L.J.M. van Soest (in press). Genetic diversity in perennial ryegrass and white clover among old Dutch grasslands as compared to cultivars and nature reserves. *Molecular Ecology*.

Vos, P., R. Hogers, M. Bleeker, M. Reijans, T. van de Lee, M. Hornes, A. Frijters, J. Pot, J. Peleman, M. Kuiper & M. Zabeau (1995). AFLP: a new technique for DNA fingerprinting. *Nucleic Acid Research*, 23, 4407-4414.

Whitkus, R., J. Doebley & J.F. Wendel (1994). Nuclear DNA markers in systematics and evolution. In: R.L. Phillips & I.K. Vasil (eds.) DNA-based Markers in Plants (Advances in Cellular and Molecular Biology of Plants, Vol. 1). Kluwer Academic Publishers, Dordrecht, the Netherlands, 116-141.

Zeven, A.C., J.M.J. de Wet (1982). Dictionary of Cultivated Plants and their Regions of Diversity. Pudoc, Wageningen, the Netherlands.

Section 1

Objectives, benefits and targets of molecular breeding

Leaves of high yielding perennial ryegrass contain less aggregated Rubisco than S23

A. Kingston-Smith and P.W. Wilkins

Institute of Grassland and Environmental Research, Plas Gogerddan, Aberystwyth, Ceredigion SY23 3EB, UK, Email: pete.wilkins@bbsrc.ac.uk

Keywords: dry matter yield, nitrogen use efficiency, plant breeding

Introduction Breeding diploid perennial ryegrass for improved dry matter yield under nitrogen-limiting conditions has reduced the nitrogen (N) concentration of the herbage (Wilkins *et al.*, 2003). Reduced N concentration in the ruminant diet is one potential way to reduce losses of N to the environment by reducing the amount of N that animals excrete. The underlying physiological basis of this increased N-use efficiency in ryegrass was investigated.

Materials and Methods Leaf samples were taken from the third harvest year (2004) of a field plot trial with 4 replicate fully randomised blocks containing two perennial ryegrass varieties that had varied consistently in N concentration throughout the first two harvest years (Wilkins *et al.*, 2003): Ba13582, which had the lowest mean N concentration over all harvests, and S23, which had the highest. Ba13582 produced significantly more dry matter than S23 in both these harvest years (2002 and 2003). Ten fully expanded leaves from each plot were frozen in liquid N_2 and stored at -80^0C. Samples were ground to a fine powder and protein was extracted by grinding in a neutral buffer (0.1 M HEPES, pH 7.5, 1 mM EDTA, 2 mM DTT, 0.1% Triton X-100, 1 mM PMSF, 1 μM E64) at a ratio of 25 ml per g dry weight. After centrifugation (5 min at $10,000g_{av}$), protein contents of the supernatants were determined (Bradford, 1977) while protein separation was achieved by denaturing gel electrophoresis (Laemmli, 1970). Gels were loaded with 10 μg protein per sample track plus molecular weight standards. They were stained with Coomassie blue and analysed by densitometry (BioRad GS710 equipped with Qantity One software, BioRad UK, Hemel Hempstead). Analyses of variance were carried out using GENSTAT.

Results Densitometric analysis of the major leaf protein bands of Ba13582 and S23 did not reveal significant differences between the varieties in concentration of the large and small subunits of Rubisco (Table 1). However, Ba13582 contained less than half the amount of high molecular weight polypeptide (~205 kDa) that was present in S23. This 205 kDa polypeptide is typical of non-heat dissociable, aggregated Rubisco subunits.

Table 1 Densitometric analysis (OD x mm^2) of Rubisco protein bands in leaf protein extracts from Ba13582 and S23 resolved by denaturing electrophoresis

Variety	Large subunit	Small subunit	Aggregated
Ba13582	31.3	10.4	2.9
S23	24.8	8.0	6.9
s.e.d.	2.23	1.18	0.56
p	NS	NS	0.006

NS, not significant at p=0.05

Conclusions Since aggregated Rubisco is unlikely to function in capturing CO_2 from the atmosphere, this result suggests a possible mechanism for the superior N-use efficiency of Ba13582. The *in vivo* significance of aggregated rubisco is unclear. It may represent an N storage pool, to which Ba13582 partitions less assimilated N than S23. Alternatively, it may indicate protein damage. In either case, Ba13582 would be predicted to achieve efficient photosynthesis with less protein N than S23. Families derived from Ba13582 are currently being used at IGER for genetic mapping. If our hypothesis is correct, it should be possible to identify quantitative trait loci that control both N-use efficiency and the amount of aggregated Rubisco.

References

Bradford, M.M (1977). A rapid and sensitive method for the quantitation of microgram quantities of protein utilizing the principle of protein-dye binding. *Analytical Biochemistry*, 72, 248-254.

Laemmli, U.K. (1970). Cleavage of structural proteins during the assembly of the head of bacteriophage T4. *Nature*, 227, 680-685.

Wilkins, P.W., J.A. Lovatt & M.L. Jones (2003). Improving annual yield of sugars and crude protein by recurrent selection within diploid ryegrass breeding populations, followed by chromosome doubling and hybridisation. *Czech Journal of Genetics and Plant Breeding*, 39 (Special Issue), 95-99.

Variability in quantity and composition of water soluble carbohydrates among Irish accessions and European varieties of perennial ryegrass

S. McGrath[1,2], S. Barth[1], A. Frohlich[1], M. Francioso[1], S.A. Lamorte[1] and T.R. Hodkinson[2]
[1]Teagasc Crops Research Centre, Oak Park, Carlow, Ireland, Email: smcgrath@oakpark.teagasc.ie,
[2]Department of Botany, University of Dublin, Trinity College, Ireland

Keywords: quality, fructose, glucose, high pressure liquid chromatography, plant breeding

Introduction The objective of this study was to identify perennial ryegrass accessions displaying high fructose and glucose contents and an improved ratio between fructose and glucose fractions across different time points throughout the year. Fructose and glucose are the main constituents of the water soluble carbohydrate (WSC) fraction in perennial ryegrass. For animal nutrition the amount of WSC is crucial as it is the primary energy source available to metabolise the intake of plant protein. The ratio between fructose and glucose fractions is important since fructosan chains, which are an excellent energy source for ruminants, are built from fructose. Furthermore the seasonal variability of WSC content in feed reflects the changing balance between protein and carbohydrates.

Materials and Methods In the summer of 2003, 33 perennial ryegrass entries were grown from true seed and planted as spaced plants in the field. Forty plants per accession were divided into 4 pools for analysis. The plant material was selected from a collection of historic indigenous Irish accessions held at Oak Park (23 entries) and current commercially grown varieties (10 entries). In 2004, at three time points during the growing season, samples were taken and processed for WSC analysis via HPLC as described by Jafari *et al.* (2003). Means and standard deviations were calculated and entries were assigned to one of four classes (1 = very good to 4 = poor) based on percentage of dryweight attributed to carbohydrates.

Results At the three time points across ecotypes and varieties a high variability was found for both the WSC content and the ratio of fructose/glucose (Table 1), *e.g.* contents of fructose ranging between 1.65 and 18.99%. Generally the material displayed wide genetic variation across the traits investigated. Among the ecotypes, several entries were superior to the commercial varieties at the third cutting time point (Table 2).

Table 1 Means (x), standard deviations (SD), minimum and maximum percentage of fructose, glucose and total water soluble carbohydrate (WSC), and ratio of fructose/glucose (ratio) across three cuts

	% fructose				% glucose				% WSC				ratio			
	X	SD	min	max	X	SD	min	max	X	SD	min	max	X	SD	min	max
cut 1	8.89	2.90	2.36	15.77	3.99	1.08	1.57	6.89	12.90	3.85	4.11	21.49	2.22	0.40	1.35	3.51
cut 2	6.34	1.96	2.47	11.29	4.39	1.24	2.04	6.68	10.73	3.13	4.54	17.37	1.44	0.19	1.04	2.29
cut 3	8.93	3.74	1.65	18.99	4.40	1.81	1.07	12.34	13.33	5.19	4.18	27.19	2.11	0.64	0.65	5.10

Table 2 Number of varieties and ecotypes within the four classes of carbohydrate content

index	fructose				glucose				WSC				ratio			
	1	2	3	4	1	2	3	4	1	2	3	4	1	2	3	4
	varieties															
cut 1	2	4	4	-	1	5	4	-	2	5	3	-	-	6	2	2
cut 2	1	6	3	-	1	6	3	-	1	6	3	-	-	7	3	-
cut 3	1	7	2	-	1	6	3	-	1	8	1	-	-	7	3	-
	ecotypes															
cut 1	1	17	5	-	2	16	5	-	1	17	5	-	4	13	4	2
cut 2	1	17	5	-	1	16	6	-	2	16	5	-	-	20	3	-
cut 3	2	15	6	-	2	2	4	-	5	15	6	-	3	15	4	1

Conclusions Perennial ryegrass genetic resource collections such as that held at Teagasc Oak Park, hold a great potential for improving the quality of ryegrass varieties. Further ryegrass traits should also be examined, *e.g.* digestibility and fatty acids. The high quality ecotypes identified in this study will be investigated further in 2005-2006.

Acknowledgements We are grateful to the grass breeding group in Oak Park for their support and to the Irish Department of Agriculture for partial funding of this study.

Reference
Jafari, A., V. Connolly, A. Frohlich, & E. J. Walsh (2003). A note on estimation of quality parameters in perennial ryegrass by near infrared reflectance spectroscopy. *Irish Journal of Agricultural & Food Research*, 42, 293-300.

Introgression breeding for improvement of winter hardiness in *Lolium /Festuca* complex using androgenenesis

T. Yamada[1], Y.D. Guo[1,2] and Y. Mizukami[3]

[1]National Agricultural Research Center for Hokkaido Region, Sapporo, 062-8555, Japan, Email: Toshihiko.Yamada@affrc.go.jp, [2]China Agricultural University, Beijing,100094, China, [3]Aichi Agricultural Research Center, Nagakute, Aichi 480-1193, Japan

Keywords: androgenesis, *Festulolium*, freezing tolerance, genomic *in situ* hybridisation, pollen fertility

Introduction Intergeneric hybrids between closely related *Lolium* and *Festuca* species are used to broaden the gene pool and provide plant breeders with options to combine complementary traits to develop robust but high quality grass varieties. Androgenesis was found to be an effective procedure for selecting *Lolium-Festuca* genotypes comprising gene combinations rarely or never recovered by conventional backcross breeding programs. Here we describe the optimisation of androgenesis in *Lolium perenne* x *Festuca pratensis*. The male fertility and freezing tolerance of the *Festulolium* microspore-derived progenies were analysed and these progenies were also analysed by using genomic *in situ* hybridisation (GISH). The object of this study is to initiate introgression breeding for the improvement of winter hardiness in *Lolium /Festuca* complex.

Materials and methods Genotypes of *Lolium perenne* x *Festuca pratensis* (*Festulolium* hybrid), 'Prior', 'Bx350' and 'Bx351'were investigated in this study. PG-96 (Guo *et al.,* 1999) with 2 mgl^{-1} 2,4-D, 0.5 mgl^{-1} kinetin was used as embryo (calli) induction media. Calli were transferred to the solid medium 190-2 (Wang & Hu, 1984) supplemented with 0.1 mgl^{-1} 2,4-D, 1.5 mgl^{-1} kinetin for green plants regeneration. The ploidy level of androgenic progenies was analysed by Partec CAII flow cytometry (Münster, Germany) with DAPI staining. GISH was carried out according to the method described by Mizukami *et al.* (1998) with some modification. Androgenic-derived plants were grown outdoor for natural hardening during autumn. Crown tissues each genotype were analysed for freezing tolerance cooled to –17°C. Male fertility was measured by staining pollen with 1% acetocarmine and counting the frequency of stainable pollen grains.

Results The calli and green plants were obtained from all three accessions, but the genotype responses differed; accessions 'Bx350' and 'Bx351' were more active than 'Prior' in androgenesis. Among microspore-derived progenies the diploids were dominant (68.2%). High levels of chromosome pairing and recombination were observed by GISH due to close homology between genomes of *L. perenne* and *F. pratensis*. These androgenic-derived *Festulolium* progenies showed a wide range of variation in freezing tolerance, 19 progenies (6.5%) exceeding that in the *F. pratensis* cv. 'Tomosakae'. More than 60% of flowing progenies produced dehiscing anthers with pollen stainability ranging from 5% to 85% in all three accessions (Table 1). The diploid progenies with both freezing tolerance and fertility potential have been crossed with *L. perenne*.

Table 1 Pollen fertility in amphidiploid *Festulolium* anther-derived progenies. F & PF, fertile and partial fertile, with pollen stainability ranging from 5% to 85%. MS, male sterile, no pollen or very few pollen (<5%) stained

B x 350				B x 351				Prior			
2n=2x=14		2n=4x=28		2n=2x=14		2n=4x=28		2n=2x=14		2n=4x=28	
F+PF	MS	F+PF	MS	F+PF	MS	F+PF	MS	F+PF	MS	F+PF	MS
25	12	8	9	43	19	11	3	26	11	4	4
46.3%	22.2%	14.8%	16.7%	56.6%	25.0%	14.5%	3.9%	57.8%	24.4%	8.9%	8.9%

Conclusions   High frequency androgenesis in *L. perenne* x *F. pratensis* was established. The diploid microspore-derived progenies with both freezing tolerance and fertility potential are promising to introduce winter hardiness of *F. pratensis* to *L. perenne* by backcrossing with *L. perenne* as introgression breeding.

References

Guo, Y.D., P. Sewón & S. Pulli (1999). Improved embryogenesis from anther culture and plant regeneration in timothy. *Plant Cell, Tissue and Organ Culture*, 57, 85-93.

Mizukami, Y., S. Sugita, N. Ohmido & K. Fukui (1998). Agronomic and cytological characterization of F$_1$ hybrids between *Lolium multiflorum* Lam. and *Festuca arundinacea* var. *glaucescens* Boiss. *Grassland Science*, 44, 14-21.

Wang, X. & H. Hu (1984). The effect of potato II medium for Triticale anther culture. *Plant Science Letter*, 36, 237-239.

A new napier grass stunting disease in Kenya associated with phytoplasma

A.B. Orodho[1], S.I. Ajanga[2], P. Jones[3] and P.O. Mudavadi[2]
[1]P.O. Box 1667 Kitale, Kenya, Email: aborodho@yahoo.com, [2]Kenya Agricultural Research Institute RRC Box 169 Kakamega, Kenya, [3]Plant Pathogens Interactions Division, Rothamsted Research, Harpenden, Herts. AL5 2JQ, UK

Keywords: blast, dairy farmers, mycoplasma

Introduction Napier grass (*Pennisetum purpureum* Schum) is a cultivated elephant grass native to Eastern and Central Africa forming the major livestock feed on East African smallholder dairy farms (Valk, 1990) as it is suitable for cut and carry zero-grazing management systems. Although several plant pathogens have been described historically they were seldom severe. However, in the 1970s there was an outbreak of snow mould fungal disease caused by *Beniowskia spheroidea* that attacked most varieties of napier grass. A napier grass variety (clone 13) was bred which is resistant to the disease. In the 1990s two major outbreaks of napier grass diseases occurred in Kenya. In Central Kenya a napier grass head smut caused by *Ustilago kamerunensis* H Sydow and Sydow in 1992 and in Western Kenya a napier grass stunting disease was first reported in Bungoma in 1997. A similar stunting disease had been reported in Uganda (Tilley, 1969), which was suspected to be a virus transmitted by insects. This new outbreak of napier grass stunting disease is of major concern as it attacks all varieties of napier grass. The main objective of this study was to survey the extent of the disease and to identify the organism causing this disease.

Materials and methods A survey was carried in 2001 of 100 farmers' fields in five districts of Western Kenya by a multidisciplinary team of scientists to assess the occurrence of napier grass stunting disease, describe its symptoms, assess its spread and collect samples for laboratory analysis. Samples collected for laboratory analysis included leaves, roots, stems and insects found feeding on the whorls of diseased plants. Individual sections of the leaves, roots and stems from healthy and diseased samples were tested for possible fungal pathogens using a method developed by Lloyd and Pillay (1980). Further samples were used to test for the presence of viruses. A third set of samples was taken to Rothamsted, U.K. to test for both virus and mycoplasma pathogens. At Rothamsted, yellowed and apparently healthy napier grass were grown under quarantine. A total DNA extraction was done from each sample for use as a template in a nested Polymerase Chain Reaction using phytoplasma 16S ribosomal DNA primers P1/P7 and R16F2n/R16R2. A band of 1250-bp rDNA product was amplified from all yellow leaves and in two of three apparently healthy leaves.

Results and discussion The survey and subsequent farm visits indicated that the disease had spread from the original district to five adjacent districts by 2001. The disease had spread to four more districts by 2003. Disease incidence ranged from 10-100%. The highest mean incidence recorded was in Bungoma and the least was in districts further away. No resistant napier grass variety was found. Cutting frequency, low soil fertility and water intensity stress intensify incidence and severity of the disease. The characteristic symptoms of the disease are yellowing of napier grass foliage, reduced leaf size, proliferation of tillers and shortening of internodes resulting in stunted growth. Laboratory analysis of external and internal morphology showed that the disease was not caused by nematodes or fungus or a virus. The DNA analysis showed that the sample grown had yellowed leaves and stunted growth and were phytoplasma positive. Restriction fragment length polymorphism (RFLP) analysis of amplimers showed similar patterns for all samples. BLAST analysis showed the phytoplasmas to be members of 16SrX1 (Rice yellow dwarf group). The highest homology (96%) was with the 16SrX1 Bermuda grass white leaf group of phytoplasma. Thrips, aphids and leaf hoppers were the main insects found feeding in the whorls of diseased plants.

Conclusion A phytoplasma similar to the rice yellow dwarf group is the casual organism of napier grass stunting disease in Western Kenya. Sugarcane and upland rice are possible hosts of this group of mycoplasma. A leaf hopper is the most probable vector. Movements of vegetatively propagated napier grass planting material provides a means for the rapid spread of the phytoplasma. There is a need to develop a napier grass variety resistant to this disease.

References
Llyod, H.L. & M. Pillay (1980). Development of an improved method for evaluating Sugarcane for resistance to smut. *Proceedings of the South African Technologists' Association*, 168-172.
Tilley, G.E.D. (1969). Elephant grass. Kawanda Agricultural Station Report, Kawanda, Uganda.
Valk, Y.S. (1990). Review report of the DEAF Survey during 1989. NDDP (M41/200) Ministry of Livestock Development, Nairobi, Kenya.

Studies of seed characteristics of ecotypes of lucerne, *Bromus* and *Agropyron* in response to *Fusarium oxysporum* and *F. solani*

M.A. Alizadeh

Research Institute of Forest and Rangeland, P.O. Box 13185-116Tehran, Iran, Email: Alizadeh@rifr-ac.ir

Keywords: *Bromus*, lucerne, *Agropyron*, *Fusarium oxysporum*, *F. solani*, germination

Introduction Vigorous seeds and seedlings are more resistant to pathogens than non-vigorous seeds and seedlings (Kim, 1994). Therefore, it is necessary to assess seed and seedling performance in response to seed borne fungi.

Material and methods Seed samples were disinfected with detergent and placed in Petri dishes and inoculated with two levels of spores of two species of *Fusarim*. The samples were germinated in a germinator at 20°C with 1000 lux light under laboratory and greenhouse conditions. The percentage and speed of germination were recorded at days 3, 6, 9, 12 and 15 according to Maguire (1962). On day 15 the shoot : root ratios of randomly selected seedlings were measured according to Lekh & Khairwal (1993). Vigour index was measured according to Abdul-baki & Anderson (1973).

Results Vigour index in the greenhouse was reduced by both *Fusaruium* spp.. Level of infection gave contradictory results, because the ecotypes responded differently. The root/shoot ratio was not affected in the greenhouse, but in the laboratory *F. oxysporum* infection significantly reduced this ratio. Speed and percentage of germination were reduced by *Fusarium* infection. (Table 1).

Table 1 Mean of the main characteristics of seeds of 13 ecotypes of t *Agropyron, Bromus* and lucerne in response to two species of *Fusarium*

Vigour index		Root length /shoot (mm)		Speed of germination.		Germination (%)		
GRH	Lab.	GRH.	Lab.	GRH.	Lab.	GRH.	Lab.	
53.11a	64.7 a	0.42 a	1.02a	13.63a	15.07a	81.33a	95.69a	Control
40.55b	56.01 b	0.45 a	1.06a	10.67b	13.61b	64.33a	84.62b	SO1
42.27b	63.07 a	0.47 a	1.03a	11.79b	14.66a	67.62a	91.28a	SO2
38.48b	58.41 b	0.42 a	0.88b	10.06b	13.89b	58.67a	88.56a	OX1
37.51b	54.54b	0.49a	0.93 b	9.76b	12.51c	58.56a	80.62b	OX2
42.38	59.34	0.45	0.97	11.8	13.95	66.1	88.15	Mean
4.8	3.76	0.044	0.05	1.28	0.77	7.09	4.44	LSD

GRH=greenhouse, Lab=laboratory, SO1, SO2 = levels 1 and 2 of spore inoculation for *Fusarium solani*. OX1 and OX2= levels 1 and 2 of spore inoculation for *Fusarium oxysporum*. Data in columns with the same are not significantly different (P≤5%)

Table 2 Compound analysis of variance for seed characteristics of 13 ecotypes of species of *Agropyron, Bromus* and lucerne in response to two species of *Fusarium* under laboratory and greenhouse conditions

Vigour index	Root length /shoot (mm)	Speed of germination	Germination %	Df	
2801.33^{**}	8465.26^{**}	746.0^{**}	4741.26^{**}	1	Condition
7134.22^{**}	375.07^{**}	121.43^{**}	2680.58^{**}	12	Ecotype
2024.02^{**}	8.6^{**}	121.10^{**}	4118.50^{**}	4	Fungi
6229.86^{**}	354.87^{**}	120.28^{**}	2689.74^{***}	12	Condition x Ecotype
267.20^{**}	8.35^{**}	14.38^{ns}	661.55^{*}	4	Condition x Fungi
246.25^{**}	5.29^{**}	13.90^{**}	367.33^{*}	48	Fungi x Ecotype
201.04^{*}	5.30^{**}	15.10^{**}	389.28^{*}	48	Condition x Ecotype x Fungi
22.62	18.61	21.66	20.46		CV

*, ** and ns: Significant at the 5%, 1% levels and non-significant respectively

References

Abdul-baki, A. A. & J. D. Anderson, (1975). Vigour determination in soybean seed by multiple criteria. *Crop Science,* 13, 630-633.

Lekh, R. & I. S Khairwal,. 1993. Evaluation of pearl millet hybrids and their parents for germinability and field emergence. Indian *Journal of Plant Pathology*, 2, 125-127.

Kim, S.H., Z.R Choe, J.H., Kang, L.O., Copeland & S.G. Elias (1994). Multiple seed vigour indices to predict field emergence and performance of barley. *Journal of Seed Science and Technology*, 22, 59-68.

Maguire, J.D. (1962): Speed of germination: aid in selection and evaluation for seedling vigour. *Crop Science*, 2, 176-177.

Genetic analysis of the interaction between the host perennial ryegrass and the crown rust pathogen (*Puccinia coronata* f.sp. *lolii*)

P.M. Dracatos[1,3], J.L. Dumsday[1,3], R.S. Olle[2,3], N.O.I. Cogan[1,3], M.P. Dobrowolski[2,3], K.F. Smith[2,3] and J.W. Forster[1,3]

[1]*Primary Industries Research Victoria, Plant Biotechnology Centre, La Trobe University, Bundoora, Victoria 3086, Australia*
[2]*Primary Industries Research Victoria, Hamilton Centre, Hamilton, Victoria 3300, Australia*
[3]*Molecular Plant Breeding Cooperative Research Centre, Australia.Email: peter.dracatos@dpi.vic.gov.au*

Keywords: crown rust, SSR, genetic diversity

Introduction Crown rust (*Puccinia coronata* f.sp *lolii*) is the most important fungal pathogen of perennial ryegrass (*L.perenne* L.). The physiological effects associated with infection include reduction of water soluble carbohydrate (WSC) reserves, causing decreased dry matter yield, digestibility and palatability for herbivores reared for meat, milk and wool production. Phenotypic variability of rust-infection in perennial ryegrass is likely to be due to environmental effects, as well as the interaction of defence and resistance genes in the grass and virulence genes in the pathogen. Classical and molecular genetic marker-based studies have previously detected both qualitative and quantitative resistance, due respectively to major genes and quantitative trait loci (QTL). In addition, evidence for physiological race variation has been demonstrated for *P. coronata* f.sp. *avenae*, the causative organisms of crown rust in oat, and has been inferred for *P. coronata* f.sp. *lolii*. Evaluation of genotypic variation in both the host and pathogen is consequently important for the analysis of the interaction.

Materials and methods Candidate gene-based genetic markers related to the plant defence response (DR genes) were added to the genetic map of the $F_1(NA_6xAU_6)$ mapping family as restriction fragment length polymorphism (RFLPs), and locus-specific single nucleotide polymorphism (SNP) markers were developed from both DR and resistance gene analogue (RGA) genes. The co-location of functionally-associated markers and QTLs for disease resistance was evaluated. Assessment of genetic variation and race structure within the pathogen is based on the use of gene-associated simple sequence repeat (SSR) markers and sampling at the single pustule level. A set of 55 unique SSR markers were developed from crown rust expressed sequence tags (ESTs) and characterised in terms of motif structure, sequence annotation, efficiency and specificity of amplification and polymorphism detection. A method based on multiple displacement amplification (MDA) was used to obtain templates from single pustules, in association with a detached leaf assay method.

Results and Conclusions In a preliminary analysis, genetic variability was detected within and between isolates from geographical locations in the UK, Japan, Australia and New Zealand (Fig 1). The sources of variation may include spore migration, mutation or sexual recombination within the pathogen. Further studies of Australian populations will provide evidence to compare between these hypotheses.

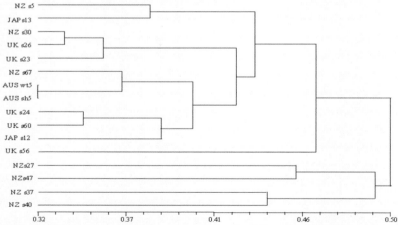

Figure 1 UPGMA dendrogram based on the analysis of variation of 12 EST-SSR markers on selected single pustule samples from UK, Japan, Australia and New Zealand

Molecular characterisation of bacterial wilt resistance in *Lolium multiflorum* Lam.

B. Studer[1], B. Boller[1], F. Widmer[1], U.K. Posselt[2], E. Bauer[2] and R. Kölliker[1]

[1] Agroscope FAL Reckenholz, Swiss Federal Research Station for Agroecology and Agriculture, CH-8046 Zurich, Switzerland

[2] State Plant Breeding Institute, University of Hohenheim, D-70593 Stuttgart, Germany

Email: bruno.studer@fal.admin.ch

Keywords: bacterial wilt resistance, Italian ryegrass, genetic mapping, QTL analysis

Introduction Italian ryegrass (*Lolium multiflorum* Lam.), a forage grass of prime importance throughout the world, is adversely affected by the pathogen *Xanthomonas translucens* pv *graminis*. Breeding for resistant cultivars is the only practicable means of disease control. However, the inheritance of bacterial wilt resistance is largely unknown. The aim of our research is to elucidate genetic control of bacterial wilt resistance using molecular technologies such as genetic linkage mapping and the analysis of quantitative trait loci (QTL).

Materials and methods A mapping population consisting of 306 F_1 progenies of a pair-cross between a susceptible and a resistant genotype of unrelated cultivars has been established. A high density genetic linkage map of dominant amplified fragment length polymorphism (AFLP) and codominant microsatellite markers has been constructed using a two way pseudo-testcross strategy (Grattapaglia & Sederoff, 1994).

Resistance to bacterial wilt of the mapping population was examined in greenhouse and field experiments by artificial inoculation using a well characterised bacterial isolate. Additionally, the susceptible and the resistant parental population was assessed in the field as reference. Disease progress was monitored using a score from 1 (no symptoms, plant is resistant) to 9 (plant is perishing and highly susceptible). The area under disease progress curve (AUDPC) as a measure of quantitative disease resistance was calculated according to the trapezoid rule (Jeger & Viljanen-Rollinson, 2001)

Results In the greenhouse, significant segregation for resistance was observed (mean disease score (MDS)=6.68, standard deviation (SD)=1.67). The skewed, bimodal frequency distribution of AUDPC disease values (Figure 1) indicates the resistance to be controlled by several genes rather than one single major gene.

In the field, the MDS of the mapping population was lower (MDS=2.76, SD=1.04) when compared to the greenhouse experiment, but differed significantly ($P<0.05$) from the MDS of the resistant and the susceptible reference population. Moreover, the results were supported by a high operative heritability of 77 % and field and greenhouse resistance data were highly correlated (r=0.67 $P<0.01$).

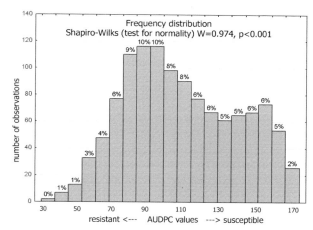

Figure 1 Frequency distribution of greenhouse AUDPC values for F_1 progenies

Conclusion Results of bacterial wilt resistance evaluation in the greenhouse and the field were highly correlated and indicate the resistance to be of quantitative nature influenced by major genes. The significant segregation for resistance within the mapping population and the genetic linkage map will serve as a valuable basis for QTL analysis and the identification of genomic regions associated with bacterial wilt resistance which are currently in progress.

References

Grattapaglia, D. & Sederoff R. (1994). Genetic linkage maps of *Eucalyptus urophylla* using a pseudo-testcross mapping strategy and RAPD markers. Genetics, 137, 1121-1137.

Jeger, M. J. & Viljanen-Rollinson, S. L. H (2001). The use of the area under the disease-progress curve (AUDPC) to assess quantitative disease resistance in crop cultivars. Theoretical and Applied Genetics, 102, 32-40.

Discriminating stay-green grasses using hyperspectral imaging and chemometrics

J. Taylor, B. Moore, J.J. Rowland*, H. Thomas and H. Ougham
*Institute of Grassland and Environmental Research, Plas Gogerddan, Aberystwyth SY23 3EB and *Department of Computer Science, University of Wales, Aberystwyth, SY23 3EB, UK. Email: janet.taylor@bbsrc.ac.uk*

Keywords: heterozygote, homozygote, machine learning, principal components analysis, trait screening

Introduction Screening of plant collections for traits can be expensive, in terms of the number of plants to be screened, the duration of the plant lifecycle and the required observations. This study describes the application of a non-invasive method, hyperspectral imaging, combined with multivariate analysis, to distinguish between homozygous wild-type (YY) *Lolium multiflorum* and *Lolium multiflorum* F2 back cross plants heterozygous for y, a recessive *Festuca pratensis* stay-green gene (Thomas *et al.*, 1997).

Materials and methods Plants were maintained and propagated in a frost-free greenhouse as vegetative tillers grown in soil-based compost. Four comparable leaf segments were imaged from 4 biological replicates of homozygous and 4 heterozygous plants, on the same day and under identical conditions. Images were acquired in the range 450nm to 720nm at a resolution of 2nm. The hyperspectral imaging system comprised a SenSys CCD camera coupled to a CRI VariSpec LCTF fitted with a hot mirror, with a standard F-mount 35mm Nikon lens. Images were processed using Matlab 6.5.1. Spectra were extracted from individual pixels in each leaf image and corrected for background reflectance using a Spectralon™ reflectance standard. Individual pixel spectra were used in the modelling, to utilise spatial and well as spectral information. One biological replicate of each of the homozygous and heterozygous samples were removed from the dataset to form an independent test set (13500 spectra). Remaining replicates were further combined into a training data set (40740 spectra).

Results Unsupervised learning was carried out using Principal Components Analysis (Jackson, 1991) (Matlab 6.5.1, The Mathworks, (www.mathworks.com), running under Windows XP). The training data set was used to form the PCA model. Data were centred and scaled prior to analysis. The first four PCs accounted for 87.3%, 3.1%, 0.9% and 0.8% of the total variance, respectively. The remaining independent test samples were projected into the original PC space; these are shown in Figure 1. While there are two distinct groups in PC space, there is some overlap of the groups using unsupervised learning. A java implementation (WEKA, www.cs.waikato.ac.nz/ml/weka/) of a supervised decision tree algorithm (C4.5, Quinlan, 1993) was applied to the training set, using 3 fold cross validation. The model produced only misclassified 90 out of 40740 spectra, an accuracy of over 99%. The resultant confusion matrix is shown in Table 1.

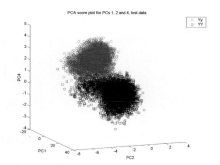

Table 1 Confusion Matrix for C4.5 decision tree

Training set (cross validated)		
Yy	YY	← classified as
17221	34	Yy
45	23440	YY

Figure 1 PCA Scores for components 1, 2 and 4

Conclusions Determination of the allele status of the F2 backcross population is currently verified by a further test cross, which can add up to a year to a breeding programme or genetic study. The combination of hyperspectral imaging, together with chemometric methods indicate that a rapid screening for this trait may be developed. The next stage step is to screen a wider number of plants to further validate the models described.

References
Thomas, H *et al.* (1997) Introgression, tagging and expression of a leaf senescence gene in Festulolium. New Phytologist 137: 29-34.
Jackson, J.E. (1991). A Users Guide to Principal Components. John Wiley and Sons, New York.
Quinlan, J.Ross (1993). C4.5 Programs for Machine Learning. Morgan Kaufmann, San Mateo, CA

Non-destructive assessment of quality and yield for grass-breeding

A.G.T. Schut[1], M.J.J. Pustjens[2], P. Wilkins[3], J. Meuleman[4], P. Reyns[2], A. Lovatt[3] and G.W.A.M. van der Heijden[1]
[1]Plant Research International, P.O. Box 16, 6700 AA, Wageningen, The Netherlands. Email: tom.schut@wur.nl
[2]Advanta Seeds B.V., P.O. Box 1, 4410AA Rilland, The Netherlands
[3]Institute of Grassland & Environmental Research, Plas Gogerddan, Aberystwyth, SY23 3EB Wales, UK
[4]Agrotechnology & Food Innovations B.V., P.O. Box 17, 6700 AA, Wageningen, The Netherlands

Keywords: grass, NIRS, measurement, spectroscopy, image analysis

Introduction Selection of cultivars has, until now, been based mainly on dry matter (DM) yields because of the high costs of sampling and chemical analysis. Imaging spectroscopy could reduce costs by limiting sampling and harvesting of individual plots to reference samples (Schut et al., accepted). In this study, the prediction accuracy of DM yields and chemical composition with imaging spectroscopy is evaluated for cultivar selection purposes.

Materials and methods A total of 13 experiments for cultivar selection were used with 6 experiments in Rilland, 3 experiments in Moerstraten (The Netherlands) and 4 experiments in Aberystwyth (Wales, UK). There were 3 and 4 replicates in experiments in the Netherlands and Wales respectively. Plot sizes varied from 3-5 m^2. From these experiments, the July harvest in Rilland and the July and September harvests in Moerstraten and Aberystwyth were used. The Imspector Mobile was used to record 2D images and hyper-spectral image-lines (Molema et al., 2003). This instrument combines a high spatial (1mm^2) and high spectral (9-13nm) resolution with a wavelength range from 450-1680nm. Images were analysed and ground coverage and reflectance spectra of grass leaves were calculated. After screening of the spectra and laboratory data, calibration models were built, with a leave-n-out partial least squares procedure. The root mean squared errors of cross validation were calculated with observations excluded from the calibration set. Q^2 values can be interpreted as the fraction of variation explained by the model. After recording 8-12 images per sensor on each plot, plots were harvested with a Haldrup and samples were taken for laboratory analysis. Samples taken in Wales were gravimetrically and chemically analysed for DM content and for concentrations of water soluble carbohydrate (sugars) and N. The samples taken in the Netherlands were analysed with Near Infrared Spectroscopy for crude protein (CP), total sugars and acid detergent fibre (ADF).

Results In Table 1 prediction results are presented. The statistical models explained a large amount of variation (Q^2 values between 0.55 and 0.91). Q^2 values in the Moerstraten experiments were slightly lower than in the Rilland or Aberystwyth experiments, partly due to limited variation. Relative errors were below 10% for all variables, except for DM yield in Aberystwyth.

Table 1 Mean Q^2 values, root mean squared error of cross validation (SECV) and relative error (RE, calculated as SECV / mean) per harvest, averaged over replicates.

Variable / Location	Rilland (N=130)			Moerstraten (N=214)			Aberystwyth (N=68)		
	Q^2	SECV	RE, %	Q^2	SECV	RE, %	Q^2	SECV	RE, %
DM yield (kg DM/ha)	0.93	146	6.7	0.69	166	7.2	0.86	111	11.5
DM content (%)	0.76	0.94	3.7	0.76	0.79	3.7	0.76	0.94	7.5
CP (g/kg DM)	0.91	8.43	5.5	0.55	6.10	3.9	*	*	*
Sugars (g/kg DM)	0.88	12.1	6.3	0.89	11.1	7.3	*	*	*
ADF (g/kg DM)	0.81	10.1	5.9	0.90	11.4	5.2	-	-	-

*Results of laboratory analysis not yet available.

Conclusions Currently, breeding and evaluation of forage grasses is expensive, largely because of the high costs of sampling for dry matter and chemical analysis of several cuts each year. Imaging spectroscopy may provide a cheaper and much faster means to measure chemical composition and dry matter yield. This study shows that DM yield and CP, sugar and ADF content of ryegrasses can be measured accurately with imaging spectroscopy and confirms results of earlier work (Schut et al., accepted). Imaging spectroscopy is a very promising method for selecting candidate varieties based on the productivity and quality of herbage.

References

Molema G.J., J. Meuleman, J.G. Kornet, A.G.T. Schut & J.J.M.H. Ketelaars, 2003. A mobile imaging spectroscopy system as tool for crop characterization in agriculture. In: A. Werner&A. Jarfe (Eds), 4th European conference on precision agriculture, Berlin, pp.499-500
Schut A.G.T., C. Lokhorst, M.M.W.B. Hendriks, J.G. Kornet & G.J. Kasper, Accepted. Potential of imaging spectroscopy as tool for pasture management. Grass and Forage Science

Root senescence in red clover (*Trifolium pratense* L.)

K.J. Webb, E. Tuck and S. Heywood
*Institute of Grassland and Environmental Research, Plas Gogerddan, Aberystwyth, Ceredigion SY23 3EB, UK,
Email: judith.webb@bbsrc.ac.uk*

Keywords: red clover, roots, senescence

Introduction Legume root systems form a mosaic of living, ageing and dead roots and nodules. The balance between these stages alters during plant development. Stressful events (drought, temperature change, reduced carbon supply, etc.) disturb the balance (Butler *et al*., 1959). Effects of root and nodule death on soil structure, composition and leaching and on plant persistency are understood poorly. Plants with differing senescence patterns are useful tools to study these effects. Molecular studies of root senescence need detailed knowledge of the process and timing of root senescence and death. Biochemical and histochemical markers of senescence were used to generate preliminary results of the effects of reduced carbon input, temporary (by defoliation, D) or permanent (by defoliation and shading, DS) on red clover shoot survival and root death.

Materials and methods In a controlled environment (20°C/15°C light/dark; 16h photoperiod; 400uM/m/sec; nitrogen-free nutrients), nodulated clover seedlings cv. Milvus were grown for 7 weeks in Agsorb. At day 0, plants were either defoliated (D), defoliated and heavily shaded with black polythene (DS), or left intact (control). Shoot re-growth, root death index (RDI, based on quantitative Evan's Blue staining (Baker & Mock, 1994)) and root catalase activities (Doulis *et al*., 1997) were measured (3 replicates/treatment) at 0, 7, 14 and 21 days. After 21 days, DS plants were grown for a further 14 days in the light before assessment of RDI. Statistical analysis was by ANOVA (Genstat).

Results Red clover plants recovered from temporary (D) but not permanent (DS) reduction in carbon supply (Figure 1A). Quantitative differences in RDI were not significant between any of the treatments for the first 14 days, indicating that similar proportions of cells of these roots were still alive (Figure 1B). By 21 days, DS roots were significantly more strongly stained ($p < 0.05$) indicating a higher level of cell death. Root catalase activity showed a similar pattern (Figure 1C); there were no significant differences in catalase activity between control, D and DS plants at 7 or 14 days. Root catalase activity of DS plants increased about 8-fold by day 21 ($p < 0.05$). After a further 14 days growing in the light, all roots from DS plants stained strongly with Evan's Blue, resulting in OD similar to dead, control roots (Figure 1B).

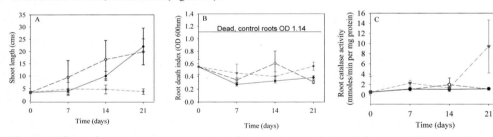

Figure 1 Effects of temporary or permanent reduction in carbon supply in red clover cv. Milvus over time. **A** Shoot growth. **B** Root death index. **C** Root catalase activity. —O— Control; —●— Defoliated (D); —▼— Defoliated and shaded (DS).

Conclusions Red clover roots survived temporary reduction in carbon supply (D) and had no change in root senescence. Permanent reduction in carbon supply (DS) caused plant death by 21 days; shoots failed to re-grow when returned to light. These data show that defoliation alone is not enough to trigger root death and provide a temporal framework for studies on differential gene expression during root senescence.

References
Baker, C. J. & N. M. Mock (1994). An Improved Method for Monitoring Cell-Death in Cell-Suspension and Leaf Disc Assays Using Evans Blue. *Plant Cell Tissue and Organ Culture*, 39, 7-12.
Butler, G. W., R. M. Greenwood & K. Soper (1959). Effects of shading and defoliation on the turnover of root and nodule tissue of plants of Trifolium repens, Trifolium pratense and Lotus uliginosus. *New Zealand Journal of Agricultural Research*, 2, 415-426.
Doulis, A. G., N. Debian, A. H. Kingston Smith & C. H. Foyer (1997). Differential localization of antioxidants in maize leaves. *Plant Physiology*, 114, 1031-1037.

Tropical vine legume-maize mixtures for enhanced silage in temperate climates

H. Riday
US Dairy Forage Research Center, USDA-ARS, Madison, WI 53706, USA   Email: riday@wisc.edu

Keywords: silage, legume, maize, intercropping, vines

Introduction Maize silage (*Zea mays*) comprises an increasing proportion of US Midwestern dairy cow rations. A weakness of maize silage is its lower protein levels; however, maize silage protein and biomass yield might be enhanced by intercropping with legumes. Maize vine legume intercropping has been examined in tropical areas in indigenous agricultural systems and for silage (Schaaffhausen, 1963; Solomon and Flores, 1994). Our initial study examined forage quality and dry matter per hectare of tropical vine legumes intercropped with maize for silage in the temperate climate of Wisconsin, USA. Plant introductions of various vine legumes were also evaluated for biomass production under a maize canopy. Existing vine legume molecular maps offer the possibility of mapping QTL for increased biomass accumulation in maize mixtures.

Materials and methods Five legume species-maize mixtures were grown in Wisconsin, USA: common bean (*Phaseolus vulgaris*), lablab, scarlet runner bean, sunn hemp (*Crotalaria juncea*), and velvet bean (*Mucuna pruriens*). Maize was planted May 3rd at standard density and spacing. Two legumes rows were planted between two maize rows on May 17th. Plant height was measured weekly throughout the growing season. During silage harvest, plot dry matter was determined. Two silage subsamples per plot were taken for moisture content, ensilability, and forage quality analysis. Plot components (legume, maize stover, and maize grain) were based on manual separation of additional subsamples.

Results Increased legume silage fractions resulted in increased crude protein but decreased maize grain (Table 1). No significant differences were determined for total dry matter yields or NDF between these five mixtures and the pure maize control. Additionally, moisture levels were acceptable for ensiling (Table 1). In most mixtures maize plants were able to support the legume vine; common bean was an exception. One week after pollen shed common bean vines reached the maize tassels and began to break them. One month after pollen shed common bean vine weight began to break the tops of maize stalks (Figure 1). In the common bean mixture, particularly, grain yield was lost to competition (Table 1). Scarlet runner bean initially grew vigorously but was severely affected by white mold and potato leafhoppers (*Empoasca fabae*). After maize plants dried and the canopy opened, many legume species displayed renewed growth (Figure 1).

Table 1 Silage mixture dry matter, legume, maize stover, and maize grain fractions

Mixture	Silage Dry M.	Legume	Maize Stover	Maize Grain	Crude Protein
	---------------- fraction ----------------				- % -
Common Bean	0.299	0.249	0.427	0.324	9.734
Lablab	0.356	0.045	0.448	0.508	7.537
S. Runner Bean	0.345	0.042	0.449	0.509	7.440
Sunn Hemp	0.363	0.026	0.451	0.524	6.791
Velvet Bean	0.361	0.034	0.449	0.517	7.274
Pure Maize	0.364	0.000	0.465	0.535	6.581
LSD (0.05)	0.029	0.051	0.018	0.044	0.373

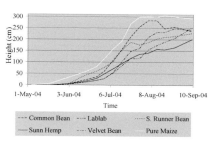

Figure 1 Plant height

Conclusions Intercropping vine legumes with maize for enhanced silage is feasible. In future studies increasing legume density in intercropped silage mixtures may increase the legume fraction. Additionally, using lodge resistant corn varieties may allow increased legume support. In conjunction with this study, improved scarlet runner bean and lablab germplasm were selected for use in a temperate maize silage intercropping system. For many of the vine legumes, mapping populations have been established. Finding QTL for important intercropping traits such as shade tolerance and rapid growth in existing mapping populations would be advantageous. Discovered QTL could be used to screen germplasm and possibly enhance field selection.

References
Schaaffhausen, R.V., (1963). Economical methods for using the legume *Dolichos lablab* for soil improvement, food, and feed. Turrialba. 13:172-179.
Solomon, T., and M. Flores, (1994). Intercropping corn and frijol chinapopo (*Phaseolus coccineus*). Tegucigalpa MDC, Honduras: CIDDICO.

Section 2

Linkage/physical mapping and map-based cloning of genes

Development of simple sequence repeat (SSR) markers and their use to assess genetic diversity in apomictic Guineagrass (*Panicum maximum* Jacq.)

M. Ebina, K. Kouki, S. Tsuruta, M. Takahara, M. Kobayashi, T. Yamamoto, K. Nakajima and H. Nakagawa
Okinawa Prefectural Livestock Experimental Station, 2009-5 Shoshi, Nakijin, Okinawa 905-0426, Japan,
Email:ebinamsm@pref.okinawa.jp

Keywords: apomixis, guineagrass, phylogeny, SSR

Introduction Guineagrass is an important and widely grown tropical forage grass. Despite its importance and increasing popularity, only little is known about its genetic diversity (Ebina *et al.*, 2001). Such information is useful for the selection of diverse parents in breeding programmes. Moreover, no simple sequence repeat (SSR) markers have been reported in any apomixis species. In this study SSR markers were developed and used to investigate genetic diversity in germplasm of apomictic guineagrass.

Materials and methods For the development of SSR markers, genomic DNA was isolated from the cultivar 'Natukaze'. An enriched SSR library was developed according to Yamamoto *et al.* (2003). EST-derived SSR markers were also used in this study. The ESTs were derived from a cDNA library of immature flowers of guineagrass. The SSR amplification method was according to the protocol of the Maize Genome Database (http://www.maizegdb.org). SSR banding patterns were detected by an ABI 310 (Applied Biosystems). The phylogenetic analysis was conducted by Diversity Database software provided by Bio-RAD laboratories.

Results 14 primers from the enriched library and 21 primers from the EST-derived SSR primers were designed successfully. Among these, 20 primers indicated polymorphic banding patterns in the first 12 guineagrass accessions. Results for 13 out of the 20 SSR primers are summarized in Table 1. Allelic differences among 96 accessions for each of the SSR markers ranged from 6 to 39. Genotype variations ranged from 6 to 54 among the 96 accessions. Therefore, the PD (Power of Discrimination) values were also high ranging from 0.485 to 0.943. The phylogenetic analysis was carried out using data obtained from the same experiment as Table 1 (data not shown).

Table 1 Summaries of SSR primers, allelic divergences, genotype variations, Ho, He and PD

Primer name	sample no.	allele no.	genotype no.	size range	Ho	He	PD
GNK01-3	94	11	36	125-138	0.693	0.833	0.943
GNK01-4	96	6	12	144-173	0.142	0.418	0.576
GNK01-5	96	17	43	208-261	0.572	0.851	0.944
GNK02-2	92	39	54	140-228	0.710	0.942	0.947
GNK02-3	96	19	54	174-221	0.655	0.899	0.943
GNK02-4	95	20	49	166-193	0.685	0.934	0.939
GNK04-2	96	8	10	94-162	0.336	0.478	0.740
GNK05-1	96	15	35	99-140	0.387	0.738	0.888
GNK03-e2	96	9	21	153-178	0.718	0.739	0.769
GNK03-e4	96	13	29	88-110	0.731	0.780	0.903
GNK03-e19	96	17	53	135-171	0.657	0.905	0.905
GNK03-e23	96	13	27	135-171	0.395	0.830	0.924
GNK03-e47	96	7	6	145-159	0.185	0.289	0.485

[#] Ho; Observed Heterozygosity, He; Expected Heterozygosity, PD; Power of Discrimination.
[#] e contained in Primer name indicates EST-derived SSR marker.

Conclusions Guineagrass is a faculitve apomictic species, however, a large amount of allelic and genotypic variation was observed in this study. The accessions, which were collected from all over native Guineagrass areas in east and south Africa, have wide genetic diversity. This indicates that sexual propagation has occurred and has resulted in considerable diversity in apomictic guineagrass in nature.

References
Yamamoto *et al.* (2002) Microsatellite markers in peach [Prunus persica (L.)Batsch] derived from an enriched genomic and cDNA libraries. *Molecular Ecology Notes,* 2: 298-301.
Ebina, M. and H. Nakagawa(2001).RAPD Analysis of Apomictic and Sexual line in Guineagrass (*Panicum maximum* Jacq.). *Grassland Science,* 47,251-255.

Construction of microsatellite-enriched libraries for tropical forage species and characterization of the repetitive sequences found in *Brachiaria brizantha*

L. Jungmann[1,2], C.B. do Valle[1], P.R. Laborda[2], R.M.S. Resende[1], L. Jank[1] and A.P. de Souza[2]
[1]*Plant Biotechnology Laboratory, Embrapa Beef Cattle, CP 154, CEP 79.002-970. Campo Grande, MS, Brazil*
[2]*CBMEG, Universidade Esdatual de Campinas, CP 6010, CEP 13.083-875, Campinas, SP, Brazil*
E-mail: jungmann@cnpgc.embrapa.br

Keywords: tropical forages, molecular markers, microsatellites, breeding program

Introduction The Brazilian cattle herd comprises 185 million animals fed with about 177 million hectares of native and cultivated pastures (IBGE, 2002). Of the grass species used for forage in Brazil, the African genus *Brachiaria* is the most widely planted, followed by *Panicum*, which also has an African origin. Legumes of the *Stylosanthes* genus, native to the South America, have emerged in the last few years as potential forage species for use with the grasses. These forage species have been bred at Embrapa Beef Cattle and the breeding programs have shown the need for more genetic information including the use of molecular markers. The objectives of this work were to construct microsatellite-enriched genomic libraries for 5 species of *Brachiaria* (*B. brizantha, B. decumbens, B. dictioneura, B. humidicola and B. ruziziensis*), for *P. maximum* and for *S. capitata*, and to characterize the microsatellites found in *B. brizantha*.

Materials and methods One genotype from the germplasm collection of EMBRAPA for each species mentioned above was used to develop libraries enriched for (AG)n and (AC)n dinucleotides. Construction of genomic libraries was carried out as described previously by Billotte *et al.* (1999). Basically this involves (1) genomic DNA digestion with a restriction enzyme, (2) ligation of oligonucleotide adaptors to fragments, followed by PRC amplification, (3) hybridization of microsatellite containing fragments with (CT)8 and (GT)8 biotinylated oligonucleotides, (4) recovery of hybridized fragments with streptavidin coated magnetic beads, followed by PCR amplification and (5) cloning of amplified fragments in the pGEM-T vector. The presence of microsatellite containing fragments in the library was confirmed after sequencing 20 clones from each library. Sequences were analyzed using the Simple Sequence Repeat Identification Tool-SSRIT (Temnykh *et al.*, 2001) to identify those containing simple perfect repetitive motifs.

Results A total of seven libraries enriched for both AC and AG were constructed, one for each species. Libraries had an average insert size of 700 bp. Sequencing of 20 clones per library showed that about 80% had repetitive motifs. Analysis of 212 *B. brizantha* clones carried out using the SSRIT (which searches only for simple and perfect motifs) found 360 repetitive motifs, 298 (82.7%) being dinucleotides and 62 (17.22%) being tri, tetra and pentanucleotides. No hexanucleotides containing perfect motifs was found. Figure 1 shows the frequency of the dinucleotides motifs. As expected due to the enrichment procedure, AC/GT, CA/TG, AG/CT and GA/TC were the most frequent motifs found. Searches for compound and/or imperfect motifs were not carried out. From all the simple and perfect motifs found, only 13 could be considered as microsatellites, based on the number of repetitions of the motifs. From these, 9 were located in regions appropriate for microsatellite amplification primer design.

Conclusions The results demonstrate that SSR enriched libraries for 5 species of *Brachiaria*, and for *Panicum maximum* and *Stylosanthes capitata* can be obtained. Analysis of *B. brizantha* repetitive motifs showed that the most common dinucleotide motifs were AC/GT and AG/CT. From all the repetitive motifs found, only 9 could be used for microsatellite primer design which means that more clones should be sequenced and analyzed.

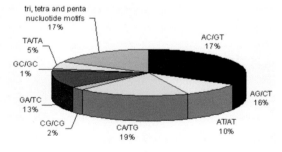

Figure 1 Frequency of repetitive motifs in the *B. brizantha* genome

References
Billotte N, Lagoda PJ, Risterucci A-M & Baurens F-C (1999). Microsatellite-enriched libraries: applied methodology for the development of SSR markers in tropical crops. Fruits 54:277-288.
IBGE (Geography and Statistics Brazilian Institute) (2002) Censo agropecuário. www.ibge.gov.br
Temnykh S, DeClerck G, Lukashova A, Lipovich L, Cartinhour S & McCouch S (2001). Computational and experimental analysis of microsatellites in rice (Oryza sativa L.): frequency, length variation, transposon associations, and genetic marker potential. Genome Research 11:1441-1452.

Isolation of SSR markers from Zoysiagrass

H. Cai, N. Yuyama and M. Inoue

Japan Grassland Farming & Forage Seed Association, Forage Crop Research Institute, 388-5, Higashiakada, Nasushiobara, Tochigi, 329-2742 Japan. Email: hcai@jfsass.or.jp

Keywords: molecular marker, SSR-enriched library, linkage map, zoysiagrass

Introduction The genus *Zoysia* consists of 16 species that are naturally distributed on sea coasts and grasslands around the Pacific. In Japan, five species of natural zoysiagrasses have been identified from southern Hokkaido to the southwest islands. Of these, *Z. japonica* Steud. and *Z. matrella* Merr. have been utilized extensively as turf in Japan and other countries in East Asia. Linkage maps based on RFLP and AFLP markers have been reported in *Zoysia* (Yaneshita *et al.*, 1999, Cai *et al.*, 2004). Simple sequence repeat (SSR) markers have the advantages of being PCR-based and multiallelic. They are highly polymorphic compared to other types of markers such as RFLPs and AFLPs, and are widely used in linkage map construction, gene tagging and QTL mapping. However, only few SSR markers from zoysiagrass have been reported. The objectives of this study were to develop zoysiagrass SSRs in larger numbers and to map them on to an AFLP-based linkage map.

Materials and methods An individual zoysiagrass plant, F08 (progeny of *Z. japonica* X *Z. matrella*), was used to construct an SSR-enriched genomic library. The repeat motifs used were $(CA)_n$, $(GA)_n$, $(AAG)_n$, $(AAT)_n$. The methods of sequencing and primer design etc. were the same as those described by Cai *et al.* (2003). To screen the working primer pairs, a panel consisting of 8 varieties including both *Z. japonica* and *Z. matrella* species was used.

Results A total of 4,000 clones (1,000 clones from each of four libraries) were sequenced. Of these, 768 unique SSR clones which could be used to design primers were identified and about half of the unique SSR clones were from the B library (GA/CT motif) (Table 1). All of the four libraries contained perfect clones with high frequencies, ranging from 67.9 % to 96.0 % and the A and C libraries contained 27.9% and 11.6% compound clones, respectively (data not shown). So far it has been found that out of the 144 primer pairs tested using the screening panel, 132 primer pairs (91.7%) could amplify polymorphic SSR products and two primer pairs amplified PCR products but gave no polymorphism in the eight varieties used (Figure 1). The rest of the 10 primer pairs were considered to be unsuitable for use, because they amplified multi-copy products or amplified no bands.

Table 1 Efficacy of SSR isolation from *Zoysia* libraries

Library	Motif	Unique SSR	Perfect SSR(%)
A	CA/GT	140	95(67.9%)
B	GA/CT	408	371(90.9%)
C	AAG/TTC	125	120(96.0%)
D	AAT/TTA	95	74(77.9%)
Total		768	

Figure 1 PCR products amplified by the ZSSR17 locus in a screening panel including 8 *Zoysia* varieties

Future work After screening the working primers, we will map the markers in a segregating population derived from the selfed progenies of a *Z. japonica* clone, 'F02' (see Cai *et al.*, 2004).

References

Cai, H. W., M. Inoue, N. Yuyama & S. Nakayama (2004). An AFLP-based linkage map of Zoysiagrass (*Zoysia japonica* Steud.). Plant Breeding, (in press).

Cai, H., N. Yuyama, H. Tamaki & A. Yoshizawa (2003). Isolation and characterization of simple sequence repeat markers in the hexaploid forage grass timothy (*Phleum pretense* L.). Theoretical and Applied Genetics, 107, 1337-1349.

Yaneshita, M., S. Kaneko, T. Sasakuma, 1999: Allotetraploidy of Zoysia species with 2n=40 based on a RFLP genetic map. Theoretical and Applied Genetics, 98, 751–756.

Development of SSR markers for variety identification in Italian ryegrass (*Lolium multiflorum* Lam.)

M. Inoue, N. Yuyama and H. Cai

Japan Grassland Farming & Forage Seed Association, Forage Crop Research Institute, 388-5, Higashiakada, Nasushiobara, Tochigi, 329-2742 Japan. Email: minoue@jfsass.or.jp

Keywords: molecular marker, SSR-enriched library, variety identification, Italian ryegrass

Introduction Italian ryegrass (IRG, *Lolium multiflorum* Lam.) is one of the most important cool-season forage grasses in the world, and is the most widely cultivated annual forage grass in Japan. Simple sequence repeat (SSR) markers have the advantages of being PCR-based, multiallelic and possessing high levels of polymorphism. They are very suitable for variety identification, especially for out-crossing species including IRG. The objective of this study was the development of SSR markers for variety identification in IRG.

Materials and methods An individual of IRG variety, Waseaoba, was used to construct the SSR-enriched genomic library. The repeat motifs used were $(CA)_n$, $(GA)_n$, $(AAG)_n$, $(AAT)_n$. The methods of sequencing and primer design etc. were same as that described by Cai *et al.* (2003). To screen the working primer pairs, a panel consisting of five IRG individuals randomly selected from five different varieties and three individuals representing the closely related species perennial ryegrass (PRG), meadow fescue (MF) and tall fescue (TF) were used.

Results A total of 4,000 clones (1,000 clones from each of four libraries) were sequenced. Of these, 796 unique SSR clones which could be used to design primers were identified and all of the four libraries contained perfect clones with very high frequencies, ranging from 88.6 % to 96.4 % (Table 1). So far, it has been found that out of the 140 primer pairs tested using the screening panel, 96 primer pairs (68.6 %) could amplify polymorphic SSR products and 23 primer pairs amplified PCR products but no polymorphisms in the five IRG individuals used (Figure 1). The rest of the 21 primer pairs were considered to be unsuitable for use because they amplified multi-copy products or amplified no bands.

Table 1 Efficacy of SSR isolation from IRG SSR libraries

Library	Motif	Clones sequenced	Unique SSR clones
A	CA/GT	1,000	151
B	GA/CT	1,000	305
C	AAG/TTC	1,000	148
D	AAT/TTA	1,000	192
Total			796

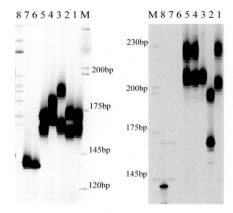

Figure 1 PCR products amplified by the loci B1H16 (left) and B1G14 (right) in the screening panel. Lane 1, IRG 1; lane 2, IRG 2; lane 3, IRG 3; lane 4, IRG 4; lane 5, IRG 5; lane 6, PRG; lane 7, MF; lane 8, TF, M, size marker

Future work After screening the working primers, we will use them to find specific SSR markers to distinguish the most frequently used 12 varieties of IRG in Japan.

References

Cai, H., N. Yuyama, H. Tamaki & A. Yoshizawa (2003). Isolation and characterization of simple sequence repeat markers in the hexaploid forage grass timothy (*phleum pretense* L.). Theoretical and Applied Genetics, 107, 1337-1349.

Development of EST-derived simple sequence repeat (SSR) markers in Italian ryegrass (*Lolium multiflorum* Lam.)

M. Hirata[1], Y. Miura[1], T. Takamizo[2] and M. Fujimori[2]
[1]*Japan Grassland Farming & Forage Seed Association, Forage Crop Research Institute, 388-5 Higashiakada, Nasushiobara, Tochigi 329-2742, Japan, Email: hirata@jfsass.or.jp*
[2]*National Institute of Livestock and Grassland Science, 768 Senbonmatsu, Nasushiobara, Tochigi 329-2793, Japan*

Keywords: simple sequence repeat, EST, Italian ryegrass, mapping

Introduction Italian ryegrass (IRG, *Lolium multiflorum* Lam.) is an important grass species used for agricultural purposes in Japan. Linkage maps in IRG have been constructed using molecular markers including RFLPs, AFLPs (Inoue *et al.*, 2003) and genomic SSRs. The SSR marker is a useful tool for genome analysis because it is PCR-based with a relatively high level of polymorphism, co-dominantly expressed. The EST-SSR markers are expected to be more conserved and have a higher rate of transferability than genomic SSR markers (Scott *et al.*, 2000), and the gene functions of ESTs may be predicted. Ikeda *et al.* (2004) have developed an IRG EST database consisting of 54,811 sequences. The aim of this study was to develop EST-SSR markers using this EST database and assign them into a reference map of IRG.

Materials and methods Clones containing a SSR motif (dinucleotide or trinucleotide repeats; ≥8 repeats) were selected from the IRG EST database. Primers were designed with PRIMER0.5 software. Primer combinations that gave a good signal in the parents were used for analysis. Sixty individuals of the reference population produced by a cross between 11S2 and 11F3 were used for mapping. Linkage analysis was performed with JoinMap software (LOD=4.0) using EST-SSR markers, selected AFLP markers and genomic SSR markers.

Results The efficiency of EST-SSRs development is given in Table 1. Of the 54,811 ESTs used, 103 (0.2%) contained SSR motifs. Among them, 78 clones were unique and 67 clones could be used to design primers. Forty-nine clones amplified products of the expected size in the parents. Of them, 38 clones were dinucleotide repeats(19(GA)n, 12(CA)n, 6(TA)n, 1(GC)n) and 11 clones were trinucleotide repeats (4(CGG)n, 3(CTT)n, 1(CAA)n, 1(CAC)n, 1(CAG)n, 1(CTA)n). The average (maximum) number of dinucleotide and trinucleotide repeats were 12.5(43) and 9.1(15), respectively. Twenty-nine clones detected polymorphism among F_1 individuals. In the linkage analysis, EST-SSR markers were distributed across 7 linkage groups and 2-6 EST-SSR markers were mapped into each of the linkage groups (Figure 1).

Table 1 Efficiency of EST-SSRs development

	No. of clones	%
ESTs	54811	
Clones containing an SSR motif	103	0.2
Unique clones	78	76
Clones for primer design	67	86
Amplified clones in parents	49	73
Polymorphic clones in population	29	59
Mapped clones into reference map	21	72

Figure 1 Linkage map of SSR markers and selected AFLP markers in the 11S2 × 11F3 2-way pseudo-testcross F1 population. (only EST-SSRs are shown)

Conclusions We developed 49 EST-SSRs using the IRG EST database. Twenty-nine clones could detect polymorphism among F_1 individuals. Finally 25 markers generated from 21clones were mapped into 7 linkage groups with 2-6 EST-SSR markers per group. The SSR markers developed in this study will be useful as core markers in genome analysis of IRG and related species.

References
Ikeda S, Takahashi W, Oishi H. (2004). Generation of expressed sequence tags from cDNA libraries of Italian ryegrass (*Lolium multiflorum* Lam.). Grassland Science 49, 593-598.
Inoue M, Gao ZS, Hirata M, Fujimori M, Cai HW (2004) Construction of a high-density linkage map of Italian ryegrass (*Lolium multiflorum* Lam.) using restriction fragment length polymorphism, amplified fragment length polymorphism, and telomeric repeat associated sequence markers. Genome 47:57–65
Scott KD, Eggler P, Seaton G, Rossetto M, Ablett EM, Lee LS, Henry RJ (2000) Analysis of SSRs derived from grape ESTs. Theoretical and Applied Genetics 100, 723-726

Development of a microsatellite library in *Lolium perenne*

J. King[1], I.P. King[1], D. Thorogood[1], L. Roberts[1], K. Skøt[1] and K. Elborough[2]
[1]IGER, Plas Gogerddan, Aberystwyth, Ceredigion, SY23 3EB, UK Email: julie.king@bbsrc.ac.uk
[2]Vialactia Biosciences, PO Box 109-185, Aukland, New Zealand

Keywords: microsatellite, *Lolium perenne,* mapping, linkage group, trait

Introduction *Lolium perenne*, as one of the most important forage grasses of temperate regions, combines a number of very useful characteristics, e.g., good seedling establishment, with a low resistance to drought and limited winter hardiness. Trait selection and introgression can be greatly enhanced by the use of molecular markers in a genetic linkage map. The aim of this project was the generation of a genomic microsatellite library which when combined with microsatellites developed from a Genethresher database would give good genome coverage coupled to high levels of marker polymorphism.

Materials and methods Six microsatellite enriched libraries of a *L. perenne* genotype (Liprior) were constructed using the procedure of Edwards *et al*. (1996). The libraries were based on three enzymes (*Rsa* I, *Alu* I and *Hae* III) with filter selection for two dinucleotide repeats (CA and CT). DNA fragments were cloned and transformed and microsatellite containing clones selected by P_{32} labelling. After sequencing with M13 forward and reverse primers, microsatellite containing sequences were entered into a local database to screen for redundancy. Non redundant primers were used to PCR genomic DNA from two mapping populations based on a forage/amenity cross:

Forage x Amenity = Hybrid   Hybrid x Forage = Forage BC2
                                   Hybrid x Amenity = Amenity BC2

Products were run on a denaturing polyacrylamide gel with bands scored as present or absent. Mapping was performed using JOINMAP 2.0. The two populations were also screened with selected microsatellite primers from the Vialactia Genethresher database. Primers selected were those already mapped to the Vialactia genetic linkage map of *L. perenne* and to the IGER WSC map (Winz *et al*., 2003).

Results After microsatellite selection and redundancy screening, primers to 230 microsatellites were produced. One hundred (45%) proved polymorphic in one or both mapping populations. Maps of seven linkage groups were produced with most microsatellites mapping to the same linkage group in both mapping populations. Screening of the mapping populations for a variety of traits has shown association for heading date to some of the markers on linkage group 4.

Conclusions The genomic microsatellites, whilst quite complex to use because of complicated banding patterns, proved consistent by mapping to the same linkage groups in both mapping populations. Genome coverage varied between chromosomes with, for example, group 7 having very good coverage but linkage group 5 possibly having regions marker free. The combination of the 100 genomic microsatellites produced at IGER with the 400 produced and mapped from the Genethresher database at Vialactia should prove a valuable resource for the analysis of traits throughout the *L perenne* genome. For example, the microsatellites shown to associate with the QTL for heading date on linkage group 4 will prove useful for the selection of this trait.

References
Edwards K.J., J.H.A. Barker, A. Daly, C. Jones and A. Karp (1996). Microsatellite libraries enriched for several microsatellite sequences in plants. *BioTechniques,* 20, 758-760
Winz, R. , Wilcox, P. L. , Echt, C. E. , Armstead, I. P. , Turner, L. B. , Humphreys, M. O. (2003) A *Lolium perenne* framework map based on microsatellite marker loci. Online at: http://www.register-for.com/mbft/ AbstractView.Asp?AbstractId=Abstract2baf Molecular Breeding of Forage and Turf, Third International Symposium, Dallas, Texas, USA, 18-22 May 2003

Tall fescue expressed sequence tag and simple sequence repeats: important resources for grass species

M.C. Saha[1], J.C. Zwonitzer[1], K. Chekhovskiy[1] and M.A.R. Mian[1,2]
[1]The Samuel Roberts Noble Foundation, 2510 Sam Noble Parkway, Ardmore, OK 73401 USA
[2]USDA-ARS, 1680 Madison Avenue, Wooster, OH 44691 USA, E-mail: mcsaha@noble.org

Keywords: tall fescue, ESTs, SSRs, polymorphism, genetic diversity

Introduction Expressed sequence tag (EST) databases have been growing exponentially. The simple sequence repeat (SSR) has become one of the most useful molecular marker systems in plant breeding and is widely used in cultivar fingerprinting, genetic diversity assessment, molecular mapping and marker-assisted selection. ESTs are a potential source for SSRs. The EST-SSR markers are of high quality and have versatile applications in molecular breeding (Bughrara *et al.*, 2003; Saha *et al.*, 2004 a, b, c). Here, we present an overview of our efforts to develop SSRs from tall fescue ESTs and their application for the genetic improvement of forage and turf grass species.

Materials and methods A total of 157 SSRs were developed from the first 20,000 tall fescue ESTs. Amplification, polymorphism and transferability across several grass species were evaluated. The EST-SSRs were mapped in a "two-way pseudo-testcross" tall fescue population, a three generation annual x perennial ryegrass population, and in creeping bentgrass. These have also been used for genetic diversity analysis in darnel ryegrass, fescue species, and Canada and Virginia wildrye populations. EST-SSRs were also used to track the *Festuca* genome introgression in *Festuca* x *Lolium* hybrids and backcross populations.

Results Among the tall fescue EST-SSRs, trinucleotide motifs were the most abundant (70%) type followed by dinucleotides (20%). A higher percent (92%) of the primers produced characteristic bands in at least one species (Saha *et al.*, 2004b). The polymorphism rate was fairly high in tall fescue and ryegrass (66%) and substantially lower in rice (43%) and wheat (38%). We developed a set of EST-SSR markers for 12 grass species covering eight genera, four tribes and two sub-families of the Poaceae family. A tall fescue linkage map was constructed with AFLP and SSR markers (Saha *et al.*, 2004c). Tall fescue EST-SSRs were present in all linkage groups and were the major source of SSR loci. In a ryegrass genetic map, one of the EST-SSR markers (TF21-230) was within 2 cM of the *SOD* locus, which discriminates the annual and perennial types (Warnke *et al.*, 2004). A creeping bentgrass population was also mapped with these loci (Zhao *et al.*, 2003). These EST-SSRs were successfully used to assess the genetic diversity within darnel ryegrass accessions and their relationship with *Festuca* species (Zwonitzer *et al.*, 2004), and variability within and among accessions of Canada and Virginia wildrye (Saha *et al.*, 2004a). Tall fescue EST-SSRs were successfully used to discriminate the *Festuca* and *Lolium* genomes and track the introgression in hybrids and subsequent generations which can reduce the number of backcross generations (Bughrara *et al.*, 2003).

Conclusions Tall fescue EST-SSR markers are valuable genetic markers for the *Festuca* and *Lolium* genera and can also be extended to other forage grass species. A set of markers have been developed for different grass species in which there is limited or no marker data available. The potential for using these markers in tracking genome introgression, genetic diversity, molecular mapping and trait association was established.

References
Bughrara, S., M.A.R. Mian, J. Wang, and M.C. Saha. 2003. Genetic analysis of a serial of *Festuca-Lolium* complex using RAPD and SSR markers. Proceedings of the 3rd MBFT symposium, Dallas, TX p.115
Saha, M.C., M.A.R. Mian, K. Chekhovskiy, and A.A. Hopkins. 2004a. Genetic diversity among a collection of wildrye determined by EST-SSR markers. Proceedings of the 96th annual meetings of ASA, Seattle, WA
Saha, M.C., M.A.R. Mian, I. Eujayl, J.C. Zwonitzer, L. Wang, and G.D. May. 2004b. Tall fescue EST-SSR markers with transferability across several grass species. Theor Appl Genet 109:783-791
Saha, M.C., M.A.R. Mian, J.C. Zwonitzer, K. Chekhovskiy, and A.A. Hopkins. 2004c. An SSR and AFLP based genetic linkage map of tall fescue. Theor Appl Genet (online first, Nov 19)
Warnke, S.E., R.E. Barker, G. Jung, S-C. Sim, M.A.R. Mian, M.C. Saha, L. Brilman, M.P. Dupal and J.W, Forster. 2004. Genetic linkage mapping of an annual x perennial ryegrass population. Theor Appl Genet 109:294-304
Zhao, H., S. Bughrara, M.C. Saha, and M.A.R. Mian. 2003. Genetic map in creping bentgrass based on SSRs, AFLPs and RFLPs. Proceedings of the 3rd MBFT symposium, Dallas, TX p.11
Zwonitzer J.C., M.A.R. Mian, M.C. Saha, and Z. Wang. 2004. SSR Diversity in Darnel Ryegrass and Its Relationship to Fescue Species. Proceedings of the 96th annual meetings of ASA, Seattle, WA

Development of EST and AFLP markers linked to a gene for resistance to ryegrass blast (*Pyricularia* sp.) in Italian ryegrass (*Lolium multiflorum* Lam.)

Y. Miura, C. Ding, R. Ozaki, M. Hirata, M. Fujimori, H. Cai and K. Mizuno
Forage Crop Research Institute, Japan Grassland Farming and Forage Seed Association, 388-5 Higashiakada, Nasushiobara, Tochigi 329-2742, Japan, Email:ymiura@jfsass.or.jp

Keywords: resistance gene, amplified fragment length polymorphism (AFLP), expressed sequence tag (ESTs), cleaved amplified polymorphic sequence (CAPS)

Introduction Ryegrass blast, caused by the fungus *Pyricularia* sp. is one of the most serious diseases of Italian ryegrass (*Lolium multiflorum* Lam.). However, in Italian ryegrass, except for a new cultivar 'Sachiaoba', no resistant cultivars were found in Japan (Mizuno *et al.*, 2003). For these reasons, improving the disease resistance of Italian ryegrass is one of the most important goals in the breeding programs. The aim of this study was to identify genes for resistance to ryegrass blast and to develop EST and AFLP markers linked tightly to the resistance genes.

Materials and methods To develop genetic mapping populations, we generated a segregating F_1 population from a cross between two heterozygous individuals: a resistant individual of cv. 'Sachiaoba' and a susceptible individual of cv. 'Minamiaoba'. Three-week-old F_1 plants were inoculated with a suspension of spores at a concentration of 5×10^4 spores/ml. The location of the resistance gene was determined from the marker order in the linkage group and the phenotypic data of the F_1 population by interval mapping in MapQTL. The position of the gene was estimated at the maximum LOD score with a 1-LOD support interval.

Results The phenotypes of the F_1 population consisting of 161 individuals showed 81 resistant individuals and 80 susceptible individuals. From the set of 512 AFLP primer combinations, we screened 25 combinations with polymorphic bands. All of the screened AFLP markers segregated in a 1:1 (present:absent) ratio in the F_1 population (Figure 1a). Linkage analysis revealed that all 25 markers formed one linkage group with one resistance gene (*LmPi1*) from the resistant parent (Figure 2). Of 30 EST-CAPS markers mapped on a reference population for Italian ryegrass, one linked EST marker, p56, was found in the segregating F_1 population and the parents. The restriction patterns of p56 amplification with *Hha*I in the F_1 population showed two short fragments (Figure 1b). One fragment of 289 bp, derived from the resistant parent, segregated in a 1:1 (present:absent) ratio in the F_1 population and the fragment was corresponding to the resistant allele at the *LmPi1* locus. In our previous study, p56 was mapped on linkage group 5 (lg5) of the reference population. Therefore, the *LmPi1* locus is also located on lg5 in Italian ryegrass.

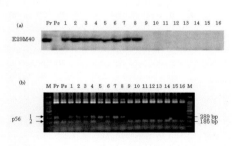

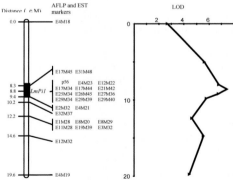

Figure 1 Segregation pattern of markers linked tightly to the resistance gene *LmPi1* in the F_1 population. The first two lanes are the resistant (Pr) and susceptible (Ps) parents, followed by eight resistant (1 to 8) and eight susceptible (9 to 16) individuals of the F_1 population.

Figure 2 Linkage map showing the position of the QTL for the resistance gene *LmPi1*.

Conclusion The results indicate that the p56 marker could be used for the introduction of the *LmPi1* resistant allele into susceptible germplasm to enhance the resistance to ryegrass blast in Italian ryegrass breeding.

Reference
Mizuno K, Yokohata Y, Oda T, Fujiwara T, Hayashi K, Ozaki R, Kobashi K, Ashizawa H, Ushimi T (2003) Breeding of a new variety 'Sachiaoba' in Italian ryegrass (*Lolium multiflorum* L.) and its characteristics. (in Japanese) Bull Yamaguchi Agric Expt Stn 54:11-24

Construction and exploitation of a bacterial artificial chromosome (BAC) library for *Lolium perenne* (perennial ryegrass)

K. Farrar, A.M. Thomas, M.O. Humphreys and I.S. Donnison
Institute of Grassland & Environmental Research, Plas Gogerddan, Aberystwyth, SY23 3EB, UK
Email: *iain.donnison@bbsrc.ac.uk*

Keywords: BAC library, *Lolium perenne*, SNP analysis

Introduction BAC libraries are an important tool in genomics, enabling physical maps, genome sequencing, marker development and map based cloning strategies. A BAC library has therefore been generated for the temperate grass species *Lolium perenne* (perennial ryegrass) which compliments an existing BAC library of the closely related species, *Festuca pratensis* also generated by IGER. Moreover the *L. perenne* BAC library will provide a useful tool for grass comparative genomics to compliment the existing BAC libraries of cereal crops including rice, wheat, barley, *Sorghum* and maize. In particular it will allow a comparison of micro-synteny between this large genome forage crop species and the model small genome monocot species *Orzya sativa*.

Materials and methods High molecular weight (HMW) DNA was isolated from 20 g of leaf material from a diploid *L. perenne* (2n=2x=14) genotype. Nuclei were isolated according to the method of Zhang *et al.* (1995) and the nuclei embedded in agarose plugs. The *L. perenne* HMW DNA was partially digested using *Hind*III and then separated by pulse field gel electrophoresis before cloning into vector pIndigoBAC-5 (Epicentre) following the method adopted by O'Sullivan *et al.* (2001). In addition to filters for hybridisation based screening, DNA pools have been generated to enable a rapid PCR based screen of the library.

Results The library currently consists of > 50,000 clones, picked into 96 well plates, with an average insert size of 100 kb. Additional clones are being generated to ensure that the final library comprises 5 genome equivalents of *L. perenne*. The first half of the library (i.e. comprising 2.5 genome equivalents) has been screened using PCR primers for candidate genes for traits of interest including forage quality, nitrogen stability, control of flowering time, and sugar metabolism. Upstream regions of these genes are being sequenced and from this new PCR primers designed. These primers will be used to amplify the respective alleles from 20 *L. perenne* genotypes which will be sequenced and aligned to derive allele-specific single nucleotide polymorphism (SNP) markers.

Conclusions A BAC library has been generated for the large genome (haploid genome size is estimated at 2034 Mbp; Bennett and Smith, 1976) species *L. perenne*, and this has provided a rapid method for obtaining 5' upstream and other non-coding regions for candidate genes of forage quality traits in this important forage and turf grass species. This *L. perenne* BAC library has been produced as part of the EU Framework 5 project, GRASP (http://www.grasp-euv.dk/). As part of this project, SNPs derived from these BAC sequences will be associated to relevant QTL and the same molecular markers will be tracked in genetically diverse *L. perenne* populations undergoing a range of selection pressures.

References
Bennett M.D. & Smith J.B. (1976). Nuclear DNA amounts in angiosperms. *Philosophical Transactions of the Royal Society of London* B, 274, 227-274.
O'Sullivan D.M., Ripoll P.J., Rodgers M. & Edwards K.J. (2001). A maize bacterial artificial chromosome (BAC) library from the European flint inbred lined F2. *Theoretical & Applied Genetics*, 103, 425-432.
Zhang H.B., Zhao X.P., Ding X.L., Paterson A.H. & Wing R.A. (1995). Preparation of megabase-size DNA from plant nuclei. *Plant Journal*, 7, 175-184.

Characterisation of perennial ryegrass parental inbred lines for generating recombinant inbred lines for fine mapping and gene cloning

U.C.M. Anhalt[1,2], S. Barth[1], T. Schwarzacher[2] and J.S. Heslop-Harrison[2]
[1]Teagasc, Crops Research Centre, Oak Park, Carlow, Ireland; [2]University of Leicester, Department of Biology, Leicester LE1 7RH, UK, Email: uanhalt@oakpark.teagasc.ie

Keywords: inbred lines, SSRs, AFLP, phenotype, inter-specific markers

Introduction Intermated recombinant inbred lines (IRIs) are a powerful tool for fine mapping and cloning of genes. Such population structures have been particularly helpful for cloning of genes in the model genetic plant *Arabidopsis thaliana*. IRIs or recombinant inbred lines (RILs) would be valuable for perennial ryegrass (*Lolium perenne*), but its allogamous character means that the construction of RILs is a difficult task. The international *Lolium* community would benefit from the development of such lines. The aims of the projects are to characterise the parental lines and initial generations at the (1) phenotypic, (2) molecular and (3) molecular cytogenetic level.

Materials and methods Two contrasting highly inbred perennial ryegrass lines were used as parental material for the further construction of recombinant inbred lines. These lines originated from an inter-specific cross between meadow fescue and perennial ryegrass. This original material was backcrossed for several generations to the ryegrass parent and subsequently selfed for ten generations. Plants from two genetic backgrounds were crossed to make an F1.

DNA was isolated from the parental lines, meadow fescue, tall fescue and pure perennial ryegrass with a modified CTAB protocol.

In total, 236 SSRs markers were tested and PCR conditions optimised. The SSR marker set comprises 54 ryegrass specific SSRs markers (CRC Australia) and 25 inter-specific grass SSR markers (Warnke *et al.*, 2004). In addition 157 tall fescue SSRs were used (The Samuel Roberts Noble Foundation and Warnke *et al.*, 2004). 60 EcoRI/MseI enzyme primer combinations were tested and a subset chosen for further analyses. Furthermore some grass RFLP probes are being converted into CAPS markers to enable connection with existing ryegrass maps based on RFLPs. Ten markers are already converted. In addition phenotypic scoring for the description of the parental plant material was carried out.

Results and discussion The perennial lines displayed different phenotypic characteristics, e.g. in plant architecture, levels of disease resistance and nutrient use efficiency. The SSR markers were polymorphic and applicable to perennial ryegrass. Of the 236 used, 127 had not been applied to perennial ryegrass before. The Noble Foundation fescue SSR markers differentiated the parental lines, fescue and *Lolium* material from each other. Cornell and CRC markers were also very useful for molecular characterisation. AFLP markers were appropriate to identify polymorphism between the different parental lines and the different grass species. To improve connections with more genetic linkage maps additional RFLP-CAPS markers will be designed. Further studies at the phenotypic, molecular and molecular cytogenetic level are in progress. The work will yield a set of characterised recombinant inbred lines that will be helpful for the construction of detailed linkage maps, and for the subsequent cloning of genes. Furthermore this work will allow insights into the genome organisation of perennial ryegrass.

Acknowledgments
We are grateful for a research license for 54 SSRs markers to the CRC for Molecular Plant Breeding, Australia and for 157 fescue markers to The Samuel Roberts Noble Foundation, USA.
Especially we acknowledge the contribution of Dr. Vincent Connolly for developing the parental lines. We thank Teagasc for award to a Research Fellowship to UCMA.

References
Warnke, S.E; Baker, R.E; Jung, G; *et al.* (2004) Genetic linkage mapping of an annual x perennial ryegrass population. Theor. Appl. Genet. 109:2, 294-304

A high-density SSR linkage map of red clover and its transferability to other legumes

S. Isobe[1], S. Sato[2], E. Asamizu[2], I. Klimenko[3], N.N. Kozlov[3], K. Okumura[1] and S. Tabata[2]

[1]National Agricultural Research Centre for Hokkaido Region, Hitsujigaoka, Toyohira, Sapporo, 062-8555, Japan, E-mail:sisobe@affrc.go.jp
[2]Kazusa DNA Research Institute, Kazusa-Kamatari, Kisarazu, Chiba, 292-0818, Japan
[3]All-Russian Williams Fodder Crop Research Institute, Lugobaya, Moscow Region, 141740, Russia

Keywords: SSR, linkage map, red clover, transferability, legumes

Introduction A high-density linkage map of red clover was constructed based on SSR and RFLP markers. In order to construct a linkage map with user (breeder) friendly markers; i.e. informative and easy detection, two policies were adopted for marker development. One was that the markers should be derived from cDNA or gene-rich regions, and the other was that the SSR markers should be detected polymorphisms on agarose gels.
We also discuss the transferability of the markers on the map to other red clover germplasm and legumes. Such highly transferable markers could be used to screen anchor markers for both on a consensus map of red clover and other legume species.

Materials and methods One hundred and eighty eight mapping individuals were derived from cross progenies between "HR" (a clone from Swiss variety "Renova" x Japanese variety "Hokuseki") and "R130" (A clone from Russian variety x wild accession collected in the Archangelsk region). SSR-enriched cDNA libraries, normalized cDNA libraries, SSR-enriched genomic libraries and methyl filtrated genomic libraries of red clover were constructed to develop SSR markers. SSR regions were detected by sequencing the libraries, and then primer pairs were designed. Polymorphic bands of SSR markers were detected on 3% agarose gels. A combined genetic linkage map was calculated using JoinMap 3.0. Eighty red clover germplasms and thirteen kinds of legumes including white clover, alfalfa, *Medicago truncatula*, *Lotus japonicus* and soybean were tested by PCR amplification using SSR markers on the map.

Results and discussion A total of 7,244 SSR marker primer pairs were successfully designed from a total of 83,172 sequences derived from the red clover genomic and cDNA libraries. Twenty percent of the primer pairs exhibited polymorphisms within the mapping population on 3% agarose gels. A combined genetic linkage map was constructed with a total of 457 markers, including 338 SSR and 119 RFLP markers.

Approximately 70% of the SSR markers on the map amplified strong bands at corresponding positions in white clover DNA. Also, 33% and 27% SSR markers amplified strong bands at corresponding positions in *Medicago truncatula* and Alfalfa DNA, respectively. This result indicated that the SSRs of red clover were highly conserved between *Trifolium* and *Medicago* species, and that the SSR markers of the present map were transferable to the other forage legumes. The present map should allow us to obtain useful genetic information on red clover, other *Trifolium* species and other members of the legume family.

Conclusions A high-density linkage map of red clover based on SSR and RFLP markers was constructed. The markers were developed taking informative and easy detection into consideration for user-friendly use. The SSR markers were transferable to *Trifolium* and *Medicago* species and could be used as anchor markers in legume species.

Estimation of the coefficient of double-reduction in autotetraploid lucerne

R. Ayadi, P. Barre, C. Huyghe and B. Julier
INRA, Unité de Génétique et d'Amélioration des Plantes Fourragères, 86600 Lusignan, France
E-mail: julier@lusignan.inra.fr

Keywords: chromatid segregation, meiosis, autotetraploid, crossing over

Introduction   Polyploidy plays an important role in the evolution of species and many cultivated species, particularly in angiosperms, are polyploids (Bever and Felber, 1992; Gallais, 2003). Autopolyploid species that show a tetrasomic inheritance have complex genetics. However, some theoretical models were built for: (1) genetic mapping (Hackett *et al.*, 1998), (2) quantitative genetics (Gallais, 2003) and (3) population genetics (Ronfort *et al.*, 1998). But in practice, most data analyses ignore one essential feature of tetrasomic inheritance that is double-reduction. Indeed, in an autotetraploid species, homologous chromosomes can form tetravalents at meiosis. In this case, a double-reduction is observed if crossing-over occurs between a locus and its centromere, and if the sister chromatids migrate to the same pole at anaphase I. The gametes may, thus, carry a pair of sister alleles. Double-reduction frequency is represented by the index of separation (α) (Demarly, 1963; Mather, 1935; Mather, 1936). The parameter α is considered as a product of four probabilities: $\alpha = q\ e\ a\ s$ (Gallais, 2003) in which " q " is the probability of multivalent formation, " e " the probability of first equational division, related to the frequency of crossing-over, " a " is the probability of non-disjunction at first anaphase and "s" is the probability of having two sister chromatids in the same gamete. If separation during anaphase II is random, $s = \frac{1}{2}$. Consequently, α will be low for a gene located in the vicinity of the centromere and will increase with distance between the gene and the centromere. It was demonstrated that double-reduction events alter the rate of progression towards equilibrium under inbreeding or under random mating, modify the recombination rate between loci and also alter the rate of decay of linkage disequilibrium under random mating (Bever and Felber, 1992). Current theoretical models allow drawing genetic maps taking into account double-reduction (Luo *et al.*, 2004). Thus, it is possible to estimate α for codominant loci in tetraploid species. To date, we have few estimates of the double-reduction frequency. Haynes and Douches (1993) on potato and Julier *et al.* (2003) on lucerne found that double-reduction occurs sporadically. In both studies, the low number of progenies hampered a precise estimation of α. The aim of our study was to estimate the frequency of double-reduction in a mapping population of lucerne that includes a large number of individuals.

Materials and methods A mapping population (Julier *et al.*, 2003) containing more than 1000 F1 individuals resulting from a cross between two parental genotypes (Magali 2 and Mercedes 4-11) was obtained. Four microsatellite markers were selected with two of them (MTIC103 and MAA660870) located at the ends of chromosome 8 and two others (AFct45 and MTIC289) located in the vicinity of the centromere of chromosome 7. Theoretical models for double-reduction estimation were applied (Hackett *et al.*, 1998; Luo *et al.*, 2004).

Results and conclusions Three of the microsatellite markers (AFct45, MTIC289 and MTIC103) showed four alleles and one null allele in the population, the fourth marker (MAA660870) had five alleles and one null allele. Detailed information was obtained on the double-reduction for these markers. The results are relevant to the low number of tetravalents observed at meioses (Armstrong, 1954) and the determination of double-reduction frequency allows a better understanding of the functioning of *Medicago sativa*.

References
Armstrong JM (1954) Cytological studies in alfalfa polyploids. Can J Bot 32:531-542
Bever JD, Felber F (1992) The theoretical population genetics of autopolyploidy. Oxford Surv Evol Biol 8:185-217
Demarly Y (1963) Génétique quantitative des tétraploïdes et amélioration des plantes. Ann Amélior Plantes 13:307-400
Gallais A (2003) Quantitative genetics and breeding methods in autopolyploid plants. INRA Editions, Paris, France
Hackett CA, Bradshaw JE, Meyer RC, McNicol JW, Milbourne D, Waugh R (1998) Linkage analysis in tetraploid species: a simulation study. Genetic Research, Cambridge 71:143-154
Haynes KG, Douches DS (1993) Estimation of the coefficient of double reduction in the cultivated tetraploid potato. Theor Appl Genet 85:857-862
Julier B, Flajoulot S, Barre P, Cardinet G, Santoni S, Huguet T, Huyghe C (2003) Construction of two genetic linkage maps in cultivated tetraploid alfalfa (*Medicago sativa*) using microsatellite and AFLP markers. BMC Plant Biol 3:9
Luo ZW, Zhang RM, Kearsey MJ (2004) Theoretical basis for genetic linkage analysis in autotetraploid species. Proc Natl Acad Sci USA 101:7040-7045
Mather K (1935) Reductional and equational separation of the chromosomes in bivalents and multivalents. J Genet 30:53-78
Mather K (1936) Segregation and linkage in autotetraploids. J Genet 32:287-314
Ronfort J, Jenczewski E, Bataillon T, Rousset F (1998) Analysis of population structure in autotetraploid species. Genetics 150:921-930

A core AFLP map of aposporic tetraploid *Paspalum notatum* (Bahiagrass)

J.P.A. Ortiz[1,2], J. Stein[2], E.J. Martínez[1], S.C. Pessino[2] and C.L. Quarin[1]

[1]*Plant Research Central Laboratory, Agronomy Department, National University of Rosario, (2123) Zavalla, Santa Fe, Argentina;* [2]*Institute of Botany of the North-East, Sargento Cabral 2131, (3400) Corrientes, Argentina. Email: jortiz@unr.edu.ar*

Keywords: *Paspalum notatum*, apospory, genetic map, AFLP, molecular markers

Introduction *Paspalum notatum* (Bahiagrass) is a perennial rhizomatous species that reproduces by aposporous apomixis. Tetraploid races (2n=4x=40) are widely distributed from Central to South America and constitute one of the most valuable natural forage grasses for the subtropical areas of Paraguay, southern Brazil and north-eastern Argentina. Apospory in the species is controlled by a single locus, which exhibits a distorted segregation ratio. The objectives of this work were to develop a core genetic linkage map of the species by using AFLP markers and characterize the genomic region related to apospory.

Materials and methods A mapping population of 113 individuals was generated by crossing a tetraploid sexual plant (Q4188) with an aposporous individual from accession Q4117. Progenies were classified for their mode of reproduction (asposporous vs. non-aposporous) by molecular and cytoembryological analyses according to Martínez *et al.* (2003). AFLP markers were generated employing *Eco*RI and *Mse*I primers with 1 and 3 selective bases (Stein *et al.*, 2004). A χ^2 test was used to determine the goodness of fit between the observed and expected genotypes for each class of segregation ratio. Data from the aposporous genotype were analyzed independently using the mapping program JoinMap1.4 with LOD values between 3.0 and 5.0. Male linkage groups were constructed considering single dose (SDA) and distorted alleles. Map units in cM were derived employing the Kosambi mapping function.

Results As a result of using a tetraploid F_1 mapping population derived from non-inbred parents, different allelic configurations per locus were observed (Table 1). Classification of F_1 plants for their mode of reproduction showed 98 non-aposporous and 15 aposporous individuals.

Table 1 Segregation of AFLP markers in the mapping population

Type of marker	Number of markers in each class of segregation ratio						Total
	1:1[a]	3:1[b]	5:1[b]	3:1/5:1[b]	11:1[c]	Distorted[d]	
Paternal	166	23	14	24	-	33	260
Maternal	245	20	11	20	-	16	308
Both	-	60[c]	-	-	38	5	103

[a] single dose alleles (SDAF), [b]double dose alleles, [c]allelic bridges, [d]at p< 0.01

A total of 199 markers segregating from the male parent were used for the linkage analysis. A core AFLP map was built with 146 markers distributed in 22 linkage groups.

The total genetic map distance covered was about 1,430 cM, with an average of approximately 10 cM between markers. The linkage group carrying the locus for apospory (apo-locus) was defined by 28 loci over a distance of 32 cM. Twelve markers were completely linked to the apo-locus, while the rest showed recombination values between 0.6 and 18 cM, either side of it. The complete co-segregation of several markers and apospory confirmed the suppression of recombination in this genomic region reported by Martínez *et al.* (2003). The average distance between markers in the apo-group was about 1 cM.

Conclusions A core genetic linkage map of aposporous tetraploid *P. notatum* was built with AFLP markers. It can be employed for fundamental research as well as for localising genes of agronomic interest. The structure of the linkage group carrying apospory suggests that it could be a large chromosome segment containing several genes. A detailed molecular analysis of this segment, based on the utilisation of the molecular markers linked to apospory, could allow the identification of the gene/s controlling the trait in the species.

References

Martínez, E.J., E. Hopp, J. Stein, J.P.A. Ortiz and C.L. Quarin (2003). Genetic characterization of apospory in tetraploid *Paspalum notatum* based on the identification of linked molecular markers. Mol. Breed. 12 (4): 319-327.

Stein, J., C.L. Quarin, E.J. Martínez, S.C. Pessino and J.P.A. Ortiz (2004). Tetraploid races of *Paspalum notatum* showed polysomic inheritance and preferential chromosome paring around the apospory-controlling locus. Theor. Appl. Genet. 109: 186–191.

Repulsion-phase linkage analysis of tetraploid creeping bentgrass (*Agrostis stolonifera* L.)

S.E. Warnke[1], N. Chakraborty[2] and G. Jung[2]
[1]USDA-ARS, National Arboretum, Washington, D.C. 20002, USA E-mail: warnkes@ars.usda.gov
[2]University of Wisconsin, Department of Plant Pathology, Madison, WI USA

Keywords: linkage analysis, repulsion-phase linkage, genetic markers

Introduction Creeping bentgrass is a cool-season grass species primarily used on golf course greens, tees, and fairways because of its tolerance of low mowing heights and recuperative ability. Creeping bentgrass is tetraploid and outcrossing and has been characterized as having bivalent chromosome pairing through cytogenetic analysis (Jones, 1956) and disomic inheritance based on the inheritance of isozyme markers (Warnke *et al.*, 1998). The objective of this experiment was to evaluate the extent of repulsion-phase linkages of single dose AFLP type markers in a creeping bentgrass mapping population and to infer chromosome pairing behaviour based on the ratio of coupling to repulsion-phase linkages.

Materials and methods A creeping bentgrass mapping population of 90 plants was used to evaluate the ratio of coupling to repulsion phase linkages using the computer program MapMaker 3.0. AFLP type DNA fragments were separated on an ABI 3730XL machine and scored using the computer program Genographer. Two hundred markers segregating in a 1:1 ratio in the progeny were scored and mapped. The ratio of coupling to repulsion-phase linkages was determined as described by Qu and Hancock (2001).

Results The markers scored for each parent in the mapping population and the number of coupling and repulsion phase linkages are shown in Table 1. The ratio of coupling to repulsion phase linkages is very close to a 1:1 ratio in both parents of the mapping population indicating disomic inheritance.

Table 1 Markers scored and the ratio of coupling to repulsion-phase linkages.

	Markers scored	AFLP: S-SAP	Distorted segregations	Repulsion linkages MaxR=37.5cm	Coupling linkages MaxR=37.5cm	p-value 1:1
Female 549	149	123 26	33(22%)	299	302	0.90
Male 372	117	96 21	28(24%)	123	127	0.80

Conclusions The results of this study provide support for genome wide disomic inheritance in tetraploid creeping bentgrass. If polysomic inheritance was operating the ratio of coupling to repulsion-phase linkages would not be expected to fit the exhibited 1:1 ratio and tight repulsion phase linkages would not be present. A thorough understanding of inheritance in tetraploid creeping bentgrass will facilitate the development of new cultivars with improved biotic and abiotic stress resistance.

References

Jones, K. 1956 Species differentiation in *Agrostis* II. The significance of chromosome pairing in the tetraploid hybrids of *Agrostis canina* subsp. Montana Hartm., *A. capillaris* Sibth. and *A. stolonifera* L. J. Genet. 54:377-393.

Qu, L., and Hancock, J.F. (2001) Detecting and mapping repulsion-phase linkages in polyploids with polysomic inheritance. Theor Appl Genet 103:136-143.

Warnke, S.E., Douches, D.S., and Branham, B.E. (1998) Isozyme analysis supports allotetraploid inheritance in tetraploid creeping bentgrass (Agrostis palustris Huds). Crop Sci. 38:801-805.

Towards a genetic map in creeping bentgrass based on SSRs, AFLPs and RFLPs

H. Zhao and S. Bughrara

Department of Crop and Soil Sciences, Michigan State University, East Lansing, MI 48824, USA. Email: bughrara@msu.edu

Keywords: creeping bentgrass, molecular markers, linkage construction

Introduction Creeping bentgrass (*Agrostis palustris*) (2n=4x=28) is commonly used in golf course, putting green, tees and fairways. In spite of the importance of the species in turfgrass industry, the genetic study of the creeping bentgrass has received relatively little attention. Genetic mapping, as a new tool, helps traditional turfgrass breeding methods through the construction of linkage, identification of quantitative trait loci linked to traits of interest, and application of marker assisted selection program. Molecular markers such as AFLPs, SSRs and RFLPs have been used extensively for the preparation of linkage maps of a number of crop species. The objective of this study is to construct a genetic linkage map of creeping bentgrass.

Materials and methods A two-way pseudo-testcross F_1 population is being used to produce a map based on AFLPs (*Pst*I/*Eco*RI and *Mse*I/*Eco*RI), SSRs and RFLPs (anchor probes), to estimate the degree of orthology and colinearity between creeping bentgrass and the Triticeae genomes for all linkage groups, and to investigate the association with genetic markers of several potential quantitative traits: leaf width, flower time, gray snow mold and dollar spot resistance.

Results and discussion An F_1 progeny set comprising 184 individuals, derived from a cross between snow mold resistance and susceptibility clones, was obtained and confirmed by a physiological marker to eliminate any progeny from selfing. The data set included three different segregation patterns: 1:1 for heterozygous markers in one parent and homozygous or null in the other, 3:1 for dominant markers heterozygous in both parents, 1:1:1:1 for co-dominant multiallelic markers. For each marker, a chi-square test (p<0.05) was used to identify deviations from the expected Mendelian ratios. Linkage analysis was carried out using JoinMap v3.0 software with a minimal LOD of 4.0 and a maximum recombination fraction of 3.0 as the group criteria. A composite interval-mapping approach was used for estimate the number of QTLs, the amount of variation explained by each of them, and their position on the genetic linkage maps.

Integration of perennial ryegrass (*L. perenne*) genetic maps using gene-associated SNPs

A.C. Vecchies[1,3], R.C. Ponting[1,3], M.C. Drayton[1,3], N.O.I. Cogan[1,3], K.F. Smith[2,3], G.C. Spangenberg[1,3] and J.W. Forster[1,3]

[1]*Primary Industries Research Victoria, Plant Biotechnology Centre, La Trobe University, Bundoora, Victoria 3086, Australia* [2]*Primary Industries Research Victoria, Hamilton Centre, Hamilton, Victoria 3300, Australia* [3] *Molecular Plant Breeding Cooperative Research Centre, Australia Email: anita.vecchies@dpi.vic.gov.au*

Keywords: EST-SSR, SNP, linkage

Introduction The reference genetic map of perennial ryegrass was developed by the International *Lolium* Genome Initiative (ILGI), using the p150/112 one-way pseudo-testcross population. A selection of public domain genetic markers including RFLPs, detected by wheat, barley, oat and rice cDNA probes, and AFLPs were mapped, allowing studies of comparative relationships between perennial ryegrass and other Poaceae species. The map was enhanced through the addition of unique perennial ryegrass genomic DNA-derived SSR (LPSSR) markers, providing the basis of framework genetic mapping in other populations. In addition, a small number of RFLP loci detected by candidate genes involved in herbage quality traits were added to the map. A second-generation reference genetic mapping family was developed based on the F_1(NA$_6$ x AU$_6$) two-way pseudo-testcross family, generating two parental genetic maps. These maps were populated by genomic SSR loci, EST-RFLP loci and EST-SSR loci (corresponding to multiple functional categories of agronomic importance). A third genetic mapping population based on an interspecific cross between perennial and annual ryegrass genotypes [F_1(Andrea$_{1246}$ x Lincoln$_{1133}$)] generated a map based on LPSSR and EST-SSR markers. Linkage groups in the two latter maps were inferred using common LPSSR loci with the p150/112 genetic map.

Materials and methods Highly efficient molecular markers based on single nucleotide polymorphism (SNP) variation in selected candidate genes were discovered and validated in the F_1(NA$_6$ x AU$_6$) mapping family using the single nucleotide primer extension (SNuPe) technique. The presence of these specific SNP variants in other mapping populations was empirically determined. Common gene-associated SNPs were mapped using a minimal selection of genotypes from the three mapping populations chosen for maximal recombination.

Results and conclusions The *Lp*CCR1 SNP locus gene was assigned to a central location on LG7 of the AU$_6$ parental genetic map and the *Lp*CCR1 gene-associated RFLP locus was assigned a similar position on LG 7 in the p150/112 reference mapping population (Fig 1). The ideogram shown is a representation of 16 genotypes of the p150/112 mapping population possessing maximal non-overlapping recombination for LG7. This approach permits enhanced genome coverage of current maps and provides a more structured reference framework between maps based on common marker positions. These enhanced maps will provide a robust basis for QTL comparisons and validation as well as comparative genomics.

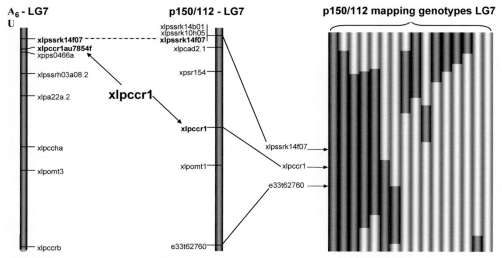

Figure 1 Linkage groups derived from analysis of the mapping populations p150/112 and F_1(NA$_6$xAU$_6$), with common markers identified.

Construction and comparison of genetic linkage maps of four F1 populations of Italian ryegrass (*Lolium multiflorum* Lam.)

M. Vandewalle

Department of Plant Genetics and Breeding, CLO-Gent, Caritasstraat 21, 9090 Melle, Belgium
Email: m.vandewalle@clo.fgov.be

Keywords: mapping, *Lolium multiflorum*, Italian ryegrass

Introduction The objective of this work was to integrate genetic maps and QTL-positions within 4 two-way pseudo-testcross F1 populations (each 110 genotypes) of Italian ryegrass (*Lolium multiflorum*) To integrate genetic maps and QTL-positions within the 4 populations and to the ILGI reference map, the maps used for initial QTL-analysis and marker assisted selection for yield and for different quality traits were further evaluated.

Materials and methods Four two-way pseudo-testcross F1 populations (each 110 genotypes) of Italian ryegrass (*Lolium multiflorum*) were generated by pair-crosses between non-related promising clones of our breeding pool. Genetic linkage maps based on microsatellites, STS and AFLPs were constructed for each population. These preliminary maps, generated by straightforward mapping of all types of markers under the "cross-pollinator" option of Joinmap, were used for QTL-analysis and marker assisted selection for yield and for different quality traits. Preliminary results will be available after evaluation in yield plots in 2005. This strategy was used due to meet the strict time schedule for plant selection, selfing and installation of polycrosses. To integrate genetic maps and QTL-positions within the 4 populations and to the ILGI reference map, the maps used for the initial analysis were evaluated further. The construction of parental maps followed by their integration was performed for each of the four crosses.

Results and discussion The integrated maps were aligned with the original maps. Most aligned linkage groups did not show major changes in length, number of markers or markers order. Some linkage groups presented changes in order and number of markers. Other groups could not be aligned. The integration or not of the less informative AFLP markers segregating as <hkxhk> has also been evaluated. Without the <hkxhk> type of markers, parental maps could not be aligned to build an integrated map within each cross. When mapping together markers of both parental origins using the cross-pollinator option in Joinmap, a map without the less informative markers could be calculated. Alignment to the original maps was possible showing in most groups a reduction of marker numbers. Other linkage groups appeared divided while they were originally grouped or the other way around. A new integrated map is being constructed.

Section 3

QTL analysis and trait dissection

QTLs for morphogenetic traits in *Medicago truncatula*

B. Julier[1], T. Huguet[2], J.M. Prosperi[3], P. Barre[1], G. Cardinet[2] and C. Huyghe[1]
[1]INRA, Unité de Génétique et d'Amélioration des Plantes Fourragères, 86600 Lusignan, France, [2]Laboratory of Plant-Microbe Interactions, UMR CNRS-INRA 2594/441, BP27, 31326 Castanet Tolosan Cedex, France, [3]INRA, Station de Génétique et d'Amélioration des Plantes, Domaine de Melgueil, 34130 Mauguio, France, E-mail: julier@lusignan.inra.fr

Keywords: growth, model species, legume

Introduction Plant morphogenesis that includes growth, development and flowering date, drives a large number of agronomical important traits in both grain and forage crops. Quantitative trait locus (QTL) mapping is a way to locate zones of the genome that are involved in the variations observed in a segregating population. Co-location of QTLs and candidate genes is an indication of the involvement of the genes in the variation. The objective of this study was to analyse segregation of aerial morphogenetic traits in a mapping population of recombinant inbred lines of the model legume species *M. truncatula* , to locate QTLs and candidate genes.

Materials and methods A population of 93 recombinant inbred lines (RILs) from the F6 generation was obtained from the cross Jemalong-6 x DZA315-16. The map was built with EST-SSR markers and anonymous markers (T. Huguet, unpublished). The lines were sown four times in greenhouses in France: in 2000 at INRA, Montpellier, in 2001 at CNRS-INRA, Toulouse, in 2002 and 2003 at INRA, Lusignan. At Montpellier and Toulouse, the flowering date was individually recorded. At Lusignan, the flowering date was recorded, and transformed to degree-days. Through the growing period, the length of the first two emerging primary branches was measured 3 times a week in 2002 and twice a week in 2003. The curve of stem elongation as a function of degree-days showed a short lag phase, followed by a linear phase. It was modelled by the slope of the linear phase, named as growth rate, and the length of the stem after one week of growth in the greenhouse, named as initial length. Analyses of variance were carried out to test the effects of lines and replicates. Correlations among traits were calculated on mean values. Broad sense heritability was calculated. QTL mapping was performed using QTLCartographer with the simple interval mapping (IM) procedure. Candidate gene position was determined from data-mining of *M. truncatula* BAC sequencing or gene mapping. Co-location between QTLs and genes was analysed.

Results Large variation was recorded for all traits in each year, with several transgressive lines, when compared to the value of the parental lines. The broad sense heritabilities were high. Stem length was positively correlated to the number of internodes, the stem diameter, the initial length and the growth rate. The correlation between stem length and number of branches was low or non significant. Dry weight was correlated to stem length and diameter in both years. Leaf to stem ratio was negatively correlated to stem length, initial length and growth rate. The flowering date was strongly negatively correlated to stem length and initial length. It was not correlated to the dry weight. QTLs were found for most traits. The R^2 for each QTL varied from 13.0 to 62.4 %. On chromosome 7, a major QTL for the flowering date was found. This region of chromosome 7 was also involved in variation for stem length in 2002 and 2003, initial length in 2002 and 2003, number of branches in 2003, leaf to stem ratio in 2003 and stem growth rate in 2003. A gene showing homology to Constans mapped in this region. The lower part of chromosome 2 was also involved in the variation of many traits: stem length in 2002, growth rate in 2002 and 2003, dry weight in 2002, leaf to stem ratio in 2002 and 2003. A gene known to be involved in dwarfism in pea (Le) mapped close to this QTL. The top of chromosome 5 was implied in stem diameter in 2002 and 2003, dry weight in 2002 and 2003. Bottom of chromosome 4 was involved in the number of internodes in 2002, stem diameter, stem radius in 2002 and 2003, dry weight in 2002 and 2003. In the middle of chromosome 1, a QTL was found for the dry weight in 2002, stem radius in 2002 and 2003. No candidate gene was mapped in the region of the QTL on chromosome 1, 4 and 5.

Conclusions QTL analysis is a powerful step to identify zones of the genome involved in the variation of traits. In a single cross in *M. truncatula*, a large variation was observed for morphogenetic traits, and QTLs with a large R^2 were found. The current progress in genome sequencing in this species progressively offers the possibility to find candidate genes in QTL regions.

A *Medicago truncatula* population segregating for aluminum tolerance

M. Sledge, B. Narasimhamoorthy and G. Jiang
The Samuel Roberts Noble Foundation, 2510 Sam Noble Parkway, Ardmore, OK, 73401, USA
Email: mksledge@noble.org

Keywords: aluminium tolerance, *Medicago truncatula*, alfalfa

Introduction Aluminium (Al) toxicity, manifested in inhibition of root elongation and reduced plant growth, is a major cause of poor crop yields on acid soils, which comprise up to 40% of the world's arable land. Al toxicity associated with acid soils has been a major obstacle in alfalfa (*Medicago sativa*) production in the USA, as well as in tropical areas of the world. Recent molecular marker mapping studies indicate that the genomes of *M. truncatula* and *M. sativa* are highly similar (Choi *et al.*, 2004). Thus, *M. truncatula* could be used as a source of genes that could be used to improve Al tolerance of cultivated alfalfa. The objective of this study is to identify QTL for Al tolerance in *M. truncatula*, using *M. truncatula* EST-SSR markers and a population from a cross between the Al sensitive Jemalong A17 and an Al tolerant USDA plant introduction, PI 566890 (Sledge *et al.*, 2004), with the long term goal of cloning Al tolerance genes to improve cultivated alfalfa for Al tolerance.

Materials and methods Jemalong A17 was crossed to PI 566890 using a dissecting microscope. Flowers were opened at an early bud stage, vacuum emasculated by attaching a micropipette to a vacuum hose, and pollen was applied from another flower, using a pair of fine forceps. The flower was then placed into a 50 ml plastic centrifuge tube with a few drops of water to prevent desiccation, and the tube was stopped with cotton. After 5-7 days, the developing seed pod was removed from the centrifuge tube and allowed to develop. Seed pods were harvested when completely dried and brown in colour. An F_1 individual was obtained, and was allowed to self-pollinate to produce the F_2 mapping population. Development of EST-SSR markers from the *M. truncatula* EST-databases has been previously described (Eujayl *et al.*, 2003). Forward and reverse primers were custom synthesized with the additional 18 nucleotides from the M13 universal primer appended to the 5' end of the forward primer. PCR reactions were prepared according to the protocol of Schuelke (2000) with the following modifications. The total reaction volume of 10µl contained 20ng of template DNA, 2.5mM $MgCl_2$, 1X PCR buffer II (Perkin-Elmer), 0.15mM dNTPs, 1 pmol of each reverse and M13 universal primer, 0.1pmol of the forward primer, and 0.5 U Ampli Taq Gold DNA polymerase (Perkin Elmer). The M13 universal primer was labeled either with blue (6-FAM), green (VIC), yellow (NED) or red (PET) fluorescent tags. PCR reactions were run in a 384 well plate format, and PCR products with different fluorescent labels and with different fragment sizes were pooled for detection. PCR products (1.6µl) were combined with 10µl of deionized formamide and 0.5µl of GeneScan-500 LIZ internal size standard and analyzed on the ABI3730 Capillary Genetic Analyzer (PE Applied Biosystems). The SSR fragments were visualized and scored using GeneMapper software (PE Applied Biosystems). A linkage map is being constructed using JoinMap 3.0 software (Van Ooijen, 2001). The population was analyzed as a cross-pollinated (CP) population, using the Kosambi mapping function. Expected segregation ratios of 1:2:1 were verified by χ^2 analysis.

Results A hydroponics study was used to identify Al tolerant and Al sensitive genotypes of *M. truncatula*. Jemalong A17 was selected as the sensitive parent, and PI 566890 as the tolerant parent. Two genotypes from these accessions were crossed, and the hybrid nature of the F_1 was confirmed with EST-SSR markers. Preliminary results indicate that approximately 30% of all primers tested in the population are polymorphic, making this a suitable mapping population. Linkage groups are being constructed.

References

Choi, H.K., D.J. Kim, T. Uhm, E. Limpens, H.Lim, J.H. Mun,P. Kalo, R.V. Penmetsa, A. Seres, O. Kulikova, B.A. Roe, T. Bisseling, G.B. Kiss, and D. R. Cook. 2004. A sequence-based genetic map of *Medicago truncatula* and comparison of marker colinearity with *M. sativa*. Genetics 166: 1463–1502.

Eujayl, I., M.K. Sledge, L. Wang, G.D. May, K. Chekhovskiy, J.C. Zwonitzer, M.A.R. Mian. 2003. *Medicago truncatula* EST-SSRs reveal cross-species genetic markers for *Medicago* ssp. and related legumes. Theor. Appl. Genet. 108:414-422.

Schuelke, M. 2000. An economic method for the fluorescent labeling of PCR fragments. Nature Biotech 18: 233-234.

Sledge, M.K., P. Pechter, and M.E. Payton. 2004. Aluminum tolerance in Medicago truncatula germplasm. Crop Sci. In Press.

Van Ooijen, J.W. and V.E. Voorrips. 2001. JoinMap 3.0, Software for the calculation of genetic linkage maps. Plant Research International, Wageningen, the Netherlands.

Genetic mapping in tetraploid alfalfa: Results and prospects

E.C. Brummer, J.G. Robins, B. Alarcón Zúñiga and D. Luth

Raymond F. Baker Center for Plant Breeding, Department of Agronomy, Iowa State University, Ames, IA 50011 USA, Email: brummer@iastate.edu

Keywords: QTL, yield, genetic mapping, forage breeding, winter hardiness

Introduction Among the difficulties of improving forages is their perennial nature, which necessarily requires long selection cycles to fully evaluate genotypes. Further, traits of particular importance—yield and winter hardiness—are difficult to assess on single plants, necessitating evaluation of progeny, which is both time consuming and expensive. Because of this, yield of many forages, and particularly alfalfa, has not improved substantially over the past 25 years (Riday and Brummer, 2002). Winter hardiness often has a negative correlation with autumn growth, although some evidence suggests this is not always true (Brummer *et al.*, 2000). One way to overcome some of these limitations may be through the use of genetic markers to help select desirable genotypes. The objective of this experiment was to test the hypothesis that quantitative trait loci (QTL) for complex agronomic traits could be identified in a segregating tetraploid alfalfa population.

Materials and methods A segregating F_1 alfalfa population consisting of 200 genotypes was developed by crossing an elite genotype ABI408 with a semi-improved genotype WISFAL-6 (Brummer *et al.*, 2000). The population was clonally propagated and planted into field trials at Ames and Nashua, IA between 1998 and 2002. Biomass yield, autumn plant height, winter survival, root mass, and a suite of physiological constituents of roots (sugars, fatty acids, starch, protein, and total nonstructural carbohydrates) were measured (Robins *et al.*, 2005; Alarcón Zúñiga, *et al.*, 2004). Genetic maps based on SSR and RFLP markers were constructed of both parents of the mapping population using the computer program TetraploidMap (Hackett *et al.*, 2003). Phenotypic data were used to map QTL based on single factor analysis of variance, and multiple regression models were built based on significant markers.

Results QTL were identified on all eight linkage groups of alfalfa, and at least one QTL was identified for each trait. Partial R^2 values for particular QTL, based on multiple regression models, ranged from 4 to 39%. A substantial proportion of the population variation could be explained by marker-trait associations for many traits averaged across environments, including biomass yield (36%), winter survival (33%), root soluble protein concentration (63%), root linoleic acid concentration (47%) and autumn plant height (69%). Although several regions of the genome included QTL for multiple traits, Linkage Group 7 (which corresponds to LG 7 in *Medicago truncatula*) is particularly noteworthy. It contains a gene associated with cold tolerance (MsaciB) and that locus is either directly associated or linked to a marker that is associated with biomass yield, autumn plant height, winter survival, and root concentrations of glucose, sucrose, linoleic acid, starch, and soluble protein.

Discussion Genetic mapping of complex traits is possible in complex autotetraploid populations. That this would be the case is not intuitively obvious, because in order for a marker to have a significant association with a trait, it needs to overcome the effects of other alleles within a given parent as well as those effects of alleles from the other parent. Dissecting the physiological basis of traits like winter survival may be possible by successively evaluating its various components. Genome locations with QTL for multiple traits, particularly those, such as linoleic acid and glucose, that are implicated in the same overarching trait (e.g., cold tolerance), are areas warranting further investigation. In particular the region on LG 7 is a prime candidate for loci controlling many important traits. How this information can be used productively in an alfalfa breeding program needs further investigation.

References

Alarcón-Zúñiga, B., E. C. Brummer, P. Scott, K. Moore, and D. Luth. 2004. Quantitative trait loci mapping of winter hardiness metabolites in autotetraploid alfalfa. In: Wang, Z.-Y. (ed.) Molecular breeding of forage and turf. Kluwer, Dordreht, The Netherlands.

Brummer, E.C. (1999) Capturing heterosis in forage crop cultivar development. *Crop Sci.*, 39, 943-945.

Brummer, E.C., M.M. Shah, and D. Luth (2000) Re-examining the relationship between fall dormancy and winter hardiness in alfalfa. Crop Sci. 40:971-977.

Hackett, C. A. and Z. W. Luo. 2003. TetraploidMap: construction of a linkage map in autotetraploid species. J. Hered. 94:358-359.

Riday, H. and E.C. Brummer (2002) Forage yield heterosis in alfalfa. *Crop Sci.*, 42, 716-723.

Robins, J.G., D. Luth, T.A. Campbell, G.R. Bauchan, C. He, D.R. Viands, J.L. Hansen, and E.C. Brummer (2005) Mapping biomass production in tetraploid alfalfa (*Medicago sativa* L.). [In review].

Quantitative trait locus analysis of morphogenetic and developmental traits in an SSR- and AFLP-based genetic map of white clover (*Trifolium repens* L.)

M.T Abberton[1], N.O.I. Cogan[2], K.F. Smith[3], G. Kearney[3], A.H. Marshall[1], A. Williams[1], T.P.T. Michaelson-Yeates[1], C. Bowen[1], E.S. Jones[4], A.C. Vecchies[2] and J.W. Forster[2]

[1]*Legume Breeding and Genetics, Institute of Grassland and Environmental Research, SY 23 3EB, UK,*
Email: michael.abberton@bbsrc.ac.uk [2]*Primary Industries Research Victoria, Plant Biotechnology Centre, La Trobe University, Bundoora, Victoria 3086, and Molecular Plant Breeding Cooperative Research Centre, Australia,* [3]*Primary Industries Research Victoria, Hamilton Centre, Mount Napier Road, Hamilton, Victoria 3300, and Molecular Plant Breeding Cooperative Research Centre, Australia,* [4]*Crop Genetics, Pioneer Hi-Bred International, 7300 NW 62nd Avenue, Johnston, Iowa 50131-1004, USA*

Overview Molecular marker-assisted plant breeding is a key target for the temperate legume pasture crop white clover (*Trifolium repens* L.). The first genetic linkage map of white clover has been constructed using self-fertile mutants to derive an intercross based fourth and fifth generation inbred parental genotypes (F_2[I.4R x I.5J]). The framework map was constructed using simple sequence repeat (TRSSR) and amplified fragment length polymorphism (AFLP) markers. Eighteen linkage groups (LG) corresponding to the anticipated 16 chromosomes of white clover (2n = 4x = 32), with a total map length of 825 cM were derived from a total of 135 markers (78 TRSSR loci and 57 AFLP loci). The F_2(I.4R x I.5J) family has been subjected to intensive phenotypic analysis for a range of morphogenetic and developmental traits over several years at IGER, Aberystwyth, Wales and East Craigs, near Edinburgh, Scotland. The resulting phenotypic data were analysed independently to identify QTL (quantitative trait loci) for the various traits, using single marker regression (SMR), interval mapping (IM) and composite interval mapping (CIM) techniques. Multiple coincident QTL regions were identified from the different years and different sites for the same or related traits. The data were reanalysed using a meta-analysis across years and sites and Best Linear Unbiased Estimates (BLUEs) were derived for the plant spread, petiole length, leaf width, leaf length, leaf area, internode length, plant height and flowering date traits. A total of 24 QTLs were identified on 10 of the linkage groups. Three regions on LGs 2, 7 and 12 all demonstrated overlapping QTLs for multiple traits (Figure 1). A meta-analysis approach can quickly identify regions of the genome that control the trait in a robust predictable manner across multiple spatial and temporal replication for rapid targeted genetic enhancement via marker-assisted breeding. This first genetic dissection of agronomic traits in white clover provides the basis for comparative trait-mapping studies and the enhanced development and implementation of marker-assisted breeding strategies.

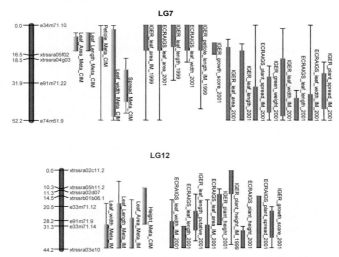

Figure 1 Linkage groups 7 and 12 from genetic map of the F_2(I.4R x I.5J) mapping population, with QTL regions identified. The QTL identified using the meta-analysis are indicated, along with comparison to QTLs from the separate datasets.

Acknowledgements

The authors are grateful for financial support from (UK) Defra and BBSRC and (Australia) Victorian DPI, DA, MLA, GGDF and MPB CRC.

Changes in gene expression during acclimation to cold temperatures in white clover (*Trifolium repens* L.)

M. Lowe, R.P. Collins and M.T. Abberton

Legume Breeding and Genetics Team, Institute of Grassland and Environmental Research, Plas Gogerddan, Aberystwyth, Ceredigion, SY 23 3EB, UK. Email matthew.lowe@bbsrc.ac.uk

Keywords: abiotic stress, acclimation, cDNA-AFLP

Introduction White clover is an important component of many temperate pastures and improved winter hardiness is a major objective of breeding programmes in many countries. Exposure to cold and fluctuations in temperature are components of winter stress and although some studies have investigated the agronomic and physiological mechanisms of cold tolerance, little research has been carried out to identify the genes involved. We are complementing mapping of quantitative trait loci (QTL) responsible for cold tolerance with studies of variation in gene expression between plants growing at different temperatures. In particular we are initially focusing on the process of acclimation by analysing plants subjected to low but above zero temperatures.

Materials and methods Three sets of 25 plants each of the variety Olwen were used. The first set (A) was kept at 2C for 24days before RNA was extracted. The second set (B) was kept at 2C for 24days then at a regime of 20C day/10C night with a 12hr day length for 14days. The final set (C) was kept at 20C/10C day/night for 38days with a 12hr day length before RNA was extracted. RNA from leaf and stolon tissue was separately extracted from all plants using TRIzol ^{TM} from Invitrogen. The purification process was carried out according to the manufacturer's instructions (RNAse free kit from Ambion ^{TM}). Comparisons of gene expression were made using the cDNA-AFLP technique. Initial comparisons were carried out on RNA from each treatment separately bulked from leaf and stolons. To confirm the pattern of polymorphisms between plants from the different treatments subsequent analysis with the same primers was carried out on individual genotypes, again comparing stolon and leaf samples separately.

Results A large number of bands were obtained in this analysis. For the bulked leaf samples, for instance, more than 100 bands from six primer combinations were seen. Similar numbers were observed for all tissue/ treatment combinations. Polymorphisms were observed for both tissue types between all treatments. These represented a small proportion of the total, facilitating further analysis. Bands present in the bulks were present in individuals and polymorphisms observed in the bulks are now being identified in the individual plants. Bands which were observed in Treatment A only are being sequenced and already some interesting similarities have been observed e .g. with a vernalisation control gene in *Arabidopsis thaliana*.

Discussion Despite its preliminary nature, this work has demonstrated the utility of cDNA-AFLP to identify changes in gene expression, consistent across a range of genotypes, during acclimation. In particular, the bands scored in the bulks represented only approx. 10% of those present when all individuals were analysed separately. Thus polymorphisms identified in bulks represented a suitable means of reducing the overall differences to a more manageable number which could subsequently be confirmed on individual genotypes and sequenced. This work is being extended in a number of directions. Clones of the genotypes used have been subjected to a test of survival under sub-zero condition and these results are being related to changes in gene expression during acclimation. Fatty acid profiles, implicated in cold tolerance in a range of species, are also being analysed in the same plants. Candidate genes identified after further analysis will be tested for co-location to QTL for cold tolerance. We will also confirm differences in expression by RT-PCR and carry out, in conjunction with the laboratory of G. Spangenberg, PBC, Victoria, Australia, microarray analysis to complement the results obtained by cDNA-AFLP.

Acknowledgements
We are grateful for the financial support of the Biotechnological and Biological Sciences Research Council and the Department for Environment, Food and Rural Affairs.

References
Abberton, M.T., Cogan, N., Smith, K. F. Marshall, A.H., Williams, T.A, Michaelson-Yeates TPT, Bowen C, Jones, E.S., Vecchies, A and Forster JW (2004) QTL analysis of morphogenetic and developmental traits in an SSR based genetic map of white clover (*Trifolium repens* L.) Forages Workshop. Plant and Animal Genome X11 San Diego, Jan 2004.
Jones, E.S, L. J. Hughes, M. C. Drayton, M. T. Abberton, T. P.T. Michaelson-Yeates, C. Bowen and J. W. Forster (2003) An SSR and AFLP molecular marker-based genetic map of white clover (*Trifolium repens* L.) *Plant Science* 165, 531-539.

QTL analysis of mineral content and grass tetany potential in *Leymus* wildryes

S.R. Larson[1] and H.F. Mayland[2]
[1]*USDA-ARS Forage and Range Research Laboratory, Utah State University, Logan, Utah 84322-6300 USA, Email: stlarson@cc.usu.edu*
[2]*USDA-ARS Northwest Irrigation and Soils Research Laboratory, Kimberly, Idaho 83341 USA*

Keywords: QTLs, potassium, calcium, magnesium, grass tetany

Introduction Grass tetany is a metabolic ailment in ruminants, occurring when animals graze rapidly growing C_3 grasses with a K/(Mg+Ca) ratio (KRAT) greater than 2.2. High KRAT values have been documented in several forage grasses including diploid Russian wildrye (Jefferson *et al.*, 2001). The objective of this experiment was to identify quantitative trait loci (QTLs) controlling KRAT in allotetraploid wildryes.

Materials and methods Full-sib mapping populations, TTC1 (164 genotypes) and TTC2 (170 genotypes), were derived from crosses of two *L. cinereus* x *L. triticoides* F_1 hybrids (TC1 and TC2) backcrossed with one common *L. triticoides* tester plant (T-tester). The F_1 hybrids were derived from crosses of the Acc:636 *L. cinereus* and Acc:641 *L. triticoides* accessions. The linkage maps include 1583 AFLP markers and 50 anchor loci in 14 linkage groups (LGs) (Wu *et al.*, 2003). Concentrations (% dry weight) of Mg, Ca, K, and other minerals were evaluated by ICP-OES of acidified forage dry ash samples harvested May 28, 2003 from clonally replicated plants on 2-m centres in randomized complete blocks (2 reps) at the Utah Agriculture Experiment Station, Richmond. Defined using equivalent units; KRAT = (%K)(0.0257) / [(%Ca)(0.0499) + (%Mg)(0.0823)]. A log of the odds (LOD) threshold of 3 was used to declare significant QTLs. Possible pleiotropy effects and correspondence between populations (homologies) were identified where QTLs overlap with LOD > 2.

Results The Acc:636 accession displayed greater %K ($p<0.001$), less %Ca ($p<0.0001$), and less %Mg ($p<0.01$) relative to Acc:641 (Table 1). Thus, Acc:636 displayed substantially higher KRAT values ($p<0.0001$), relative to the Acc:641 (Table 1). Likewise, the TC1 and TC2 hybrids also showed less %Ca ($p<0.005$) and greater KRAT ($p<0.05$) than the *L. triticoides* T-tester (Table 1). Correlations (*r*) among TTC1 clones were 0.42, 0.61, 0.45, and 0.52 for K, Ca, Mg, and KRAT respectively; and 0.45, 0.47, 0.46, and 0.39 respectively among the TTC2 clones. The range of values (averages of 2 clones per genotype) for K, Ca, Mg, and KRAT varied 1.6-, 2.3-, 2.0-, and 2.3-fold respectively in the TTC1 population; and 1.5-, 2.5-, 1.9-, and 2.1-fold in the TTC2 population.

	%K	%Ca	%Mg	KRAT	
Acc:636 (13 genotypes)	3.58 (0.50)	0.225 (0.057)	0.157 (0.018)	3.85 (0.63)	**Table 1** Means
Acc:641 (20 genotypes)	2.90 (0.32)	0.362 (0.097)	0.182 (0.029)	2.30 (0.44)	(standard deviations)
TC1 (13 clones)	3.16 (0.42)	0.281 (0.040)	0.202 (0.035)	2.68 (0.49)	for mineral
TC2 (12 clones)	3.16 (0.25)	0.273 (0.052)	0.217 (0.023)	2.61 (0.43)	concentration and
T-tester (12 clones)	3.28 (0.13)	0.370 (0.072)	0.225 (0.026)	2.29 (0.20)	grass tetany potential
TTC1 (164 genotypes[#])	3.36 (0.32)	0.364 (0.060)	0.185 (0.026)	2.64 (0.41)	[#]Averages of two
TTC2 (170 genotypes[#])	3.32 (0.24)	0.367 (0.058)	0.201 (0.024)	2.49 (0.34)	clones per genotype

Significant %K QTLs were detected on TTC1 LG2a and LG3a; TTC2 LG1b and LG2b; and corresponding regions of TTC1 and TTC2 LG1a. Significant %Ca QTLs were detected on TTC1 LG1b and LG2a; TTC2 LG2a and LG6a; and corresponding regions of TTC1 and TTC2 LG3b. Significant %Mg QTLs were detected on TTC1 LG5x and LG7a in addition to TTC2 LG1b, LG3a, and LG3b. Significant KRAT QTLs were detected on TTC1 LG2a; TTC2 LG7b; and corresponding regions of TTC1 and TTC2 LG3b. Possible pleiotropy for %Ca and KRAT on TTC1 LG2a; %K and %Mg on TTC2 LG1b; and %Ca and KRAT on TTC1 and TTC2 LG3b.

Conclusions This experiment identified QTLs for all three minerals (i.e. K, Ca, and Mg) contributing to grass tetany potential, including a major QTL effect for %Ca and KRAT on LG3b in both populations. Interestingly, the LG3b region and homoeologous regions of LG3a also have major effects on rhizome proliferation (results unpublished). Like Russian wildrye, *L. cinereus* is a tall caespitose grass with relatively high grass tetany potential whereas *L. triticoides* is a strongly rhizomatous grass with relatively low grass tetany potential. These evaluations identified plant materials and methods to reduce grass tetany potential in perennial wildryes.

References

Jefferson, P.G., H.F. Mayland, K.H. Asay, & J.D. Berdahl (2001). Variation in mineral concentration and grass tetany potential among Russian wildrye accessions. *Crop Science*, 41, 543-548.

Wu, X-L., S.R. Larson, Z-M. Hu, A.J. Palazzo, T.A. Jones, R.R-C. Wang, K.B. Jensen, & N.J. Chatterton (2003). Molecular genetic linkage maps for allotetraploid *Leymus* (Triticeae). *Genome*, 46, 627-646.

QTL analysis of mineral content in perennial ryegrass (*Lolium perenne* L.)

N.O.I. Cogan[1,4], A.C. Vecchies[1,4], T. Yamada[3], K.F. Smith[2,4] and J.W. Forster[1,4]

[1]*Primary Industries Research Victoria, Plant Biotechnology Centre, La Trobe University, Bundoora, Victoria 3086, Australia* [2]*Primary Industries Research Victoria, Hamilton Centre, Hamilton, Victoria 3300, Australia* [3]*National Agricultural Research Centre for Hokkaido Region, National Agriculture and Bio-orientated Research Organisation, Hitsujigaoka, Sapporo, 062-8555, Japan*[4]*Molecular Plant Breeding Cooperative Research Centre, Australia, Email: noel.cogan@dpi.vic.gov.au*

Keywords: hypomagnesemia, mineral content, QTL

Introduction Variation in mineral content of grasses can be strongly influenced by genetic factors. Grass tetany (hypomagnesemia) of cattle and sheep is due to disturbances in serum magnesium levels. In Southern Australia, resultant levels of mortality in cattle vary between 0.5-1.5% of total stock numbers. Serum magnesium variation may be due to feed deficits, or dietary imbalances that interfere with magnesium metabolism. High levels of potassium appear to exert negative effects on the levels of magnesium in the blood. Italian ryegrass genotypes with high levels of magnesium can alleviate the incidence of grass tetany. The genetic control of mineral content, including magnesium, in perennial ryegrass has been investigated using molecular marker-based analysis.

Material and methods The p150/112 population, which formed the basis for reference map construction through the International Lolium Genome Initiative (ILGI), and the second generation reference mapping population [F_1(NA_6xAU_6)] were used for phenotypic analysis. The F_1(NA_6 x AU_6) mapping population is based on a two-way pseudo-testcross structure, generating two parental genetic maps, and is largely populated by gene-associated markers. The maps of each population have been aligned through common simple sequence repeat (SSR) markers. Mineral content in each population was assessed through the use of inductively-coupled plasma mass spectroscopy (ICP-MS). The p150/112 population was sampled in Hokkaido Japan, while the F_1(NA_6 x AU_6) population was sampled in Hamilton, Australia.

Results A total of 14 mineral content traits were examined and heritability values were calculated varying from 0.19 to 0.75. QTL analysis based on simple interval mapping (SIM) and composite interval mapping (CIM) permitted the identification of a total of 68 QTLs. These QTL identified coincident regions of the genome associated with content level variation for several different mineral traits, with a number of common regions apparent across the three different genetic maps (Fig 1). This analysis provides the basis for targeted marker-assisted breeding for mineral content traits in pasture grass improvement.

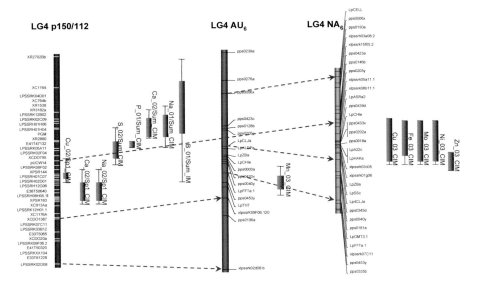

Figure 1 Selected linkage groups from the mapping populations p150/112 and F_1(NA_6xAU_6), with QTL regions controlling mineral content indicated. Common markers are shown for comparison of co-locations.

A glucanase gene cosegregates with a QTL for crown rust resistance in *L. perenne*

H. Muylle[1], J. Baert[1], E. Van Bockstaele[1,2] and I. Roldán-Ruiz[1]

[1]*Department of plant genetics and breeding, Centre of Agricultural Research, Caritasstraat 21, 9090 Melle, Belgium;* [2] *Faculty of Bioscience Engineering, University of Ghent, Coupure Links 653, 9000 Gent, Belgium. Email: h.muylle@clo.fgov.be*

Keywords: *Lolium*, crown rust, STS marker, candidate gene approach

Introduction An important disease in *Lolium* spp. is crown rust caused by the fungal pathogen *Puccinia coronata*. In order to study the genetic background of crown rust resistance in *L. perenne,* a mapping study was carried out and is discussed below. To identify genomic regions or genes involved in resistance, STS markers are extremely useful. This candidate gene approach was applied in the present study.

Material and methods In order to study the genetic background of crown rust resistance in *L. perenne,* an F_1 population segregating for crown rust resistance was created by crossing a resistant and a susceptible parent. The parents were heterozygous plants chosen among breeding materials. Phenotypic analysis was done as described by Adams *et al.* (2000). Mapping and QTL studies were performed using AFLP, SSR, STS and RFLP markers and using Joinmap3.0 and MapQTLv4 (Van Ooijen & Voorrips, 2001).

STS markers were developed from a cDNA library constructed from leaf tissue. 130 cDNAs were sequenced. 58 cDNA sequences showed interesting homologies with DNA sequences with known gene function. For 44 cDNA sequences, it was possible to design primer pairs that span an intron at the genomic DNA-level.

Results Phenotypic analysis revealed that resistance was oligogenic in this segregating F_1 population. Mapping studies using AFLP, SSR, STS and RFLP markers resulted in a genetic map of 833 cM. QTL analysis revealed 4 genomic regions involved in crown rust resistance explaining 45 % of the variance in the population.

Out of the developed STS marker set, one marker, showing homology to a glucanase gene, could be mapped in the F_1 population segregating for crown rust resistance. It coincides with a QTL on LG1 explaining 6.4 % of variance for crown rust resistance in this population.

Conclusion With the candidate gene approach, we could map a STS marker, showing homology to glucanase genes, in a genomic region involved in crown rust resistance. Glucanase genes are known to be involved in defence responses of the plant against biotic stress. Expression profiling will have to confirm these results.

References
Adams, E., I. Roldán-Ruiz, A. Depicker, E. Van Bockstaele & M. De Loose, 2000. A maternal factor conferring resistance to crown rust in *Lolium multiflorum* cv. 'Axis'. Plant Breeding 119: 182-184.
Van Ooijen, J.W. & R.E. Voorrips, 2001. Joinmap® 3.0, software for the calculation of genetic linkage maps. Plant Research International, Wageningen, The Netherlands.

Mapping water-soluble carbohydrate content in perennial ryegrass

L. Turner, J. Gallagher, I. Armstead, A. Cairns and M. Humphreys
Institute of Grassland and Environmental research (IGER), Aberystwyth, Wales SY23 3EB, UK
E-mail lesley.turner@bbsrc.ac.uk

Keywords: perennial ryegrass, water-soluble carbohydrate, fructan, QTL, candidate genes

Introduction Perennial ryegrass (*Lolium perenne* L.) is the main species used in UK agriculture and shows considerable genetic variation for water-soluble carbohydrate (WSC) content (Humphreys, 1989, Turner *et al.,* 2001, 2002). High-sugar grasses have already proved useful in UK livestock production (Miller *et al.,* 2001), but can be unpredictable in the field. Increased understanding of carbon partitioning in ryegrass would benefit future breeding programmes.

Materials and methods A perennial ryegrass mapping family derived from a high WSC x low WSC cross was used to map components of WSC comprising fructan polymers, fructan oligomers, sucrose, glucose and fructose. In order to identify regions of the genome which control basic carbohydrate metabolism, a strategy to maximise G-effects and minimise GxE and E-effects was employed. Data were replicated over years (ie one replicate was collected each year for several years), and QTL reproducible over years were characterised. In this way confounding effects from short-term environmental variation were minimised. Two different tissues (leaves and tiller bases) and two sampling times (spring and autumn) were included to give some information on tissues with different roles and carbon status. Carbohydrates were measured by HPLC. Interval and composite mapping were carried out with MapQTL.

Results Most traits showed considerable variation within the family. Tiller bases always had higher WSC than leaves and autumn samples had more WSC than spring samples. In most of the tissues analysed high molecular weight fructan constituted the major part of the WSC pool. Correlations between traits didn't always lead to corresponding clusters of QTL and some traits had no reproducible QTL. Many QTL were observed in only one year's data and were disregarded. Reproducible QTL were found in only a few regions of the genome and tended to form clusters. Leaf and tiller base QTL did not coincide; tiller base QTL were identified on linkage groups 1 and 5 and leaf QTL on linkage groups 2 and 6. The QTL explained between 8 and 59% of the variation in the traits. The QTL do not currently overlie the positions of those fructosyltransferases that have been mapped. These have been located to linkage groups 3 and 7. However, some of the QTL on linkage group 6 do overlie invertase genes.

Conclusions Environment has a strong influence on carbohydrate content, but some QTL are reproducible over years. These QTL are located in regions that have previously been identified as important in analyses of single replicates (Humphreys *et al.,* 2003). However the QTL do not currently correspond with mapped candidate genes for fructan synthesis. Although fructan breakdown genes could be involved, it is also possible that these QTL are determined by regulatory genes.

References

Humphreys, MO. (1989) Water-soluble carbohydrates in perennial ryegrass breeding. I. Genetic differences among cultivars and hybrid progeny grown as spaced plants. Grass For Sci, 44, 231-236.
Humphreys, M, Turner, L, Skøt, L, Humphreys, M, King, I, Armstead, I & Wilkins, P. (2003). The Use of Genetic Markers in Grass Breeding. Czech Journal of Genetics and Plant Breeding, 39, 112-119. Biodiversity and Genetic Resources as the Bases for Future Breeding. 25th Meeting of the Fodder Crops and Amenity Grasses Section of Eucarpia: Brno, Czech Republic
Miller, LA, Moorby, JM, Davies, DR, Humphreys, MO, Scollan, ND, Macrae, JC, & Theodorou MK. (2001) Increased concentration of water-soluble carbohydrate in perennial ryegrass (*Lolium perenne* L.). Milk production from late-lactation dairy cows Grass and Forage Science, 56, 383-394.
Turner, LB, Humphreys, MO, Cairns, AJ & Pollock, C.J. (2001). Comparison of growth and carbohydrate accumulation in seedlings of two varieties of Lolium perenne. Journal of Plant Physiology, 158, 891-897.
Turner, LB, Humphreys, MO, Cairns, AJ & Pollock, C.J. (2002). Carbon assimilation and partitioning into non-structural carbohydrate in contrasting varieties of Lolium perenne. Journal of Plant Physiology, 159, 257-263.

Quantitative trait loci for vegetative traits in perennial ryegrass (*Lolium perenne* L.)

A.M. Sartie[1], H.S. Easton[1], M.J. Faville[1] and C. Matthew[2]

[1]*AgResearch Ltd., Grasslands Research Centre, Private Bag 11008, Palmerston North, New Zealand*
[2]*Institute of Natural Resources, Massey University, Private Bag 11222, Palmerston North, New Zealand*
Email: alieu.sartie@agresearch.co.nz

Keywords: physiology, linkage map, PCA, QTL, ryegrass

Introduction Physiological (EP) research in forage grasses relates traits such as leaf elongation rate (LER), leaf elongation duration (LED), and leaf appearance interval (ALf), to forage yield (Chapman & Lemaire, 1993). This paper reveals preliminary quantitative trait locus (QTL) discovery for eight EP traits in perennial ryegrass. It also investigates the potential role of multivariate analyses such as principal component analysis (PCA) in QTL analysis of EP data.

Materials and methods A full-sib mapping population (n=200) created by pair-crossing plants from cv. 'Grasslands Impact' and cv. 'Grasslands Samson' (Faville *et al*. this volume) was replicated three times in a randomised complete block design glasshouse experiment. From April to June 2003, ALf, ligule appearance interval (ALg) and leaf length (LL) were recorded. Tiller number (TN) and plant dry weight (DW) above 25 mm were measured at the end of the experiment. Mean tiller weight (TW), and a productivity index (PI, Hernández Garay *et al*., 1999), were derived from DW and TN. LER and LED were also calculated. Data were subjected to PCA. EP traits and PCA scores (PCs) were used for QTL discovery, by interval mapping (MapQTL 4.0).

Results and discussion Plants were significantly (p≤0.001) different for all traits measured. Twenty-two QTL were discovered involving all EP traits except TN (Table 1). The majority of the QTL occur on linkage groups (LG) 1, 3, 4 and 6, with single QTL on LG 2, 5 and 7. Plants varied significantly for the first three PCs yielded by PCA of EP traits (83% phenotypic variation explained, PVE). Generally QTL for PC's are well defined and co-locate with QTL for some but not all correlated traits. For example, PC2 (25% PVE) correlates most strongly with LL (R=0.73) and ALf (R=0.60), while its strongest QTL co-locates with a QTL for TW, and PC3 correlates most strongly with DW (R=0.80) but co-locates with a QTL for LER. Future work will determine whether PC's, accounting in three parameters for most of the variation in nine EP traits, and located with well defined QTL, are repeatable over experimental conditions and between populations, and can be used in marker-assisted selection.

Table 1 Linkage groups, QTL peaks (cM), LOD scores and PVE for EP traits and PC scores

LG	QTL peak (cM)	EP Trait (LOD; PVE)	PC (LOD; PVE)
	0.0	PI (3.0; 7.6)	PC2 (4.9; 7.7)
1	7.2	LER (3.7; 9.4), LL (7.4; 19.7)	
	22.5	ALg (3.7; 10.1)	PC2 (4.5; 10.0)
2	9.4	TW (3.2; 12.5)	
	36.1	ALf (3.4; 8.8), ALg (3.7; 9.3)	
3	45.7	LER (3.0; 8.0)	
	88.1	TW (6.2; 19.0)	PC2 (6.4; 20.5)
4	40.3	LED (7.9; 21.8), ALf (10.5; 26.4), LL (3.6; 9.3), ALg (5.8;15.6)	PC2 (4.7; 12.6)
	43.5		PC1 (6.1; 15.0)
5	60.9	ALf (2.8; 10.4)	
	0.0	LL (4.0; 11.0)	PC2 (3.6; 8.6)
6	3.2	TW (3.2; 10.7)	
	16.4	ALf (3.6; 9.5), ALg (4.1; 10.5)	
	43.5	ALf (3.3; 10.4), ALg (3.9; 11.7), DW (3.9; 8.6)	PC1 (6.9; 12.5)
7	17.1	LER (4.0; 10.1)	PC3 (3.4;8.2)

References
Chapman, D.F. & G. Lemaire (1993) Morphogenic and structural determinants of plant regrowth after defoliation. Proceedings XVII International Grassland Congress: 95 – 104.
Hernandez Garay, A., C. Matthew & J. Hodgson (1999) Tiller size-density compensation in ryegrass miniature swards subject to differing defoliation heights and a proposed productivity index. Grass and Forage Science 54: 347-356.

Approaches for associating molecular polymorphisms with phenotypic traits based on linkage disequilibrium in natural populations of *Lolium perenne*

L. Skøt, J. Humphreys, I.P. Armstead, M.O. Humphreys, J.A. Gallagher and I.D. Thomas
Plant Genetics and Breeding Department, Institute of Grassland and Environmental Research, Aberystwyth, Ceredigion SY23 3EB, UK. E-mail: leif.skot@bbsrc.ac.uk

Keywords: alkaline invertase, heading date, linkage disequilibrium, *Lolium perenne*, soluble carbohydrates

Introduction Association mapping relies on linkage disequilibrium (LD) between haplotypes and quantitative trait loci (QTL). The level of LD in a genome determines the resolution of this approach. In out-breeding species, LD is expected to decay rapidly, thus allowing for high-resolution mapping. It has been most extensively used in human genetics, but recent work with maize populations has demonstrated its potential in plants (Thornsberry *et al.*, 2001; Wilson *et al.*, 2004), and used in *L. perenne* to identify AFLP markers associated with a major QTL for heading date on linkage group 7 (Skøt *et al.*, 2004). The objective of the present work is to associate allelic variation in candidate genes for heading date and water soluble carbohydrates (WSC) in natural populations of *L. perenne* with phenotypic variation. Both these traits are important breeding targets in ryegrass.

Materials and methods 100 genotypes from each of 9 *L. perenne* populations with a wide range of variation in heading date, were grown in pots, vernalised during the winter, prior to determining the heading date in days after March 1st for emergence of the third inflorescence. The plants were then cut back before taking tillers for planting two replicates of each genotype as spaced plants in the field in a completely randomised design. The above ground biomass was later harvested from the remaining tillers in the pots, and ground for analysis of water soluble carbohydrates, total nitrogen and dry matter digestibility using Near Infrared Spectroscopy. DNA was extracted using a QIAGEN kit, and sequencing performed with an ABI 3100 Genetic Analyzer.

Results In *L. perenne* the orthologue of *Hd1* in rice is a potential candidate gene for involvement in the control of photoperiod dependent flowering. A 7.3 kb region containing the putative *Hd1* orthologue was sequenced from a BAC clone, and identification of single nucleotide polymorphisms (SNP) in this region is in progress. An alkaline invertase gene located on linkage group 6 co-locates with a QTL for WSC and is thus a candidate for involvement in this phenotypic trait. A 6238 bp genomic clone of this gene was sequenced (Figure 1), and SNPs were identified in 24 genotypes representing the variation of the populations being investigated. One SNP per 28 nucleotide was found. The average nucleotide diversity was 0.0039, approximately 5 times higher than in *Sorghum*, but similar to maize. LD decays to 0.1 within 2-3 kb (Figure 2). These data indicate a high level of diversity. Association analysis of the SNPs identified in all the genotypes for both the alkaline invertase and the *Hd1* orthologue is being related to phenotypic data for WSC and heading date.

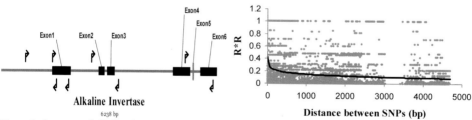

Figure 1 Structure of alkaline invertase gene in *L. perenne*   **Figure 2** Pattern of LD within alkaline invertase

Conclusions The high level of nucleotide diversity, and the rapid decay of LD within the alkaline invertase gene is consistent with *L. perenne* being a self-incompatible species. Association mapping has potential value in candidate gene approaches for identifying allelic variants contributing to phenotypic variation.

References
Skøt, L., Humphreys, M.O., Armstead, I.P., Heywood, S., Skøt, K.P., Sanderson, R., Thomas, I.D., Chorlton, K.H. and Sackville Hamilton, N.R. (2004). An association mapping approach to identify flowering time genes in natural populations of *Lolium perenne* (L.). *Molecular Breeding* (in press).
Thornsberry, J.M., Goodman, M.M., Doebley, J., Kresovich, S., Nielsen, D. and Buckler IV, E.S. (2001) *Dwarf8* polymorphisms associate with variation in flowering time. *Nature Genetics*, 28, 286-289.
Wilson, L.M., Whitt, S.R., Ibáñez, A.M., Rocheford, T.R., Goodman, M.M. and Buckler, E.S. (2004) Dissection of maize kernel composition and starch production by candidate gene association. *Plant Cell*, 16, 2719-2733.

Identification of quantitative trait loci for flowering time in a field-grown *Lolium perenne x Lolium multiflorum* mapping population

R.N. Brown[1], R.E. Barker[2], S.E. Warnke[3], L.A. Brilman[4], M.A.R. Mian[5], S.C. Sim[6] and G. Jung[6]
[1]*University of Rhode Island, Kingston RI 02881 USA;* [2]*National Forage Seed Production Research Center, USDA-ARS, Corvallis OR 97331 USA;* [3]*National Arboretum, USDA-ARS, Washington DC 20002 USA;* [4]*Seed Research Oregon, 27630 Llewellyn Rd., Corvallis OR 97333 USA;* [5]*USDA-ARS, Wooster OH 44691 USA;* [6]*University of Wisconsin, Madison WI 53706 USA Email: brownreb@onid.orst.edu*

Keywords: flowering time, ryegrass, molecular markers, mapping

Introduction Perennial ryegrass (*L. perenne*) and annual, or Italian, ryegrass (*L. multiflorum*) are considered to be separate species by the seed trade, and are used and bred for distinct purposes. However, the two species are cross-fertile. Seed producers rely on the different flowering times of the two species to produce pure seed. Flowering times can overlap, leading to genetic mixing. Contamination of perennial ryegrass seed lots with annual types is an expensive problem for grass seed producers in western Oregon, USA.

Materials and methods The mapping population is a pseudo-F_2 derived from *L. perenne* 'Manhattan' and *L. multiflorum* 'Floregon' The population of 186 individuals was planted in two replications in each of two sites near Corvallis, Oregon, USA, (OSU and SRO) in March 2000. Four replications were planted in Ardmore, Oklahoma, USA (NF) in October 2000. Days to anthesis, defined as three tillers shedding pollen, was recorded for each plant. Plants were also scored on whether they flowered each year. Data was collected in 2000 and 2001 in Oregon, and in 2001 in Oklahoma. None of the plants were vernalised before being transplanted to the field in 2000. QTLs were mapped using both non-parametric analysis and interval analysis. Only those QTLs which were significant (LOD $\geq$ 3.0, $\alpha \leq$ .005) in both analyses are reported. The two or four replications were pooled for a given year and location.

Results In 2000 58% of the plants flowered and 42% remained vegetative. In 2001 all of the plants flowered at all three locations. The locations and LOD scores for the significant QTLs are given in Table 1. The wheat vernalisation genes *Vrn1* and *Vrn2* are located on chromosomes syntenic to *Lolium* linkage group (LG) 4 (Sim *et al.*, 2005), and a homologue of the wheat *Vrn1* gene has been mapped to LG4 in *L. perenne* (Jensen *et al.*, 2005). Growth chamber studies with this population identified vernalisation and photoperiod QTLs on LG1 and LG7. Genes on these linkage groups account for much of the variation among plants whose vernalisation needs have been met. The QTLs on LG3 and LG6 may reflect genes for earliness *per se*, or they may reflect stress response genes.

Conclusions Linkage groups four and seven are most important in determining flowering time in this population. Additional mapping efforts focused on these linkage groups would be useful to generate accurate markers for distinguishing between annual and perennial types in *Lolium*, and for developing cultivars with more defined flowering periods.

Table 1 QTL data for flowering time and vernalisation requirement

Trait	LG	Peak Marker	LOD	Position	Variation
Flower in	4	CDO1196	3.9	54 cM	15%
2000	4	TF45-420MFB	3.1	87 cM	16%
SRO 2000	4	CDO1387	3.0	25 cM	28%
	7	CDO595	3.4	76 cM	30%
OSU 2000	1	CDO105.2	7.0	27 cM	66%
SRO 2001	2	CDO1376	3.5	38 cM	14%
	3	CDO281	3.35	46 cM	15%
OSU 2001	7	CDO464	3.4	53 cM	27%
	7	BCD938	3.5	71 cM	18%
NF 2001	1	CDO105.1	3.7	54 cM	17%
	3	CDO460	3.6	89 cM	16%
	4	CDO541	3.3	35 cM	21%
	6	CDO1380	3.9	69 cM	15%
	7	TF58-292MFA	7.2	22 cM	37%

References

Jensen, L.B., J.R. Andersen, U. Frei, Y. Xing, C. Taylor, P.B. Holm, and T. Lübberstadt. 2005. QTL mapping of vernalisation response in perennial ryegrass reveals co-location with an orthologue of wheat *VRN1*. Theoretical and Applied Genetics (In press, published online Dec. 24, 2004).

Sim, S., T. Chang, J. Curley, S. Warnke, R. Barker, and G. Jung. 2005. RFLP marker based analysis of large-scale chromosomal rearrangements differentiating the *Lolium* genome from other Poaceae species. Theoretical and Applied Genetics (In press).

Crown rust resistance of Italian ryegrass cultivar 'Axis' to an isolate from Japan

T. Kiyoshi[1], M. Hirata[2], T. Takamizo[1], H. Sato[1], Y. Mano[1] and M. Fujimori[1]
[1]National Institute of Livestock and Grassland Science, 768 Senbonmatsu, Nishinasuno, Tochigi 329-2747, Japan, Email: tkiyoshi@affrc.go.jp
[2]Japan Grassland Farming & Forage Seed Association, Forage Crop Research Institute, 388-5 Higashiakada, Nishinasuno, Tochigi 329-2742, Japan

Keywords: major gene, linkage analysis, pseudo-testcross

Introduction Crown rust (*Puccinia coronata*) is one the most serious diseases of Italian ryegrass. Crown rust resistance genes in Italian ryegrass have been identified from 'Yamaiku130' and 'Harukaze'. The aim of this study was to identify novel major resistance genes for gene pyramiding in order to develop cultivars with high levels of durable resistance to crown rust.

Materials and methods The Italian ryegrass cultivar 'Axis' provided basic material for this study. Following artificial inoculation, we selected 8 individuals with high level of resistance (no visible symptoms). In order to produce F_1 populations for analysis, the 8 resistant individuals were crossed with susceptible individuals of the cultivar 'Waseaoba'. In order to evaluate resistance, seedlings were grown in plastic trays and inoculated with a crown rust isolate from the Yamaguchi Agricultural Experiment Station in Japan. Two weeks after inoculation, the resistance of individuals was scored according to the following criteria, i: no visible symptom of infection, 0: tiny necrotic and chlorotic flecks without pustules, 1: tiny necrotic and chlorotic flecks with small sized pustules, 2; chlorotic flecks with medium to large sized pustules. The Yamaguchi isolate has been used in our previous work including identification of major resistance gene from Yamaiku130 and Harukaze.

Results The degree of resistance of Axis and the 8 F_1 populations are shown in Fig1. 28 (50%) out of 56 individuals of Axis showed complete resistance (i) and a total of 39 individuals (70%) showed some resistance to crown rust after artificial inoculation (Fig1). Among the 8 F_1 populations, there was no significant difference between reciprocal crosses, indicating that a cytoplasmic factor did not contribute resistance to the Yamaguchi isolate. Four out of the 8 F_1 populations segregated in a 1:1 ratio for resistance and susceptibility (Axis2, Axis3, Axis4, Axis5). Two populations in particular (Axis4 and Axis5) segregated in a 1:1 ratio without partial resistance, suggesting that a resistance gene with major effect was segregating in these populations. Linkage analysis using these populations is efficient in identifying major resistance genes. A large portion of partially resistant plants in populations such as Axis1 and Axis7 suggests that these populations contain resistance genes.

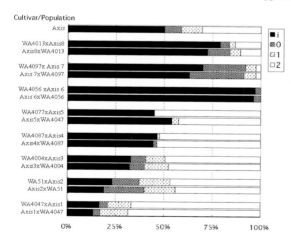

Figure 1 Frequency distribution of the degree of resistance to crown rust in Axis and 8 F_1 populations produced by crossing resistant individuals from Axis with susceptible individuals from Waseaoba. The rust scores are explained in the materials and methods

Conclusions Our data indicate that cytoplasmic factors did not contribute resistance to the Yamaguchi isolate in our populations. This is different from the result obtained by Adams *et al.* (2000). One possible explanation is a different reaction of the Axis cytoplasm to physiological races. The genetic analysis indicated that Axis possesses at least one dominant resistance gene. In order to identify major resistance genes, we will select 4 populations for linkage analysis in this study.

Reference
Adams E, Roldan-Ruiz I, Depicker A, van Bockstaele E, de Loose M (2000) A maternal factor conferring resistance to crown rust in *Lolium multiflorum* cv. 'Axis'. Plant Breeding 119 (2): 182-184.

Locating, and utilising *Festuca pratensis* genes for winter hardiness for the future development of more persistent high quality *Lolium* cultivars

M.W. Humphreys[1], D. Gasior[2], A. Kosmala[3], O.A. Rognli[4], Z. Zwierzykowski[3] and M. Rapacz[2]

[1]*Institute of Grassland and Environmental Research, Plas Gogerddan, Aberystwyth, Ceredigion SY23 3EB, UK;* [2]*Agricultural University of Craców, Department of Plant Physiology, Faculty of Agriculture, Podłużna 3, Kraków, Poland;* [3]*Institute of Plant Genetics Polish Academy of Sciences, Strzeszyńska 34, 60-479 Poznań, Poland.Agricultural* [4]*The University of Norway, Department of Chemistry and Biotechnology, P.O. Box 5040, 1432 Ås, Norway; Email: mike.humphreys@bbsrc.ac.uk*

Keywords: *Festulolium*, introgression-mapping, QTL-mapping, cold-hardening, photosystem II (PSII)

Summary Genes for freezing-tolerance and winter hardiness were located in *Festuca pratensis* by QTL analysis and introgression-mapping. QTL for freezing-tolerance on *F. pratensis* chromosome 4 were orthologous to rice chromosome 3, and Triticeae chromosome 5. Increased energy dissipation during the autumn through a lower maximum quantum yield of photosystem II (PSII) was correlated with improved winter survival. Freezing-tolerance in *Lolium* was achieved by the transfer and subsequent expression of *F. pratensis* genes from chromosome 4 that govern the expression of a non-photochemical (NPQ) mechanism for the dissipation of excess light energy under low temperature.

Introduction We describe outcomes from field tests for winter hardiness in Norway and Poland and simulated freeze testing and cold acclimation studies for adaptations to low temperatures. Quantitative trait loci (QTL) for winter hardiness were assigned to chromosomes and the first QTL map of *Festuca pratensis* constructed. Introgression-mapping in *Lolium* x *Festuca* hybrids is a particularly efficient mechanism for trait "dissection" as it harnesses the high frequencies of chromosome recombination inherent in these hybrids with the capability to locate sites of chromosome introgression by genomic in situ hybridisation (GISH). This has led to the location of *F. pratensis* genes associated with winter survival and freezing-tolerance.

Materials and methods Plant populations used were: a *F. pratensis* mapping family (Alm *et al.*, 2003); backcross populations involving *Lolium perenne* x *F. pratensis* and *L. multiflorum* x *F. pratensis* hybrids; a series of monosomic chromosome addition lines where a *F. pratensis* chromosome replaces its *L. perenne* homoeologue (King *et al.*, 2002). GISH techniques were used to locate sites of *Festuca* introgressions on *Lolium* chromosomes (e.g. King *et al.*, 2002). Simulated freeze-testing involved use of a climatic cooler (Cambridge Scientific UK, Ltd.) and followed cold acclimation for 2w at 2°C, 8h light, 400μmol m^{-2}s^{-1}. Chlorophyll fluorescence measurements were as in Rapacz *et al.* (2004).

Results and discussion Two major QTL for freezing tolerance and four QTL for winter survival in the field were located. QTL for freezing tolerance, *Frf4* on chromosome 4 and *Frf5* on chromosome 5 explained a total of 34.3% of the total phenotypic variation. *Frf4* was orthologous to the frost tolerance *Fr1* and *Fr2* loci in wheat which are also associated closely with the *Vrn-1* (vernalization) gene and 2 regulatory genes *Rcg1* and *Rcg2* known to have a major role in expression of cold-responsive (*cor*) genes. A monosomic chromosome addition line for *F. pratensis* chromosome 4, or introgression lines of *Lolium spp* with *F. pratensis* genes from chromosome 4, displayed cold acclimatory adaptations of PSII typical of *F. pratensis*. The presence of *Festuca*-derived NPQ mechanisms for protection against high-light induced inactivation of PSII at low temperature, in *Lolium* led to improved freezing-tolerance. Norwegian and Polish field trials and simulated freezing-trials led to the selection of *Lolium*-like diploid (2n = 2x = 14) backcross derivatives with good winter hardiness and freezing-tolerance. Plants from the Norwegian field study were found repeatedly with introgressed *F. pratensis* genes at the same location on *Lolium* chromosomes 3 and 4. A *F. pratensis* introgression on chromosome 2 of *Lolium* led to the breakdown of mechanisms required for effective cold acclimation.

References

Alm V., C. Fang, C.S. Busso, K.M. Devos, K. Vollan, Z. Grieg, & O.A. Rognli (2003). A linkage map of meadow fescue (*Festuca pratensis* Huds.) and comparative mapping with other Poaceae species. *Theoretical and Applied Genetics*, 108:25-40

King J., I.P. Armstead , I.S. Donnison, H.M. Thomas, R.N. Jones, M.J. Kearsey, L.A. Roberts, A. Thomas, W.G. Morgan, & I.P. King (2002). Physical and genetic mapping in the grasses *Lolium perenne* and *Festuca pratensis*. *Genetics*, 161:315-324

Rapacz M., D. Gasior, Z. Zwierzykowski, A. Lesniewska-Bocianowska, M.W. Humphreys, & A.P. Gay (2004). Changes in cold tolerance and the mechanisms of cold acclimation of photosystem II to cold hardening generated by anther culture of *Festuca pratensis* x *Lolium multiflorum* cultivars. *New Phyt.*, 162:105-114

QTL analysis of vernalisation requirement and heading traits in *Festuca pratensis* Huds.

Å. Ergon[1], C. Fang[1], Ø. Jørgensen[1], T.S. Aamlid[2] and O.A. Rognli[1]

[1]Norwegian University of Life Sciences, Dept. of Plant and Environmental Sciences, P.O.Box 5003, N-1432 Ås, Norway [2]Norwegian Crop Research Institute, Apelsvoll Research Centre, Division Landvik, Reddalsvn. 215, 4886 Grimstad, Norway, E-mail: ashild.ergon@umb.no

Keywords flowering time, heading time, induction of flowering, QTL

Introduction The transition from the vegetative phase to the reproductive phase occurs as a result of environmental and endogenous stimuli. In *Festuca pratensis*, low temperature and/or short days over a certain period (primary induction) followed by long days (secondary induction) will lead to heading and flowering (Heide, 1988). We present results from QTL mapping of vernalisation requirement and heading traits and mapping of the Vrn-1 ortholog in *F. pratensis*.

Materials and methods A mapping family of *F. pratensis* consisting of 138 progeny from a cross between Norwegian genotype HF2/7 and Yugoslavian genotype BF14/16 were characterized for vernalisation requirement and heading traits. In a greenhouse experiment, plants were vernalized at 6 °C and 8h photoperiod for 12, 9, 6 or 0 weeks and then transferred to warmer temperatures and long days to stimulate heading. Vernalisation requirement was recorded, together with heading date, final number of panicles and final number of panicles per tiller after 12 weeks of vernalisation. In a field experiment, non-vernalized plants were planted at two locations in the spring and the number of panicles produced during the summer season was recorded. The data was subjected to QTL analysis using previously obtained genotype data (Alm *et al.*, 2003) and the MapQTL software (van Ooijen and Maliepaard, 1996). The presence of cereal anchor probes on the map allows the comparison with genes and QTLs for similar traits mapped in other species. A segment of a gene with high similarity to the homoeoallelic series of Vrn-1 genes controlling vernalization requirement in cereals (Snape *et al.*, 2001; Yan *et al.*, 2003) was PCR-amplified and sequenced. A single nucleotide polymorphism identified two different alleles in the mapping population. Individuals in the mapping population were characterized for these alleles and coupling analysis was performed to determine the posistion of the Vrn-1 gene.

Results A large portion of linkage group (LG) 4 was strongly associated with vernalisation requirement in both greenhouse and field experiments. The upper half of LG4 had the strongest effect on the phenotype, however, to our knowledge no genes or QTLs relating to vernalisation requirement or heading time are reported in comparable regions of other genomes. Strong QTLs were also found in the lower part of the LG, as well as near Xwg644, a marker which is closely linked to several interesting genes; the Vrn-1-series in cereals and grasses, Tack2a/Hd6 (casein kinase 2α) in wheat/rice and Phytochrome C in barley. The sequence of the *F. pratensis* Vrn-1 gene segment (most of its MADS-box domain) was 100% identical to that of Vrn-1 in wheat and ryegrass species at the amino acid level and it was closely linked to Xwg644. It therefore represents a member of the Vrn-1-series and is likely to affect vernalisation requirement in *F. pratensis*. QTLs for heading time were found on LG1 and LG5, whereas QTLs for number of panicles were found on LG1, LG4, LG5, LG6 and LG7. The QTL for heading time on LG5 is located in a region comparable to that containing the rice flowering time QTL FLT2 and the QTL for number of panicles on the same LG is located near Xwg364a, which is linked to the barley cold-binding factor Hvcbf3. The QTLs on LG7 were found in a region known to contain genes in rice, barley and ryegrass with similarity to CONSTANS, an important gene in the photoperiod pathway in arabidopsis.

References

Alm, V., Fang, C., Busso, C.S., Devos, K.M., Vollan, K., Grieg, Z., and Rognli, O.A. (2003). A linkage map of meadow fescue (Festuca pratensis Huds.) and comparative mapping with other Poaceae species. *Theoretical and Applied Genetics* 108, 25-40.

Heide, O.M. (1988). Flowering requirements of Scandinavian Festuca pratensis. *Physiologia Plantarum* 74, 487-492.

Snape, J.W., Butterworth, K., Whitechurch, E. and Worland, A.J. (2001) Waiting for fine times: genetics of flowering time in wheat. *Euphytica* 119, 185-190.

Van Ooijen, J.W. and Maliepaard, C. (1996). MapQTL (tm) version 3.0: Software for the calculation of QTL positions on genetic maps. CPRO-DLO, Wageningen.

Yan, L., Loukoianov, A., Tranquili, G., Helguera, M., Fahima, T. and Dubcovsky, J. (2003). Positional cloning of the wheat vernalization gene VRN1. *Proceedings of the National Acadmy of Sciences* 100, 6263-6268.

Consistency of QTL for dollar spot resistance between greenhouse and field inoculations, multiple locations, and different population sizes in creeping bentgrass

N. Chakraborty[1], J. Bae[1], J. Curley[1], S. Warnke[2], M. Casler[3], S. Bughrara[4] and G. Jung[1]
[1]Dept. of Plant Pathology, Univ. of Wisconsin, Madison, WI53706 USA Email:jung@plantpath.wisc.edu
[2]USDA-ARS, National Arboretum, Washington D.C. 20002, USA, [3]USDA-ARS, U.S. Dairy Forage Research Center, 1925 Linden Dr., Madison, WI 53706, USA, [4]Dept. of Crop and Soil Sciences, Michigan State University, East Lansing, MI 48824, USA

Keywords: linkage map, QTL, dollar spot resistance, allotetraploid, creeping bentgrass

Introduction Dollar spot caused by *Sclerotinia homoeocarpa* F. T. Bennett is the most economically important turf disease in North America. Previous work indicated differences among cultivars in their susceptibility to dollar spot (Bonos *et al.*, 2003). Studies have indicated that dollar spot resistance might be quantitatively inherited (Bonos *et al.*, 2003) but the number, location and effect of genomic regions conferring resistance is still not known. Therefore the objective of this research is to understand the effect of population size, inoculation assays, and field locations on QTL for dollar spot resistance in creeping bentgrass.

Materials and methods A full-sib genetic mapping population consisting of 697 progeny was developed from a cross between the out-crossing clones named 372 and 549. A dense linkage map with fourteen linkage groups has been constructed using various markers. Ninety-three of the RFLP markers are from sequenced probes from cDNA libraries, constructed from leaf tissue of clones 372 and 549.

Results The linkage map covers 1144 cM with 227 RFLPs, 177 RAPDs, and 73 AFLPs. One major QTL on linkage group 7 is consistent over multiple ratings as the disease progressed and also across locations and experiments in WI. Greenhouse inoculation, rating methods, and field location-specific QTL with minor effect were also detected on various linkage groups. Effect of population size (94-697) on the significance of the QTLs was also examined.

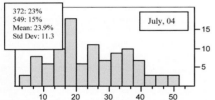 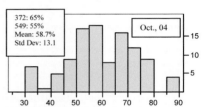

Figure 1 Frequency distribution of disease severity (%) at south central WI field plot inoculated with one virulent isolate at different rating dates (July and October, 2004).

Conclusions Continuous frequency distribution for dollar spot resistance among the progeny indicated the quantitative nature of resistance. Transgressive segregation was detected which suggests contribution of resistance alleles from both parents. The consistency of the major QTL on linkage group 7 across locations and disease progress indicated that it is stable over environments, while the minor QTLs were influenced by inoculation assay, rating methods and also rating time during the disease progress. Markers tightly linked to the QTL which are stable under environmental factors can be applied for MAS in future dollar spot resistant creeping bentgrass-breeding programs.

Reference
Bonos, S. A., Casler M. D., and Meyer W. A. 2003. Inheritance of dollar spot resistance in creeping bentgrass. Crop Sci. 43:2189-2196.

Section 4

Genomics, model species, gene discovery and functional analysis

Structural and functional genomic research in model legume plants: The National BioResource Project (NBRP) in Japan

S. Tsuruta, M. Hashiguchi and R Akashi

Faculty of Agriculture, University of Miyazaki, Miyazaki 889-2192, Japan, E-mail: rakashi@cc.miyazaki-u.ac.jp

Keywords: legume base, *Lotus japonicus*, *Glycine max*

Introduction *Lotus japonicus* is a wild perennial plant with a small genome and a short life cycle. This plant is expected to play a role as the model organism of leguminous plants, which include important crop plants such as soybean (*Glycine max*). Legume Base, a resource centre for *Lotus japonicus* and *Glycine max*, was established in April 2004. The scope of Legume Base is the collection, development and conservation of the genetic resources of *L. japonicus* and *G. max* and the distribution of the material for utilization by the research community. DNA resources including genomic DNA clones will be also available through Legume Base web site.

Legume Base is supported by the National BioResource Project of Japan. The core facility of Legume Base is at Miyazaki University and the sub facility is at Hokkaido University. Some parts of the distribution work are carried out by following facilities on commission: the National Agricultural Research Centre for the Hokkaido Region, Nihon University College of Bioresource Sciences and RIKEN Plant Science Centre

Results The following genotypes are/will be made available for distribution (Items listed in bold are now available.)

Lotus japonicus	*Glycine max*
1. Miyakojima MG-20, Gifu B-125	**1. Cultivated accession lines**
2. Accession lines (collected throughout Japan)	**2. Wild accession lines (collected throughout Japan)**
3. LjMG RI lines (RI lines between Miyakojima MG-20 and Gifu B-129)	3. RI lines (RI lines between Misuzudaizu and Moshidou Gong 503)
4. Activation tag lines	4. Mutants in fatty acid composition
5. EMS mutants	
6. Root culture system (Super Roots isolated from *Lotus corniculatus*)	

For more detailed information visit Legume Base web site at http://www.legumebase.agr.miyazaki-u.ac.jp

Identification of putative *At*TT2 R2R3-MYB transcription factor orthologues in tanniferous tissues of *L. corniculatus* var. *japonicus* cv *Gifu*

D.N Bryant[1], P. Bailey[2], P. Morris[1], M. Robbins[2], C. Martin[2] and T. Wang[2]

[1]*Plant, Animal and Microbial Sciences Department, Institute of Grassland and Environmental Research, Plas Gogerddan, Aberystwyth, Ceredigion SY23 3EB, UK* [2]*John Innes Centre, Norwich Research Park, Colney, Norwich, NR4 7UH, UK. Email: David.Bryant@bbsrc.ac.uk*

Keywords: proanthocyanidin, R2R3-MYB, BHLH, *Lotus*, *Arabidopsis*

Introduction R2R3-MYB plant transcription factors are sequence–specific DNA-binding proteins, which regulate the expression of specific gene(s) following the R2R3 DNA-binding domain interacting with the corresponding promoter sequence(s). The biosynthetic pathway leading to the production of anthocyanins has been demonstrated to be under MYB transcriptional regulatory control (Cone *et al.*, 1986), while the accumulation of proanthocyanidins (PAs) in *Arabidopsis* seed coats is determined by the R2R3-MYB *At*TT2 (Nesi *et al.*, 2001). Using an informatics approach, partial sequences of putative *AtTT2* orthologues have been identified and cloned from the forage legume *Lotus corniculatus* var. *japonicus* cv *Gifu*.

Materials and methods Total RNA and cDNA were prepared from flower, stem and leaf tissue harvested from *Lotus corniculatus* var. *japonicus* cv Gifu grown under glass. 180 bp fragments were amplified using degenerate PCR primers designed to consensus sequences within the MYB DNA-binding domain (Romero *et al.*, 1998). The subsequent PCR products were cloned into *E. coli* via pGEMT easy prior to preparation for sequencing and analysis via DNA for windows and ClustalX.

Results We isolated and cloned candidate sequences *Lj*MYB38 and *Lj*MYB72 from cDNA derived from stem and flower tissues. Multiple amino acid sequence alignment of the DNA binding domain of *Lj*MYB38 and *Lj*MYB72 revealed 81% and 70% identity and 87% and 88% respective similarity to *At*TT2 Within the amino acid sequence of the *Arabidopsis* basic helix-loop-helix interaction motif, spanning helices 1 & 2 of the R3 domain, the essential residue at position 20 was Asp-20 while *Lotus* sequences differed with Lys-20 (Figure 1).

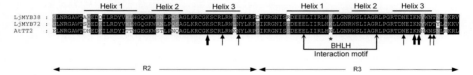

Figure 1 Multiple amino acid sequence alignment of the R2R3-MYB DNA-binding domain of *Arabidopsis* AtTT2 with R2R3-MYBs cloned from *Lotus*. Homologous regions are highlighted in black, while grey shading represents amino acids with similar physico-chemical properties. Arrows represent amino acids that interact with DNA with arrow thickness denoting stronger interaction. An essential difference in the amino acid sequence of the basic helix-loop-helix (BHLH) interaction motif is indicated by *.

Conclusions These data indicate that two putative orthologues to *At*TT2, the R2R3-MYB transcription factor required for PA biosynthesis in *Arabidopsis*, are expressed in the tanniferous stem and flower tissue of *L. corniculatus*. The presence of Lys-20 in the R3 domain of *Lj*MYB38 and *Lj*MYB72, as opposed to Asp-20 in *At*TT2 represents a significant alteration in the amino acid sequence of BHLH interaction motif (Zimmermann *et al.*, 2004). Thus, the BHLH protein with which *Lj*MYB38 and *Lj*MYB72 could interact, may be distinct from the corresponding BHLH, *At*TT8 (Nesi *et al.*, 2000), in *Arabidopsis* and might contribute to the differential tissue specific biosynthesis of PAs between these species.

References
Cone *et al.*, (1986) Molecular analysis of the maize anthocyanin regulatory locus C1. Genetics. 83, 9631–9635.
Nesi *et al.*, (2000) The *TT8* gene encodes a basic-helix-loop-helix protein required for expression of *DFR* and *BAN* genes in Arabidopsis siliques. The Plant Cell. 12, 1863-1878.
Nesi *et al.*, (2001) The Arabidopsis *TT2* gene encodes an R2R3-MYB domain protein that acts as a key determinant for proanthocyanidin accumulation in developing seed. The Plant Cell. 13, 2099-2114.
Zimmermann *et al.*, (2004) Comprehensive identification of *Arabidopsis thaliana* MYB transcription factors interacting with R/B-like BHLH proteins. The Plant Journal. 40, 22–34.
Romero *et al.*, (1998) More than 80 R2R3-MYB regulatory genes in the genome of *Arabidopsis thaliana*. The Plant Journal. 14(3), 273-284.

Foliar expression of candidate genes involved in condensed tannin biosynthesis in white clover (*Trifolium repens*)

S.N. Panter[1,2], J. Simmonds[1,2], A. Winkworth[1,2], A. Mouradov[1,2] and G.C. Spangenberg[1,2]

[1]*Primary Industries Research Victoria, Plant Biotechnology Centre, La Trobe University, Bundoora, Victoria 3086, Australia* [2]*Molecular Plant Breeding Cooperative Research Centre, Australia*
Email: stephen.panter@dpi.vic.gov.au

Keywords: transgenic white clover, condensed tannins, chalcone synthase, anthocyanidin and leucoanthocyanidin reductases, bloat safety

Introduction Bloat disease in cattle and sheep is caused by the rapid microbial degradation of protein-rich fodder in the rumen. This leads to the production of protein foams that trap gases, causing bloat, a condition that is often fatal to livestock and costly to farmers. Condensed tannins (CTs) are phenolic polymers produced by the phenylpropanoid pathway of plants (Figure 1). CTs bind to proteins under acidic to neutral conditions, such as those present in the rumen, slowing their breakdown. A diet with a CT content of between 2% and 4% by dry weight, which is provided by some pasture legumes (e.g. *Lotus corniculatus*), protects livestock against bloat and improves the absorption of amino acids from the diet. White clover (*Trifolium repens* L.), a protein rich legume widely used in temperate regions, has virtually no CTs in leaves, although they are present in flowers.

Materials and methods To test whether the foliar expression of clover candidates for genes involved in CT synthesis enhances CT production in leaves of transgenic white clover plants, a targeted EST discovery program was undertaken and cDNAs encoding white clover homologs of the genes involved in the CT biosynthesis pathway were identified.

Results and conclusions The putative protein encoded by *Tr*CHSh (1599 bp), isolated from cDNA from vegetative stolon tips, shares 84.8% amino acid identity with *Arabidopsis* chalcone synthase (TT4), an early enzyme in the phenylpropanoid pathway. Putative proteins encoded by *Tr*BANa (1309 bp) and *Tr*LARb (1551 bp), both isolated from inflorescence cDNA libraries, share 89.3% and 70.3% amino acid identity with *Medicago truncatulata* anthocyanidin reductase and *Desmodium unicinatum* leucoanthocyanidin reductase, respectively. The latter two enzymes catalyse different steps in the CT-specific branch of the phenylpropanoid pathway.

Transgenic white clover plants ectopically expressing chimeric *Tr*CHSh, *Tr*BANa and *Tr*LARb genes – individually and combined (i.e. *Tr*BANa plus *Tr*LARb; *Tr*BANa plus *Tr*LARb plus *Tr*CHSh) - under the control of constitutive and leaf-prevalent promoters were generated using *Agrobacterium*-mediated transformation. The transgenic nature of the plants recovered was demonstrated by quantitative PCR and Southern hybridisation analysis revealing integration of 1 – 5 transgene copies. Selected transformation events showing elevated levels of *Tr*CHSh, *Tr*BANa and *Tr*LARb transcripts were targeted for biochemical and metabolomic assays to identify the level and composition of CTs in leaves. This research forms part of a molecular breeding approach in white clover deploying exclusively white clover genes and promoters for the development of transgenic white clover cultivars with improved nutritional quality and bloat safety.

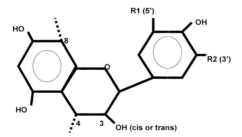

Figure 1 Schematic diagram of a condensed tannin subunit. Subunits are linked by C4 -C8 linkages. Functional groups, shown as R1 and R2, can be either H or OH, depending on precursors entering the CT-specific branch of the phenylpropanoid pathway, where individual enzymes can generate cis or trans conformations at position 3

Discovery, isolation and characterisation of promoters in white clover (*Trifolium repens*)

C.M. Labandera[1,2], Y.H. Lin[1,2], E. Ludlow[1,3], M. Emmerling[1,2], U. John[1], P.W. Sale[4], C. Pallaghy[3] and G.C. Spangenberg[1,2]

[1]Primary Industries Research Victoria, Plant Biotechnology Centre, La Trobe University, Bundoora, Victoria 3086, Australia [2]Molecular Plant Breeding Cooperative Research Centre, Australia [3]Department of Botany, La Trobe University, Bundoora, Victoria 3086, Australia [4]Department of Agriculture, La Trobe University, Bundoora, Victoria 3086, Australia Email: Marcel.Labandera@dpi.vic.gov.au

Keywords: white clover, bacterial artificial chromosomes (BACs), BAC library, promoters, reporter genes

Introduction The availability of a suite of promoters with a range of spatial, temporal and inducible expression patterns is of significant importance to control targeted expression of genes for molecular breeding in forage species. A range of resources and tools have been developed for promoter isolation and characterisation in white clover (*Trifolium repens* L.), including a comprehensive BAC library and a 15K unigene microarray.

Materials and methods Discovery, isolation and characterisation of heterologous and endogenous promoters was undertaken in white clover. Expression patterns of chimeric *gusA* reporter genes encoding bacterial β-glucuronidase (GUS) with four differentially regulated promoters from *Arabidopsis thaliana* (*atmyb32*, *adh*, *xero2* and *SAG12*) were assessed in transgenic white clover plants generated by *Agrobacterium*-mediated transformation.

Results and conclusions Molecular analysis of independent transformants confirmed the stable integration of T-DNAs containing the various promoter-*gusA* reporter genes. Histochemical staining of plant tissues and organs revealed that the *atmyb32* promoter directed *gusA* expression in leaf and root vascular tissue including lateral roots and nodules with low levels of expression in reproductive organs. Wound-response of the a*tmyb32* promoter in white clover leaves and stolons was also shown. The *adh* promoter showed anaerobic stress and dehydration stress response. The *xero2* promoter directed strong expression in roots, leaf vascular tissue, inflorescences, anther filaments and pollen grains, while the *A. thaliana SAG12* promoter resulted in senescence-associated *gusA* expression in white clover leaves. A white clover BAC library consisting of 50,302 BAC clones with 101 kb average insert size, corresponding to 6.3 genome equivalents and 99% genome coverage was established. Root-prevalent promoters were isolated from white clover following screening of the BAC library. White clover BAC clones hybridising to phosphate transporter (*TrPT1*) and iron transporter (*TrRit1*) sequences were identified (Fig 1), and corresponding 5' regulatory sequences were isolated. Transgenic white clover plants expressing a chimeric *TrPT::gfp* gene encoding green fluorescence protein (GFP) fusion were produced. They revealed fluorescence in root tissues, mainly in root-tips and root nodules. This research provides a toolbox of promoters with a range of specificities for targeted gene expression as part of a molecular breeding approach in white clover deploying exclusively white clover genes and promoters for transgenic product development.

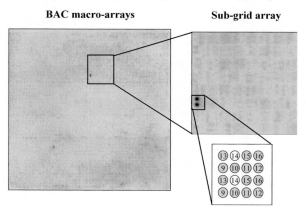

BAC macro-arrays   **Sub-grid array**

Figure 1 White clover BAC macro-array membrane after hybridisation with a *TrPT1* cDNA ^{32}P-labelled probe revealing *TrPT1*-hybridising genomic clones for *TrPT1* promoter isolation.

Application of molecular markers derived from *Medicago truncatula* in white clover (*Trifolium repens* L.)

C. Jones and M.T. Abberton

Legume Breeding and Genetics Team, Institute of Grassland and Environmental Research, Plas Gogerddan, Aberystwyth, Ceredigion, Wales SY23 3EB Email: charlotte.deakin@bbsrc.ac.uk

Keywords: genetic mapping, synteny, stolons

Introduction White clover is the major forage legume of temperate areas. Genome maps have been produced recently (Jones *et al.*, 2003; Barrett *et al.*, 2004) and the location of QTL for important agricultural traits reported (Abberton *et al.*, 2004). White clover is closely related to the model legume *Medicago truncatula* and there is likely to be considerable benefit in applying genomic resources from model to crop. However, the extent of synteny between the species must be established. Here we present preliminary results detailing progress towards this goal.

Materials and methods An F₁ mapping family of white clover has been developed. Parents were derived from material that was produced from three generations of divergent selection for stolon characteristics, variation in which largely governs persistence. The primary traits selected were stolon profuseness and thickness. Molecular markers derived from *M. truncatula* were applied to this F₁ mapping family. Simple sequence repeat (SSR) markers were used from expressed sequence tags (ESTs) of *M. truncatula* based on information kindly supplied by T. Huguet, INRA, France. DNA was extracted from 350 plants of the mapping family including the two parents using either a modified 2x CTAB or Qiagen DNA extraction kit. A total of 205 primer pairs were used with the following amplification conditions: a 25 µl reaction volume using an ABI GeneAmp PCR System 9700 of 50ng DNA, 1x buffer with 1.5µM MgCl₂ (Roche), 80µM dNTP, 50pM primers and 0.4U Taq DNA polymerase (Roche), initial denaturation 94°C, 5 min, denaturing step 94°C, 1 min, 55°C annealing for 1 min, 72°C elongation 1 min, and a final elongation step of 72°C.

Results 23 polymorphic primers (see Table 1) were re-assessed on polyacrylamide gels and 17 were chosen for further analysis. The mapping population was screened with each of these primers on an ABI 3100 Genetic Analyser. The primers detected from two to ten peaks in the white clover DNA, however the molecular weights were not always exactly those observed in *M. truncatula*.

Table 1 Structure of SSRs from *M. truncatula*, their amplification and the extent of polymorphism in the white clover F₁ mapping family.

Structure	Number	Efficient amplification (%ᵃ)	Polymorphic primer pairs (%ᵇ)
Perfect	189	169 (82)	22 (12)
Compound	12	10 (5)	1 (0.5)
Imperfect	1	1 (0.5)	0 (0)
TOTAL	202	180 (89)	23 (13)

ᵃ total number of primer pairs. ᵇ primer pairs showing efficient amplification.

Discussion Markers developed in *M. truncatula* are able to amplify products in white clover, showing a degree of conservation of microsatellite flanking sequences between the two species. The proportion of primers showing amplification was high, although polymorphism in this family was limited. The SSR data are now being supplemented by twenty AFLP markers. Alongside the molecular work, a range of characteristics were measured on each plant of the mapping family including thickness, length and number of stolons and other agronomically important traits. These measurements will be combined with the molecular data to identify the number and location of QTL for key stolon morphology traits. The framework map will also allow preliminary observations on the degree of synteny between white clover and *M. truncatula*.

Acknowledgments
This research was supported by the BBSRC and Defra

References
Abberton, M.T., Cogan, N., Smith, K.F., Marshall, A.H., Williams, T.A., Michaelson-Yeates, T.P.T., Bowen, C., Jones, E.S., Vecchies, A., and Forster, J.W. (2004) QTL analysis of morphogenetic and developmental traits in an SSR based genetic map of white clover (*Trifolium repens* L.). Forages Workshop. PAGA XII San Diego.
Barrett, B., Griffiths, A., Schreiber, M., Ellison, N., Mercer, C., Boulton J., Ong, B., Foster, J., Sawbridge, T., Spannenberg, G., Bryan, G., and Woodfield, D. (2004) A microsatellite map of white clover. *TAG* 109(3), 596-608.
Collins, R. P., Abberton, M. T., Michaelson-Yeates, T. P. T., Marshall, A. H., and Rhodes, I. (1998). Effects of divergent selection on correlations between morphological traits in white clover (*Trifolium repens* L.). *Euphytica* 101: 301-305.
Jones, E.S., Hughes, L.J., Drayton, M.C., Abberton, M.T., Michaelson-Yeates, T.P.T., Bowen, C., and Forster, J.W. (2003) An SSR and AFLP molecular marker based genetic map of white clover (*Trifolium repens* L.). *Plant Science* 165, 531-539.

Gene-associated single nucleotide polymorphism discovery in white clover (*T. repens* L.)

M.C. Drayton[1,3], R.C. Ponting[1,3], A.C. Vecchies[1,3], T.C. Wilkinson[2,3], J. George[1,3], N.O.I. Cogan[1,3], N.R. Bannan[2,3], K.F. Smith[2,3], G.C. Spangenberg[1,3] and J.W. Forster[1,3]

[1]*Primary Industries Research Victoria, Plant Biotechnology Centre, La Trobe University, Bundoora, Victoria 3086, Australia.* [2]*Primary Industries Research Victoria, Hamilton Centre, Hamilton, Victoria 3300, Australia.* [3]*Molecular Plant Breeding Cooperative Research Centre, Australia. Email: michelle.drayton@dpi.vic.gov.au*

Keywords: SNP, chalcone synthase, haplotype

Introduction Single nucleotide polymorphism (SNP) discovery permits the discovery of molecular marker variation associated with functionally-defined genes. SNP markers have been developed for the temperate pasture legume crop white clover (*Trifolium repens*) using public and proprietary genic sequences correlated with key agronomic traits of interest.

Materials and methods Several two-way pseudo-test cross white clover trait-specific genetic mapping families have been developed for the analysis of multiple agronomic traits, including abiotic stress tolerance, floral development, vegetative yield and herbage quality. These crosses include parental genotypes from the common population source LCL, a salt-tolerant selection from the Israeli cultivar Haifa. In addition to the other phenotypic traits, the progeny sets are likely to segregate for the important production trait of saline stress tolerance. The trait-specific families provide an important resource for *in vitro* SNP development and genetic map construction. Several strategies have been employed for SNP discovery. *In silico* SNP identification was performed in over 1400 contigs from white clover ESTs derived from cultivar Grasslands Huia, using the D2 clustering program and AutoSNP analysis.

Results Primers were designed to amplify the target regions and to validate the SNP loci in the parents of the mapping families. SNPs have also been identified *in vitro* from cDNA sequences of annotated genes of interest. Locus specific primers were designed for amplicons (c. 300-600 bp) spanning the transcriptional unit, and genomic sequences were amplified from parental and selected progeny genotypes. Subsequent cloning, sequencing and alignment of sequences allowed the identification of SNPs, within and between parental genotypes, followed by validation using the single nucleotide primer extension (SNuPe) assay. Validated gene-associated SNPs will be used to create genetic linkage maps in the trait specific mapping families, and co-location with QTLs for correlated traits will be assessed. Gene-length SNP haplotype structure will be compared to phenotypic variation and a strategy for confirmation of associations will be tested, based on comparison of non-synonymous SNP variation and variations in gene expression. SNP haplotypic data from diverse genotypes will be analysed for linkage disequilibrium (LD) studies. The gene-associated SNP markers provide a valuable resource supporting marker-assisted breeding for key agronomic traits.

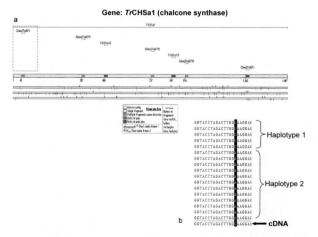

Figure 1 a) PCR primer locations spanning the *Tr*CHSa1 chalcone synthase gene (flavonoid biosynthesis), for SNP discovery and validation in white clover; b) identification of a G/A transition SNP locus in the *Tr*CHSa1 gene.

A molecular study of alfalfa megasporogenesis

D. Rosellini, S. Capomaccio and F. Veronesi

Department of Plant Biology and Agro-environmental Biotechnology, Borgo XX Giugno, 74 - 06121 Perugia (Italy). Email: roselli@unipg.it

Keywords: megasporogenesis, cDNA-AFLP, female sterility, *Medicago sativa*, RT-PCR

Introduction Our ability to control plant reproduction impacts on both seed production and plant breeding. A female sterility mutation was previously described (Rosellini *et al.*, 1998; 2003) revealing a female-specific arrest of sporogenesis associated with ectopic, massive callose deposition within the nucellus. The goal of this study is to isolate and characterize genes involved in megasporogenesis and female sterility in alfalfa.

Materials and methods In a 50-plant F_1 alfalfa population segregating for female sterility, ten plants were selected, 5 with sterility higher than 97% and 5 with sterility lower than 5%. Flower buds were excised from each plant at five flower development stages; leaves were also sampled as a control. cDNA was synthesized from total RNA of each sample; two cDNA bulks were formed using the same amounts of cDNA from each of the sterile (sterile bulk) and fertile (fertile bulk) plants and cDNA-AFLP analysis was performed. Full-length cDNAs corresponding to four selected differentially expressed transcripts were isolated by the RACE technique and sequenced. RT-PCR was performed for these four genes to confirm differential expression.

Results 96 Expressed Sequence Tags (ESTs) differentially expressed between sterile and fertile bulks were generated, and most of them published (GenBank from CB165074 to CB165159). Similarities with genes involved in the cell cycle, development and callose metabolism were found. Four were chosen for further studies: CB165076 similar to *A. thaliana* eukaryotic initiation translation factor *eIF4G III*; CB165091 similar to a soybean flower *beta 1,3-glucanase*; CB165125 similar to an *A. thaliana MAPKKK*; CB165105 similar to the *A. thaliana* transcription factor *SCARECROW* gene regulator. RT-PCR confirmed the differential expression between sterile and fertile plants for three of these genes, and flower specificity for two of them (Figure 1). Different alleles of CB165076 are probably revealed by RT-PCR performed with different primer pairs. The full length cDNA sequences (Table 1) revealed that the *eIF4G III*-like gene is much shorter than the similar published genes (not shown) and misses the MIF4G domain (RNA, DNA, protein binding).

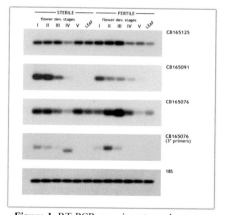

Table 1 Features of the isolated cDNAs

Clone	Similar to	cDNA nt	CDS nt
CB165076	At3g60240	4084	3468
CB165091	AAC04713	2103	1692
CB165125	CAB87658	1379	1050
CB165105	At5g66670	2320	1656

Figure 1 RT-PCR experiments on three differentially expressed genes

Conclusions Among the 96 ESTs, similarities for genes involved in the cell cycle, development and callose metabolism were found. Full-length cDNAs were cloned for four alfalfa genes whose similarity with published sequences suggests a possible involvement in the sterility trait. Our work continues with cloning and sequencing the genomic regions of these genes and with Southern and *in situ* hybridization experiments to assess gene copy number and expression patterns. This research was funded by the Italian Ministry of University and Scientific and Technological Research (project: Isolation and mapping of genes affecting sporogenesis and gametogenesis in *Medicago* spp) and is part of S. Capomaccio's *Dottorato di Ricerca* research activity.

References

Rosellini, D., F. Ferranti, P. Barone & F. Veronesi (2003). Expression of female sterility in alfalfa (*Medicago sativa* L.). Sexual Plant Reproduction 15, 271-279

Rosellini D, F. Lorenzetti & E.T. Bingham (1998). Quantitative ovule sterility in *Medicago sativa*. Theoretical and Applied Genetics 97, 1289-1295

The efficacy of GeneThresher® methylation filtering technology in the plant kingdom

U. Warek[1], J.A. Bedell[1], M.A. Budiman[1], A.N. Nunberg[1], R.W. Citek[1], D. Robbins[1], N. Lakey[1] and P.D. Rabinowicz[2]

[1]Orion Genomics LLC, 4041 Forest Park Avenue, St. Louis, MO 63108, USA.,Email: uwarek@oriongenomics.com
[2]The Institute for Genomic Research, Rockville, MD 20850, USA

Keywords: gene discovery, sequencing, SSRs, synteny

Introduction The genomes of many plants are known to be composed of a large fraction of repetitive DNA, while a small portion is dedicated to genes. The bulk of the repetitive DNA constitutes transposable elements and is heavily methylated. GeneThresher technology has been developed to take advantage of these differential methylation patterns by filtering genomic shotgun libraries to exclude methylated sequences (Rabinowicz *et al.*, 1999; Palmer *et al.*, 2003; Martienssen *et al.*, 2004). The result is a gene-enriched genomic shotgun library. Random shotgun sequencing of plant gene space, enabled by GeneThresher technology, is a rapid and cost-effective strategy for comprehensive gene discovery in agriculturally important crops.

Materials and methods We have applied GeneThresher to a number of plants that span the major branches of the plant kingdom. We have tested species representing monocots, dicots, gymnosperms, and non-vascular plants with a last common ancestor estimated at 500 million years ago.

Results Gene enrichment was achieved in all plants tested suggesting that GeneThresher will be effective across the whole plant kingdom. Genes discovered in the filtered sequences appear to be a random, unbiased representation of the gene set and represent the 5', internal, and 3' portions of genes with equal frequency. GeneThresher subclone libraries contain virtually all of the genes in a plant genome, and preferentially represent exons and introns, promoters, non-coding RNAs, and simple sequence repeats, while minimizing the representation of interspersed repeats. DNA sequence obtained from GeneThresher libraries provides a robust view of the functional parts of the genome, and enables the design of DNA microarrays bearing the complete gene set of a plant. A combination of GeneThresher sequencing with low coverage BAC sequencing recovers rare methylated genes and anchors the sequence to a physical map. This is a plant-specific modification of the strategy used to sequence the rat genome (Gibbs *et al.*, 2004).

After only a 1.1x coverage of the ~250 Mb gene space of the Sorghum bicolor genome, we have tagged more than three quarters of the Sorghum gene set using GeneThresher methylation-filtering technology (Bedell *et al.*, 2005). The Sorghum genome, as with other plants, is composed of a large percentage of methylated repetitive elements that have expanded the genome several-fold. A comparison of the Sorghum gene set to the completed Rice and *Arabidopsis* genomes enhances our understanding of the gene family structure of Sorghum and adds to our knowledge of the syntenic relationships among the grasses. In addition to allowing the rapid cataloguing of the gene sets of large plant genomes, GeneThresher sequences can enhance and complement other genome discovery efforts.

References

Bedell, J. A., M. A. Budiman, A. Nunberg, *et al.* (2005) Sorghum genome sequencing by methylation filtration. PLoS Biol 3(1): e13.

Gibbs, R. A., G. M. Weinstock, M. L. Metzker, *et al.* (2004) Genome sequence of the Brown Norway rat yields insights into mammalian evolution. Nature 428(6982): 493-521.

Martienssen, R. A., P. D. Rabinowicz, A. O'Shaughnessy and W. R. McCombie (2004) Sequencing the maize genome. Curr Opin Plant Biol 7(2): 102-7.

Palmer, L. E., P. D. Rabinowicz, A. L. O'Shaughnessy, V. S. Balija, L. U. Nascimento, S. Dike, M. de la Bastide, R. A. Martienssen and W. R. McCombie (2003) Maize genome sequencing by methylation filtration. Science 302(5653): 2115-7.

Rabinowicz, P. D., K. Schutz, N. Dedhia, C. Yordan, L. D. Parnell, L. Stein, W. R. McCombie and R. A. Martienssen (1999) Differential methylation of genes and retrotransposons facilitates shotgun sequencing of the maize genome. Nat Genet 23(3): 305-8.

Screening of perennial grasses and a mutant maize collection by Fourier-Transformed InfraRed (FTIR) spectroscopy for improved biofuel traits

S.C. Thain, P. Morris, S. Hawkins, C. Morris and I.S. Donnison
Institute of Grassland & EnvironmentalResearch. Aberystwyth, SY23 3EB, UK
Email: simon.thain@bbsrc.ac.uk

Keywords: plant cell wall, lignin, phenolics, FTIR

Introduction Currently the potential of biomass crops, including grasses, is limited because most species have not been bred for this purpose. However traits such as lignification, phenolic cross-linking and carbohydrate accessibility, which are also important for nutritive quality in forage grasses, can affect potential biofuel quality in applications such as combustion, fast-pyrolysis or fermentation. A collection of *Lolium* and *Festuca* species known to exhibit a range of lignin, cell wall phenolic and carbohydrate concentrations have been used to test optimum characteristics for biofuel processing. This collection formed a "calibration" set for subsequent high through-put FTIR chemical screening of additional plant lines: (1) A set of *Lolium-Festuca* substitution lines, in which *L. perenne* chromosomes or chromosome segments are substituted by homoeologous regions of *F. pratensis,* that provide the potential to physically map biofuel traits to an individual chromosome or chromosome segment; (2) A maize transposon (Robertson's *Mutator*) induced mutant collection, which provides the potential to identify gene sequences underlying important biochemical traits linked to biofuel as determined by FTIR analysis.

Materials and methods. High through-put biochemical analyses of plant material were undertaken using FTIR spectroscopy followed by some molecular genetic analysis. Plant material was finely ground and subjected to the Attenuated Total Reflectance (ATR) FTIR. Alternatively leaf discs or sectioned material were analysed by transmission mode FTIR. Maize PCR was performed using *Mutator* and gene specific PCR primers to screen candidate plants identified by FTIR analysis.

Results *Lolium* and *Festuca* species were analysed by FTIR to produce calibration sets, e.g. for lignin content (Figure 1). After appropriate data correction, the application of simple Principle Component Analysis (PCA) to the complete FTIR spectra indicates that different grass genotypes can be readily distinguished by these techniques. Data on the spectral "signatures" from detailed FTIR analyses of known cell wall biochemical mutants will be compared to newly characterised plants along with the available genetic characterisation data.

Conclusions The data from this study indicates that FTIR is a sensitive, robust and rapid screening method that following the application of appropriate calibrations can be used to detect differences in a broad range of cell wall biochemical components. When coupled to rapid methods for identifying the associated genes this provides a powerful tool for integrating trait analysis with the underlying molecular biology. This work has the potential to identify the biochemical and molecular bases of a number of quality traits which affect the suitability of grasses for use as livestock or biofuel based feedstocks.

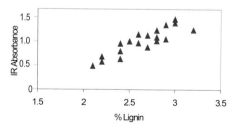

Lignin Absorbance @ 840-810cm

Figure 1 The sum FTIR absorbance values at 840 and 810cm^{-1} specific for G- and S-lignin respectively, plotted against %Lignin derived from Klason lignin analysis

A proposal for an international transcriptome initiative for forage and turf: microarray tools for expression profiling in ryegrass, clover and grass endophytes

T. Webster[1,2], N. Nguyen[1,2], C. Rhodes[1,2], S. Felitti[1,3], R. Chapman[1,2,4], D. Edwards[1,3,4] and G.C. Spangenberg[1,2,3,4]

[1]*Primary Industries Research Victoria, Plant Biotechnology Centre, La Trobe University, Bundoora, Victoria 3086, Australia* [2]*Victorian Microarray Technology Consortium, Australia* [3]*Molecular Plant Breeding Cooperative Research Centre, Australia* [4]*Victorian Bioinformatics Consortium, Australia*
Email: german.spangenberg@dpi.vic.gov.au

Keywords: microarrays, transcriptome analysis, International Transcriptome Initiative for Forage and Turf, perennial ryegrass, white clover, grass endophytes

Introduction Knowledge of the expression pattern of genes provides a valuable insight into gene function and role in determining the observed heritable phenotype. High–density cDNA and oligonucleotide microarrays represent powerful tools for transcriptome analysis to gain an understanding of gene expression patterns for thousands of genes. Internationally coordinated efforts in transcriptome analyses and sharing of microarray resources will benefit the advancement of our understanding of gene function in forage and turf species.

Materials and methods Within a joint Pasture Plant Genomics Program co-funded by Agriculture Victoria Services Pty Ltd and AgResearch Ltd. (New Zealand), we have developed high-density cDNA microarrays representing over 15 000 unique genes for each perennial ryegrass and white clover (Sawbridge *et al.*, 2003ab). Within the Molecular Plant Breeding Cooperative Research Centre, we have developed unigene microarrays allowing the interrogation of over 5 000 *Neotyphodium* and *Epichloe* genes (Nchip™ microarray and EndoChip™ microarray, Felitti *et al.*, 2005).

Results Microarrays have been applied in hybridisations with labelled total RNA isolated from a variety of genotypes, plant organs, developmental stages, and growth conditions. The collated results enable validation of functions predicted through comparative sequence annotation and suggest roles for novel genes lacking comparative sequence annotation. We have applied this data to assess the expression of genes associated with selected metabolic pathways and developmental processes to dissect these at the transcriptome level and to identify novel genes co-regulated with template genes known to be involved in these processes. Furthermore, these microarrays have enabled applications for gene and promoter discovery when used in concert with BAC libraries established for each of the target species.

Conclusions An *International Transcriptome Initiative for Forage and Turf* (ITIFT) to facilitate international efforts in microarray-based transcriptome analyses for key forage and turf plants and their endosymbionts is proposed. The Plant Biotechnology Centre will support ITIFT contributing access to and transcriptional profiling with unigene microarrays for key forage and turf species, namely perennial ryegrass (*Lolium perenne*), white clover (*Trifolium repens*) and grass endophytes (*Neotyphodium/Epichloe*), as well as access to a range of platforms for microarray spotting, hybridisation and scanning operationally integrated through a Scierra laboratory workflow system. The Plant Biotechnology Centre further proposes to develop and maintain an ITIFT database with a web based front-end portal for secure access by the research community to appropriate data and information. Anyone interested in participating in, contributing to ITIFT or accessing ryegrass, clover and endophyte microarrays, should contact german.spangenberg@dpi.vic.gov.au.

References
Sawbridge, T., Ong, E.K., Binnion, C., Emmerling, M., Meath, K., Nunan, K., O'Neill, M., O'Toole, F., Simmonds, J., Wearne, K., Winkworth, A. and Spangenberg, G. (2003a). Generation and analysis of expressed sequence tags in white clover (*Trifolium repens* L.). *Plant Science*, 165, 1077-1087.
Sawbridge, T., Ong, E.K., Binnion, C., Emmerling, M., McInnes, R., Meath, K., Nguyen, N., Nunan, K., O'Neill,
M., O'Toole, F., Rhodes, C., Simmonds, J., Tian, P., Wearne, K., Webster, T., Winkworth, A. and Spangenberg, G. (2003b). Generation and analysis of expressed sequence tags in perennial ryegrass (*Lolium perenne* L.). *Plant Science*, 165, 1089-1100.
Felitti, S.A., Tian, P., Webster, T., Edwards, D. and Spangenberg, G.C. (2005) Microarray-based transcriptome analysis of the interaction between perennial ryegrass (*Lolium perenne*) and the fungal endophyte *Neotyphodium lolii*. This volume.

Isolation and characterisation of genes encoding malate synthesis and transport determinants in the aluminium-tolerant Australian weeping-grass (*Microlaena stipoides*)

R.M. Polotnianka[1,2,3], E. Ribarev[1], L. Mackin[1], K.A. Sivakumaran[1,2], G.D. Nugent[1,3], U.P. John[1,2,3] and G.C. Spangenberg[1,2,3]

[1]*Primary Industries Research Victoria, Plant Biotechnology Centre, La Trobe University, Victoria 3086, Australia* [2]*Australian Centre for Plant Functional Genomics, Australia* [3]*Victorian Centre for Plant Functional Genomics, Australia Email: renata.polotnianka@dpi.vic.gov.au*

Keywords: weeping grass, aluminium tolerance, malate dehydrogenase, aluminium-induced malate transporter, aluminium exclusion

Introduction Acid soils cover some 40% of the Earth's arable land where they represent a major limitation to plant production. Plant growth on acid soils is primarily limited due to aluminium (Al) solubilized by acidity into toxic Al^{3+} cations which will inhibit root growth resulting in poor uptake of water and nutrients. Many important pasture species lack sufficient Al tolerance within their germplasm to allow effective breeding for this character.

Materials and methods A gene discovery and functional genomics program was undertaken in the Australian weeping grass (*Microlaena stipoides*) in order to determine the molecular basis of its enhanced tolerance to toxic Al^{3+} species associated with acidic soils, as a "Xenogenomics" approach that seeks to isolate and characterise determinants of abiotic stress tolerance in indigenous and exotic non-model and non-crop plants. *M. stipoides* has a highly efficient, rapidly induced aluminium exclusion mechanism in its roots whereby within hours of exposure to elevated Al levels at low pH it ceases to assimilate Al.

Results and conclusions EST sequencing of *M. stipoides* cDNA libraries derived from root material of plants exposed to 1mM Al revealed an enrichment of clones encoding malate dehydrogenase (MDH) orthologues (0.58%), relative to root EST collections of other grasses (< 0.1%). Northern hybridisation analysis showed that steady state levels of mRNA encoding the predominant cytoplasmic MDH (cMDH) form are not elaborated in response to Al or low pH but are enriched 4-fold in root tips relative to the mature portion of the root (Figure 1). Thus *cMDH* transcripts are preferentially localised to the region of the root most vulnerable to Al toxicity. A 1.4 kb region of upstream sequence has been isolated by adaptor PCR and cloned into promoter-reporter gene vectors for analysis in transgenic plants.

1   2   3

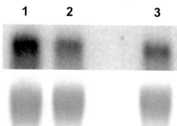

Figure 1 Northern hybridisation analysis of cytoplasmic MDH transcripts in root tips (1), mature part of root (2), and leaves (3) of *M. stipoides*. Lower panel shows rRNA as loading control.

In order to exploit malate exudation to confer aluminium tolerance, *M. stipoides* must not only be able to synthesise malate but it must efficiently secrete it. To investigate transport of malate, orthologues of the aluminium-induced malate transporter (*ALMT1*) gene of wheat that confers malate efflux and aluminium tolerance (Sasaki *et al.*, 2004), were isolated and characterised. The genome of *M. stipoides* contains a small family of *ALMT1* related sequences. Adaptor PCR has been used to isolate two distinct genomic clones encoding putative orthologues with 71.4 and 70.4% identity to wheat ALMT1. Further analysis of the expression and function of these candidate malate transporters is in progress.

Reference
Sasaki, T., Yamamoto, Y., Ezaki, B., Katsuhara, M., Ahn, S.J., Ryan, P.R., Delhaize, E., Matsumoto, H., (2004) A wheat gene encoding an aluminium-activated malate transporter. *Plant Journal*, 37,645-53.

Novel genotypes of the subtropical grass *Eragrostis curvula* for the analysis of apomixis (diplospory)

S. Cardone[1,2], P. Polci[1], J.P. Selva[1], M. Mecchia[3], S. Pessino[3], P. Voigt[4], G.C. Spangenberg[5] and V. Echenique[1,6]
[1]*Depto. de Agronomía, Universidad Nacional del Sur, San Andrés 800, 8000 Bahía Blanca, Argentina* [2]*Facultad de Agronomía, Universidad de Buenos Aires, Argentina* [3]*Facultad de Ciencias Agrarias, Universidad Nacional de Rosario, Rosario, Argentina* [4]*USDA, Agricultural Research Service, Appalachian Farming Systems Research Center, Beaver, USA* [5]*Primary Industries Research Victoria, Plant Biotechnology Centre, La Trobe University, Bundoora, Victoria 3086, Australia* [6]*CERZOS (CONICET), Bahía Blanca, Argentina*
Email: echeniq@criba.edu.ar

Keywords: *Eragrostis curvula*, diplosporous apomixis, novel genotypes, apomixis gene(s)

Introduction *Eragrostis curvula* (Schrad.) Nees is a variable grass native to Southern Africa. Its several forms, known as lovegrasses, were introduced to Australia, USA and Argentina as forage perennial grasses. Apomixis is a common trait in the genus *Eragrostis*, with diplospory being the most frequent type. Sexual reproduction also occurs in *Eragrostis*, although not frequently. Since most tetraploid *Eragrostis* lines are apomictic, the generation of a sexual tetraploid strain is a requirement for linkage analysis of the gene(s) governing the apomictic character. Furthermore, isogenic lines of the same ploidy, reproducing alternatively by sexuality or apomixes, represent an ideal system for comparative transcriptome analysis. The aim of this work was the generation and characterization of two novel genotypes of *E. curvula*: a dihaploid strain obtained *in vitro* from an apomictic cultivar and a tetraploid plant derived from the dihaploid after chromosome duplication.

Materials and methods Plants were obtained by *in vitro* culture of immature inflorescences of the apomictic *E. curvula* cultivar Tanganyka (2n = 4x = 40) on Murashige and Skoog medium supplemented with 2,4-D and BAP. Morphological and phenological traits of the regenerated plants and the original tetraploid cultivar were comparatively assessed. Chromosome number was determined in root tips by Feulgen staining. The reproductive mode was assessed by the study of megasporogenesis and embryo sac development and by the evaluation of progeny plants using RAPD amplification. Duplication of chromosome number of the dihaploid plant was done by 0.05% colchicine treatment. Progeny tests were performed by RAPD amplification on 8 individuals originated by open pollination from the dihaploid plant.

Results One out of 23 regenerated plants showed 20 chromosomes in root tip cells instead of the normal 40 chromosomes in the original tetraploid cultivar. This plant (named UNST1122) produced wider and shorter leaves than those of the other regenerated plants and the original cultivar. Other unusual characteristics of this plant were a typical pubescence on the base of the adaxial surface of leaves, a clear tendency to produce aerial tillers, a very small ligula (0.2 mm) and panicles exhibiting variation in shape and size, with rachillas radially arranged forming an angle of approximately 90° with the rachis. Anthers were pale yellow and seeds were ellipsoidal with a light brown coat, smaller and lighter than those produced by control plants (weight of 100 seeds: 23 mgs vs 40 mgs). Under greenhouse conditions, UNST1122 flowered almost all year round, except in the colder months of June and July (Southern hemisphere). However, the number of seeds per inflorescence was lower than that of the original cultivar. Studies of megasporogenesis and megagametophyte development, as well as progeny tests using RAPD amplification, showed that this dihaploid plant was sexual and self-incompatible. Colchicine treatment of seeds from the dihaploid gave rise to two plants with 40 chromosomes, UNST1112 and UNST1131. Comparison of the amplification profiles obtained for UNST1122 with those of 8 progeny individuals obtained by open pollination revealed polymorphism that ranged from 4.6% to 40% for all analyzed plants. The lack of offspring showing a maternal profile is evidence of sexual reproduction in UNST1112.

Conclusions The molecular basis underlying mechanisms of diplosporous reproduction remains largely unknown. The novel genotypes described here will allow the future establishment of mapping populations of *Eragrostis curvula* that can be used in strategies of forward genetics to map and positionally clone the determinants of the apomictic trait. They also represent an excellent system for the identification of genes involved in diplospory and/or ploidy level gene regulation by using transcriptional profiling techniques such as differential display, cDNA-AFLP or ESTs discovery.

Discovery and functional categorisation of expressed sequence tags from flowers of *Eragrostis curvula* genotypes showing different ploidy levels and reproductive modes

V. Echenique[1,2], S. Felitti[3], N. Paniego[4], L. Martelotto[5], S. Pessino[5], D. Zanazzi[4], P. Fernández[4], M. Díaz[1], P. Polci[1] and G.C. Spangenberg[3]

[1]Depto. de Agronomía, Universidad Nacional del Sur, San Andrés 800, 8000 Bahía Blanca, Argentina [2]CERZOS (CONICET), Bahía Blanca, Argentina [3]Primary Industries Research Victoria, Plant Biotechnology Centre, La Trobe University, Bundoora, Victoria 3086, Australia [4]Instituto de Biotecnología, CICVyA, INTA-Castelar, Argentina [5]Facultad de Ciencias Agrarias, Universidad Nacional de Rosario, Rosario, Argentina
Email: echeniq@criba.edu.ar

Keywords *Eragrostis curvula*, expressed sequence tags (ESTs), diplosporous apomixis, ploidy

Introduction Two novel genotypes of weeping lovegrass (*Eragrostis curvula*) - a dihaploid strain obtained *in vitro* from an apomictic cultivar and a tetraploid plant derived from the dihaploid after chromosome duplication – have recently been developed. These materials represent an excellent system for the identification, through transcriptional profiling, of genes involved in diplospory and/or ploidy level gene regulation. The aim of this work was the discovery and functional classification of expressed sequence tags (ESTs) from immature inflorescences of the apomictic *E. curvula* cultivar Tanganyika (2n=4x=40), a dihaploid sexual strain derived from it (2n=2x=20) and a tetraploid sexual strain (2n=4x=40) obtained by colchicine duplication of the dihaploid.

Materials and methods A dihaploid plant was obtained by *in vitro* culture of immature inflorescences of the apomictic *E. curvula* cultivar Tanganyka (2n = 4x = 40) on Murashige and Skoog medium supplemented with 2,4-D and BAP. Duplication of chromosome number of the dihaploid plant with 0.05% colchicine led to the recovery of two tetraploid sexual plants. Inflorescences of Tanganyka, the sexual dihaploid and one of the colchicine-duplicated tetraploid sexual plants were collected at the same developmental stage. Total RNA was extracted from flowers using the RNeasy® total RNA isolation kit (Promega). cDNAs were obtained using a the SMART® PCR synthesis kit, cloned into the pGEM®-T easy vector and used for transformation of XL10-Gold® Ultracompetent *E. coli* cells. Bacterial cultures were performed in 96-well blocks and DNA isolations using a QIAprep®96 turbo Miniprep kit were carried out on a Qiagen Biorobot 9600. 2400 randomly selected cDNA clones from each library were sequenced using a MegaBACE™ DNA sequence analyzer. Same experimental approaches were used for the generation of ESTs from a leaf cDNA library from Tanganyka. Raw EST sequence traces were processed and annotated after the analysis using the XGI pipeline (http://www.ncgr.org/xgi) for automated EST clustering. Processed ESTs were clustered into consensus sequences that were compared using BLASTX against NCBIs non-redundant protein database. Functional categories were assigned by means of gene ontology (GO) annotations.

Results and conclusions A total of 9600 randomly selected cDNA clones from 4 cDNA libraries of *E. curvula* (2400 from 01EC dihaploid flowers, 2400 from 02EC apomictic tetraploid flowers, 2400 from 03EC apomictic tetraploid leaves and 2400 from 04EC sexual tetraploid flowers) were sequenced. Cluster analysis of the corresponding ESTs allowed the identification of 2,218 unique sequences (719 for EC01, 564 for EC02, 589 for EC03 and 346 for EC04). BLASTX analysis revealed no significant similarities at an E-value greater than 10^{-6} for 54% of the non-redundant *Eragrostis* sequences. Functional categories were assigned to 402 out of 1198 annotated sequences. The EST collections from the different cDNA libraries showed a distinctive distribution among different functional categories. Furthermore, the *E. curvula* sequences were compared with those corresponding to differential bands obtained after an extensive transcript profiling of diploid and tetraploid genotypes of the aposporous pasture grass *Paspalum notatum* with different ploidies, in order to identify common patterns in relation to ploidy levels and modes of reproduction.

A comprehensive analysis of gene expression and genomic alterations in a newly formed autotetraploid of *Paspalum notatum*

L.G. Martelotto[1], J.P.A. Ortiz[1], F. Espinoza[2], C.L. Quarin[2] and S.C. Pessino[1]

Plant Research Central Laboratory, Agronomy Department, National University of Rosario, 2123 Zavalla, Santa Fe, Argentina; Institute of Botany of the North-East, Sargento Cabral 2134, 3400 Corrientes, Argentina. Email: pessino@arnet.com.ar

Keywords: ploidy changes, gene expression alterations, differential display, forage grasses

Introduction The proportion of angiosperms that have experienced one or more episodes of chromosome doubling is estimated to be of the order of 50-70%. This prominence of polyploidy probably implies some adaptive significance. The emergence of novel phenotypes could allow polyploids to enter new niches or enhance their chances of being selected for use in agriculture. Although the causes of novel variation in polyploids are not well understood, they could involve changes in gene expression through dosage-regulation, altered regulatory interactions and rapid genetic/epigenetic changes. The objective of this work was to carry out an extensive transcript profiling and genome analysis of a diploid genotype of the pasture grass *Paspalum notatum* and its tetraploid derivative obtained after colchicine treatment.

Materials and methods Several different clonal individuals of a *Paspalum notatum* diploid genotype (C4-2x, 2n = 2x = 20) and its newly synthesized autotetraploid derivative (C4-4x, 2n = 4x = 40) (Quarin *et al.*, 2001) were studied. Total RNA was obtained from flowers (at a particular developmental stage) or leaves. Differential display experiments were conducted under the general protocol reported by Liang and Pardee (1992). Differential expression was validated by reverse-Northern blot. Sequencing was done by Macrogen Inc., Korea. RAPDs reactions were conducted according to the method of Williams *et al.* (1990) and electrophoresed in polyacrylamide gels.

Results Differential display banding patterns generated from flowers of the diploid and tetraploid lines were compared in duplicated tests. From a total of 9617 bands scored, 129 (1.34%) were polymorphic between both lines. The isolated clones were subjected to reverse-Northern validation to discard false positives. Sixty four (64) were confirmed to represent up- or down-regulated genes in the autotetraploid. Sequencing showed that 42 of them were homologous to 26 different genes of known function, involved in processes related to DNA repair, protein trafficking, regulation of transcription, proteolysis, protein folding, carbohydrate and lipid metabolism and signal transduction The remaining 22 clones represented novel sequences. We also analysed the expression of around 2000 transcripts from leaves and found alterations in the profile of expression of 10 transcripts. Six of them were novel sequences, while the rest corresponded to an rRNA sequence and two mRNA of uncharacterized function. RAPD analysis was used to compare the genomic structure of both lines. Genome fingerprinting assays using 565 markers revealed the presence of a significant rate of polymorphisms (9.2%), that were mainly revealed by 4 particular decamers. Twelve polymorphic bands were isolated from the polyacrylamide gels, cloned and sequenced. Five of the isolated bands showed significant homology to a protein of uncharacterized function containing a SIS (sugar isomerase) binding domain. Two clones showed homology with a family of retroelements and a methyltransferase domain, respectively. The rest of the clones showed no significant similarities in the data banks.

Conclusions The results of the present study show that a change in the number of genomic complements modifies the expression of several genes in flowers and leaves of *Paspalum notatum*. Several general pathways (like carbohydrate and lipid metabolism, regulation of chromatine structure, ubiquitination and signal transduction) might be affected. We detected genomic modifications associated with polyploidization. Alterations in the structure of genomic sectors codifying retrotransposons and methyltransferases could be related to the well-documented epigenetic modifications occurring in recently formed polyploids.

References

Liang, P. and Pardee, A.B. (1992) Differential display of eukaryotic messenger RNA by means of the polymerase chain reaction. *Science* 257: 967-971.

Quarin, C.L., Espinoza, F., Martínez, E.J., Pessino, S.C.and Bovo, O.A. (2001) A rise of ploidy level induces the expression of apomixis in *Paspalum notatum*. *Sex Plant. Reprod.* 13: 243-249.

Williams, J.G.K., Kubelik, A.R., Livak, K.J., Rafalski, J.A. and Tingey, S.V. (1990) DNA polymorphisms amplified by arbitrary primers are useful as genetic markers. *Nucleic Acids Res.* 18: 6531-6535.

Gene discovery and molecular dissection of fructan metabolism in perennial ryegrass (*Lolium perenne*)

J. Chalmers[1,2], A. Lidgett[1,2], X. Johnson[1,2], K. Terdich[1,2], N. Cummings[1,2], Y.Y. Cao[1,2], K. Fulgueras[1,2], M. Emmerling[1,2], T. Sawbridge[1], E.K. Ong[1,3], A. Mouradov[1,2] and G.C. Spangenberg[1,2]

[1]*Primary Industries Research Victoria, Plant Biotechnology Centre, La Trobe University, Bundoora, Victoria 3086, Australia* [2]*Molecular Plant Breeding Cooperative Research Centre, Australia* [3]*Department of Botany, La Trobe University, Bundoora, Victoria 3086, Australia* Email: german.spangenberg@dpi.vic.gov.au*

Keywords: perennial ryegrass, fructan metabolism, transgenic plants, nutritive value

Introduction Fructans are the main soluble carbohydrate stored in up to a third of the vegetation of the earth, including the economically important temperate grasses. Fructans are polymers of fructose attached to a sucrose precursor. Perennial ryegrass (*L. perenne* L.) accumulates fructans of the inulin series, inulin neoseries and levan neoseries. Four enzymes are required to produce fructans of this profile: 1-SST (sucrose:sucrose 1-fructosyltransferase), 1-FFT (fructan:fructan 1-fructosyltransferase), 6G-FFT (6-glucose fructosyltransferase) and 6-FFT (fructan:fructan 6-fructosyltransferase) or 6-SFT (sucrose:fructan 6-fructosyltransferase) (Figure 1). Fructan biosynthetic enzymes have evolved from invertases and thus it is argued that fructan metabolism is an extension of sucrose metabolism. A high fructan content is a valuable resource in perennial ryegrass as it can be readily mobilised to sustain regrowth immediately after defoliation as well as adding to the nutritive value of the feed. However, the physiological role of fructans in grasses is not fully understood.

Materials and methods A targeted gene discovery program in perennial ryegrass has led to the isolation and characterisation of cDNA and genomic clones encoding enzymes involved in sucrose and fructan metabolism: sucrose phosphate synthase (SPS), invertases (INV), sucrose synthase (SS), sucrose transporter (ST), 1-SST (sucrose:sucrose 1-fructosyltransferase), 1-FFT (fructan:fructan 1-fructosyltransferase) and fructan hexohydrolase (FEH).

Results and conclusions Following sequence analyses, expression profiles (i.e. organ specificity, developmental regulation, light/dark and temperature regulation) of these sucrose and fructan metabolism genes in perennial ryegrass were determined using microarray-based and northern hybridisation analyses. Gene organisation, copy number and genetic map location using RFLPs were determined for LpSPS, LpINV, LpSS, Lp1-SST, LpFFT and LpFEH. Functional analyses of sucrose and fructan metabolism genes were undertaken in *Pichia pastoris* and *in planta*. Expression vectors were generated using sense, antisense and hpRNA technologies for overexpression, co-suppression or downregulation of target genes in transgenic perennial ryegrass plants. Transgenic perennial ryegrass plants have been generated and characterised at the molecular and biochemical level. This provided the basis for a molecular dissection of fructan metabolism to understand its physiological role in perennial ryegrass. It further underpinned the design of transgenic elite perennial ryegrass genotypes carrying modular vectors with a combinatorial modulation of key target fructan metabolism genes using exclusively ryegrass gene sequences for improved nutritive value, persistence and quality.

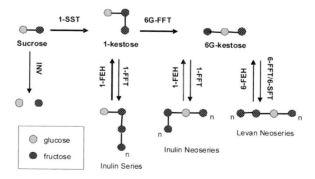

Figure 1 Hypothetical pathway of fructan metabolism in perennial ryegrass

Gene discovery and molecular dissection of lignin biosynthesis in perennial ryegrass (*Lolium perenne*)

A. Lidgett[1,2], M. Emmerling[1,2], R. Heath[1,2], R. McInnes[1,2], D. Lynch[1,2], A. Bartkowski[1,2], K. Fulgueras[1,2], T. Sawbridge[1], E.K. Ong[1,3], K.F. Smith[2,4], A. Mouradov[1,2] and G.C. Spangenberg[1,2]

[1]*Primary Industries Research Victoria, Plant Biotechnology Centre, La Trobe University, Bundoora, Victoria 3086, Australia* [2]*Molecular Plant Breeding Cooperative Research Centre, Australia* [3]*Department of Botany, La Trobe University, Bundoora, Victoria 3086, Australia* [4]*Primary Industries Research Victoria, Hamilton Centre, Hamilton, Victoria 3300, Australia Email: german.spangenberg@dpi.vic.gov.au*

Keywords: perennial ryegrass, lignin biosynthesis, transgenic plants, herbage quality

Introduction Lignification of plant cell walls has been identified as a major factor limiting forage digestibility. It limits the amount of digestible energy available to livestock, resulting in an incomplete utilisation of cellulose and hemicellulose by ruminant animals. Modification of the lignin profile of ryegrasses (*Lolium* spp.) and fescues (*Festuca* spp.) is undertaken through modulating the expression of genes encoding enzymes involved in the biosynthesis of monolignols.

Materials and methods A targeted gene discovery program in perennial ryegrass (*L. perenne* L.) has led to the isolation and characterisation of cDNA and genomic clones encoding the following enzymes involved in the biosynthesis of the monolignol precursors and their extracellular polymerisation to yield lignins: phenyl alanine ammonia lyase (PAL), cinnamate-4-hydroxylase (C4H), ferulate-5-hydroxylase (F5H), caffeic acid *O*-methyltransferase (OMT), 4-coumarate-CoA ligase (4CL), caffeoyl-CoA 3-*O*-methyltransferase (CCoAOMT), cinnamoyl-CoA reductase (CCR), cinnamyl alcohol dehydrogenase (CAD), peroxidase (PER) and laccase (LAC) (Figure 1).

Results and conclusions Following sequence analyses, expression profiles (i.e. organ specificity, developmental regulation and wound-inducibility) of these lignification genes in perennial ryegrass were determined using microarray-based and northern hybridisation analyses. Gene organisation, copy number and genetic map location using RFLPs were determined for LpPAL, LpC4H, LpF5H, LpOMT, Lp4CL, LpCCoAOMT, LpCCR, LpCAD and LpPER. Functional analysis *in planta* was undertaken following production of transgenic model (i.e. arabidopsis, tobacco) plants for overexpression of chimeric target lignification genes under control of the CaMV35S promoter and the generation of transgenic ryegrass plants for overexpression and down-regulation of the chimeric genes under control of constitutive and xylem-specific promoters. This provided the basis for a molecular dissection of the biosynthesis of monolignol precursors in perennial ryegrass to enhance the understanding of lignin monomeric composition and properties in grasses. It furthermore underpinned the design of transgenic elite perennial ryegrass genotypes carrying modular vectors with a combinatorial modulation of key target lignification genes using exclusively ryegrass gene sequences for improved forage quality and nutritive value.

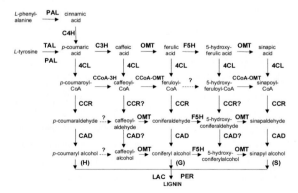

Figure 1 Hypothetical pathways of lignin biosynthesis in perennial ryegrass

An *in silico* DNA sequence comparison of the perennial ryegrass and rice genomes

M.J. Faville

AgResearch Ltd., Grasslands Research Centre, Private Bag 11008, Palmerston North, New Zealand
Email: marty.faville@agresearch.co.nz

Keywords: colinearity, EST-SSR, rice, ryegrass, synteny

Introduction Comparative mapping studies in the family Poaceae, which includes rice (a model species for this family) and perennial ryegrass (PRG) have indicated macro-colinearity of genes is generally conserved across different genomes. Genome mapping of simple sequence repeat markers derived from expressed sequence tags (EST-SSRs) for PRG (Faville *et al.*, 2004) provides a vehicle for DNA sequence-based matching of mapped PRG genes to orthologous positions in the rice genome, which can be used to establish comparative relationships between these species' genomes. We have initiated such an analysis using an EST-SSR-based PRG genome map. Our objective was to assess this *in silico* approach as a tool for candidate gene identification from rice, and for targeting markers to specified PRG genome map regions based on rice genome position.

Materials and methods Two PRG bi-parental genetic linkage maps were combined to construct a consensus genome map (JoinMap 3.0) with 167 single locus EST-SSR locations. Potential rice orthologues of mapped PRG ESTs were identified by BLASTN analysis against a Rice Gene Index (http://www.tigr.org/tigr-scripts/tgi/T_index.cgi?species=rice). Criteria used were: E value $<e^{-20}$; sequence identity >85% over >100 bp alignment length. Rice tentative consensus (TC) sequences associated with map-ordered rice BAC (bacterial artificial chromosome) and PAC (P1-derived artificial chromosome) clones were identified, providing a rice genome position for their putative PRG orthologues.

Results Of 167 EST-SSRs used to construct a consensus PRG genome map, 37% were common to both PRG component maps. Common marker orders were conserved between component maps (r^2 0.93 across linkage groups), and component maps and the consensus map (r^2 0.98 for both). Under our BLASTN criteria, potential rice orthologues were identified for 136 mapped EST-SSRs. This information facilitates an overview of macro-colinearity between the PRG and rice genomes (Figure 1), with large portions of PRG linkage groups showing homology to 1-2 rice chromosomes. These relationships endorse earlier comparisons between the two species (Jones *et al.*, 2002), and are concomitant with the wheat-rice relationship (Sorrells *et al.*, 2003). Within macro-colinear blocks, apparent rearrangements of orthologous sequences were observed (data not shown). Verification of EST-SSR positions in more than one PRG mapping pedigree will confirm that these rearrangements are genuine and not due to ambiguous marker ordering.

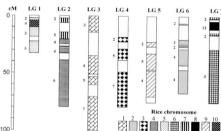

Figure 1. Perennial ryegrass (PRG) genome map-rice genome relationships. Patterned regions on the map represent macro-colinear blocks (intervals with ≥ 2 contiguous mapped PRG EST-SSRs matching rice sequences from one rice chromosome; no. matches indicated left of each block). White spaces contain EST-SSRs matches to one or no rice sequence. Macro-colinear blocks were not found for rice chromosomes 11 and 12. LG = linkage group; cM =centimorgans.

Conclusions The approach used was successful for identifying macro-colinearity between the PRG and rice genomes, and will improve by enhancing EST-SSR density and genome coverage. This information will be useful for selecting DNA sequence-derived markers targeted to broad PRG genome map regions. However, rice-PRG transfer of candidate gene information may be complicated by intrachromosomal rearrangements.

References

Faville M.J., A.C. Vecchies, M. Schreiber, M.C. Drayton, L.J. Hughes, E.S. Jones, K.M. Guthridge, K.F. Smith, T. Sawbridge, G.C. Spangenberg, G.T. Bryan & J.W. Forster. (2004) Functionally-associated molecular genetic marker map construction in perennial ryegrass (*L. perenne* L.). *Theoretical and Applied Genetics* 110: 12-32

Jones E.S., N.L. Mahoney, M.D. Hayward, I.P. Armstead, J.G. Jones, M.O. Humphreys, I.P. King, T. Kishida, T. Yamada, F. Balfourier, C. Charmet & J.W. Forster (2002). An enhanced molecular marker-based map of perennial ryegrass (*Lolium perenne* L.) reveals comparative relationships with other Poaceae species. *Genome* 45: 282-295

Sorrells, M.E., M. La Rota, C.E. Bermudez-Kandianis *et al.* (2003). Comparative DNA sequence analysis of wheat and rice genomes. *Genome Research* 13: 1818-1827

The identification of genetic synteny between *Lolium perenne* chromosome 7 and rice chromosome 6 genomic regions that have major effects on heading-date

I.P. Armstead, L.B. Turner, L. Skøt, I.S. Donnison, M.O. Humphreys and I.P. King
Plant Genetics and Breeding Dept., Institute of Grassland and Environmental Research, Aberystwyth, SY23 3EB, UK, Email: ian.armstead@bbsrc.ac.uk

Keywords *Lolium perenne*, rice, genetic synteny, heading date QTL

Introduction Comparative genetic mapping between plant species has established that there has been a conservation of genomic organisation which reflects evolutionary relationships. The genetic mapping of *L. perenne* has identified such syntenic relationships with both the Triticeae and rice. The recent publication of the complete sequence of the rice genome has allowed these relationships to be analysed more closely and has raised the possibility of using the rice genome as a template for chromosome landing-based gene identification in related non-model species.   The aim of the present work was to map particular markers and genes associated with heading-date in rice in *L. perenne* in order to test this comparative genomics approach.

Materials and methods A *L. perenne* F2 mapping family of 188 individuals constructed from parental material with early and late heading dates was established and genetically mapped with a marker set that included comparative mapping RFLP probes (Armstead *et al.*, 2004).   The family was evaluated for heading date and a QTL analysis performed. Using known comparative genetic relationships between rice and *L. perenne*, STS markers based upon rice gene sequences known to be associated with rice heading date QTL were developed and mapped in the *L. perenne* mapping population.

Results Interval and MQM mapping with MapQTL 4.0 (Van Ooijen *et al.*, 2002) identified a QTL for heading date which accounted for up to 64% of the variance and which was associated with *L. perenne* chromosome (C) 7. The trait appeared to be under the control of a single dominant gene, or a block of tightly linked dominant genes. Comparative mapping indicated that this region of *L. perenne* C7 showed a degree of conserved genetic synteny with rice C6, a region of the rice genome which contains both the Hd1 and Hd3 rice heading date QTL. The mapping of STSs based upon gene sequences associated with these QTL in rice indicated the major effect in the *L. perenne* genome in this mapping family seemed to be associated more closely with the Hd3 as opposed to the Hd1 equivalent regions.   However, a parallel heading-date linkage disequilibrium study (Skøt *et al.*, 2004) identified molecular markers which, when mapped in the same *L. perenne* mapping family, were associated with Hd1 equivalent region of *L. perenne* C7.

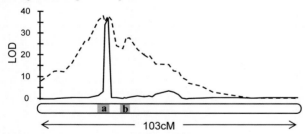

Figure 1   Position of major QTL for heading date on *L. perenne* chromosome 7; dashed line and solid line illustrate LOD profiles produced by interval and MQM mapping, respectively. **a** = positions of markers associated with the Hd3 region of rice; **b** = markers associated with the Hd1 region of rice and markers generated from the *L. perenne* linkage disequilibrium study.

Conclusions Comparative genetics indicates that *L. perenne* C7 and rice C6 share a degree of conserved synteny. Further detailed analysis indicates that this relationship extends beyond just co-linearity of molecular markers. It also implies similarities in terms of the significance of this genomic region in determining an important trait in that both the Hd3 and Hd1 chromosomal regions of rice C6 and the equivalent chromosomal regions of *L. perenne* C7 can be associated independently with heading-date determination.

References
Armstead, I.P., L.B. Turner, M. Farrell, L. Skøt, P. Gomez, T. Montoya, I.S. Donnison, I.P. King, & M.O. Humphreys (2004). Synteny between a major heading-date QTL in perennial ryegrass (*Lolium perenne* L.) and the Hd3 heading-date locus in rice. Theoretical and Applied Genetics 108: 822-828.
Van Ooijen J.W., M.P. Boer, R.C. Jansen & C. Maliepaard, 2002 Map QTL® 4.0, Software for the calculation of QTL positions on genetic maps. Plant Research International, Wageningen, the Netherlands.
Skøt, L., M.O. Humphreys, I. Armstead., S. Heywood, K.P. Skøt, R. Sanderson, I. Thomas, K. Chorlton, & R. Sackville-Hamilton (2004). An association mapping approach to identify flowering time genes in natural populations of *Lolium perenne*. Molecular Breeding in press.

Towards understanding photoperiodic response in grasses

M. Gagic[1], I. Kardailsky[1], N. Forester[1], B. Veit[1] and J. Putterill[2]
[1]AgResearch, Grasslands, Palmerston North, New Zealand[2], School of Biological Sciences, University of Auckland, New Zealand. Email: milan.gagic@agresearch.co.nz

Keywords: flowering, photoperiod candidate gene orthologues

Introduction In many plants, day length is the critical environmental parameter that controls flowering time. In long day plants, such as Arabidopsis and ryegrass (*Lolium perenne*), increasing day length in spring signals flowering, while in short day plants like rice, flowering is accelerated when days become shorter. Recently, significant progress has been made in understanding the molecular genetic mechanisms that govern this response. Most results have been obtained in the model plant *Arabidopsis* where *CONSTANS (CO)* is a critical candidate gene. Upstream of it is the *GIGANTEA (GI)* gene which is associated with the circadian clock mechanism (1). The *FT* gene is the immediate downstream genetic target of *CO*, and is a direct promoter of flowering (2). Characteristically, all three genes show circadian expression, albeit in different phases, and both the *CO* and *FT* genes are up-regulated under long-day (inductive) conditions. Work in ryegrass should help reveal both the conserved and divergent segments of the photoperiod response between different plant species.

Material and methods Putative orthologues of *Arabidopsis* (At) GI, CO, and FT genes were isolated and sequenced using a combination of different methods including a ryegrass EST library screen, multiple alignment and degenerate primers, 5' and 3' RACE-PCR, and gene walking. Gene expression was carried out using real time RT-PCR on clones of Grasslands Impact ecotype grown in a controlled environment.

Results The full-size *LpGI* gene was found to contain 11 introns and encoded a 1149 amino acid (aa) protein. Expression analysis of GI showed diurnal oscillation and different expression patterns under long day (LD) and short day (SD) conditions with the peaks in LD coinciding with the light period of the day. The *LpCO* gene encoded a 365 aa protein whose sequence contained two zinc-finger domains, a feature specific for the *CO*–like genes, as well as a CCT region near the carboxy domain similar to that observed in the barley *CO* and rice *Hd1* gene. In addition, the intron region contained a Dof2 transcription factor binding domain whose role is yet to be elucidated. Phylogenetically, the *LpCO* groups align closely with *OsHd1*, *HvCO1*, and *AtCO* (Fig 1). We have also isolated 3 ryegrass FT-like genes with *LpFT3* showing the highest similarity with rice *Hd3a*. Expression pattern analysis using real-time RT-PCR revealed a 30-fold increase under LD conditions which is consistent with previous findings in *Arabidopsis*.

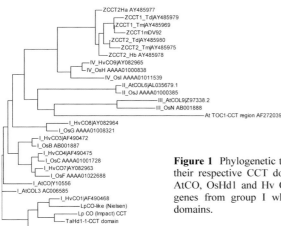

Figure 1 Phylogenetic tree of the CO and COL genes deduced from their respective CCT domains. The LpCO groups closely with the AtCO, OsHd1 and Hv CO1 genes as well as the rest of the COL genes from group I which all contain two functional zinc-finger domains.

Conclusion Initial results reveal that the ryegrass photoperiod pathway genes show high similarity to their wheat and rice orthologues. The CO gene contains two complete B-box regions as indicated by the presence of His and Cys residues with conserved positions in agreement with the general consensus. Expression analysis under LD and SD conditions showed that GI and FT cycle in the same manner as the *Arabidopsis* orthologues.

References
Fowler, S. *et al. Embo J* 18, 4679-88 (1999)
Kardailsky, I. *et al. Science* 286, 1962-5 (1999)

Controlled flowering project for *Lolium perenne* at Agresearch: an overview

I. Kardailsky, B. Veit, N. Forester, M. Gagic, K. Richardson, M. Faville and G. Bryan
AgResearch, Grasslands, Palmerston North, New Zealand. Email: igor.kardailsky@agresearch.co.nz

Keywords: flowering, photoperiod, vernalisation, candidate gene orthologues

Introduction Ryegrass (*Lolium perenne*) is an important forage crop in New Zealand. The work presented here has the goal of developing a system for complete and arbitrary control of the transition from vegetative to floral development. For this, we have pursued an integrated approach utilising genomics with both forward and reverse genetics. Like other model plants, photoperiodic and vernalization pathways are presumed to be operating in ryegrass and control the activity of the meristem identity/floral patterning genes. The candidate gene approach targeting the photoperiodic pathway is described in an accompanying abstract (Gagic *et al.*). Other candidate genes include the meristem identity gene LEAFY and a range of the MADS box transcription factors. Relevant expression profiles are established for these genes, i.e. vernalization time course at weekly intervals, and daily and circadian collections during the secondary induction. A detailed genetic map of ryegrass has been developed at AgResearch (see abstract by Faville *et al.*) which we are using to map candidate genes. We are also conducting detailed phenotypic analysis of the flowering behaviour variation within this population in an effort to isolate relevant QTLs. Ryegrass transformation has been used to ascertain functions of the candidate genes and to manipulate flowering time control directly. We are developing a universal switch to turn on the flowering that consists of a cassette of the arabidopsis genes under a control of a chemically inducible promoter.

Materials and methods For the candidate gene isolation, a combination of approaches involving screening of our ryegrass EST library and direct amplification using degenerate oligonucleotides derived from conserved sequences, followed by additional screening to obtain full-size cDNA and genomic sequences (5' and 3'-RACE, genome walking) have been used. Gene expression was assayed using real-time RT-PCR analysis from samples of cloned material grown in a controlled environment. To map the genes, SNPs were identified and their segregation scored using direct sequencing of PCR products from genomic DNA. For the QTL mapping, sets of clones of the mapping population were exposed to variable vernalization and then either induced under artificial long days, or naturally. Dates of the 1st and 3rd head emergence were scored and the number of total and floral tillers counted at the end of the experiment. Traits reflecting flowering time, uniformity of induction and percentage of induced tillers were derived from these measurements, as well as the effect of vernalization on the traits across treatments. For transformation, micro projectile bombardment of leaf explants/young inflorescences was used with subsequent callus formation and plant regeneration in tissue culture. For the floral induction switch, we used a derivative of the pHTOP/LhG4::GR (Moore *et al.*, 1998) system in which expression is controlled with dexmethasone.

Results Candidate genes were isolated that are putative orthologues of the arabidopsis GI, CO, LFY, FT, TFL, MFT, AP1, SOC1, SVP, and wheat VRN1. GI, FT and CO expressions have circadian cycles (see Gagic *et al.*). LFY, GI, SVP, FT and CO have been mapped and the others are in progress (see Faville *et al.*); the VRN1 mapping has been published (Jensen *et al.*, 2004). Significant QTLs have been found on LG1, 2 and 4. Some of these may overlap with other published QTLs (Armstead, *et al.*, 2004) and the position of candidate genes. Direct over-expression of the *Arabidopsis* FT gene in ryegrass led to an expected acceleration of flowering. The inducible flowering switch works very efficiently in *Arabidopsis* and its utility in other species including ryegrass is being investigated.

Conclusion A complex approach toward understanding the genetics of flowering in ryegrass is an efficient way to isolate relevant genes and provides a foundation for the manipulation of flowering behaviour using both transgenic and conventional breeding strategies.

References
Armstead, I. P. *et al*. Theor Appl Genet 108, 822-8 (2004).
Jensen, L. B. *et al*. Theor Appl Genet (2004).
Moore, I., Galweiler, L., Grosskopf, D., Schell, J. & Palme, K. Proc Natl Acad Sci U S A 95, 376-81 (1998).

The investigation of flowering control in late/rare flowering *Lolium perenne*

S. Byrne[1], I. Donnison[2], L.J. Mur[3] and E. Guiney[1]
[1]*Teagasc Crops Research Centre, Oak Park, Carlow, Ireland.* [2]*Institute of Grassland and Environmental Research, Aberystywth.* [3]*University of Wales, Aberystwyth. Email: sbyrne@oakpark.teagasc.ie*

Keywords: perennial ryegrass, flowering control, gene expression

Introduction Flowering in *Lolium perenne* (perennial ryegrass) results in reduced digestibility and its inhibition would enhance forage quality. Flowering regulation has been well studied in *Arabidopsis thaliana* (Simpson and Dean, 2002) and orthologs of *Arabidopsis* flowering genes underlying heading date Quantitative Trait Loci (QTL) have been identified in rice (Yano, M *et al.,* 2000). However it is not clear yet how universally applicable such studies are to *Lolium*. The project goals are to characterise the gene expression profiles of late/rare flowering *L. perenne* plants to determine factors affecting flowering and to map the genes involved in the flowering process. Initial studies, reported here, have focussed on the ability of 6 plant lines from the Oak Park breeding programme, previously identified as rare or non-flowering under natural day length conditions, to flower in controlled environments.

Materials and methods The 6 lines, representing 3 families, together with controls from each family (plants with normal heading pattern) and an early flowering variety (Moy) were vernalised. 5 tillered plants from each line were then transferred to growth chambers at 18^0C with continuous light for 21 days, after which they were monitored under greenhouse conditions for days to heading with day 0 taken as the start of secondary induction (Table 1). A large number of plants from each line was also observed in pots under outdoor conditions (Data not shown).

Results All 6 test lines headed under controlled conditions in 2004 except X15 (Table 1). The severely stunted X15 line did produce immature inflorescence in 2004 but they did not fully emerge. This was also observed in outdoor pot conditions (data not shown) in direct contrast to previous years. The 3 controls and Moy headed at the same time under controlled conditions, a situation not seen under natural day length conditions (controls headed as much as 18 days apart). This indicates saturation of the long day requirement. The J family plant lines, flowered as many as 8 days later than other families under controlled conditions, indicating a greater long day requirement. Within the J family, J43(c)and the later flowering J51 had the greatest difference in days to heading in both controlled conditions (8 days) and outdoor pot conditions (25 days).

Table 1 Average days to heading in controlled environments (c) = controls

L. perenne Line	Days to heading
X15	No Heading
X19 (c)	21
J74	28.6
J51	29.2
J54	26
J47	26.4
J43 (c)	21.2
V16	22.6
V13 (c)	21
Moy	21

Conclusion On the basis of these results J43 (c) and J51 will have their shoot apical meristem gene expression profiles compared. The comparison will be made after vernalisation and during secondary induction using Suppression Subtractive Hybridisation PCR, which allows the identification of differentially expressed genes. The expression of selected flowering genes that have been identified in other monocots will also be investigated in the subtracted library. J43(c) and J51 have been cross pollinated and a genetic linkage map will be constructed from the resulting population to identify possible heading date QTL .

References

Simpson, G. and Dean, C., (2002). Arabidopsis, the rosetta stone of flowering time? Science, 296, 285-289.
Yano, M., Katayose, Y., Ashikari, M., Yamamouchi, U., Monna, L., Fuse, T., Baba, T., Yamamoto, K., Umehara, Y., Nagamura, Y. and Sasaki, T. (2000). *Hd1*, a major photoperiod sensitivity quantitative trait locus in rice, is closely related to the Arabidopsis flowering time gene *CONSTANS*. The Plant Cell, 12, 2473-2483.

Isolation of candidate genes involved in cold temperature response in *Festuca pratensis* Huds., using suppression subtractive hybridisation and microarray approaches

H. Rudi[1], V. Alm[1,3], L. Opseth[1], A. Larsen[2] and O.A. Rognli[1]
[1]*The Agricultural University of Norway, Department of Plant and Environmental Sciences, N-1432 Ås, Norway,*
Email: heidi.rudi@ipm.nlh.no
[2]*The Norwegian Crop Research Insitute, Vågønes Research Station, N-8010 Bodø, Norway*
[3]*Present address: Department of Molecular Biosciences, Univ. of Oslo, N-0315 Oslo, Norway*

Keywords: cold-acclimation, suppression subtractive hybridisation, expression profiling

Introduction The objective of this work was to isolate candidate genes which are differentially expressed following cold-acclimation and develop SNPs to test for associations between candidate genes and frost tolerance. The ability to develop sufficient levels of tolerance against freezing temperatures through cold-acclimation (hardening) is crucial for survival of grasses and winter cereals in temperate climate. Meadow fescue (*Festauca pratensis* Huds.) is one of the most important forage grass species in Northern Europe. The preference of *Festuca* instead of *Lolium* in Norway is due to its superior combination of winter hardiness and forage quality.

Materials and methods Cold-acclimated and non-treated *Festuca* crown tissue of plants with low and high frost tolerance were used. RNA was isolated and cDNA libraries were prepared using suppression subtractive hybridisation (Clontech PCR-Select cDNA subtractive kit). 559 ESTs from the cDNA libraries were sequenced and spotted on a microarray. Expression profiling using mRNA pools isolated from individual plants of meadow fescue populations selected for high and low frost tolerance was conducted.

Results The genome of *Festuca pratensis* is not sequenced and in order to get useful information from our cDNA libraries we performed Blast searches of the 559 ESTs against the NCBI database. Through these searches we identified 242 homologous sequences in other species, 131 of our ESTs were homologous to putative proteins in other species, 78 to sequenced clones with no predicted protein, 81 with no significant similarities in the database and 27 homologous to unknown proteins in other species. We found homologues involved in cold stress but also other stress related genes encoding enzymes involved in several metabolic pathways, ribosomal proteins, histones, ubiquitin, elongation factors and others. The GeneSpring expression analysis software was used to analyze the expression profiling data. In the high frost tolerant hardened (Hi-H)/high frost tolerant non-hardened (Hi-NH) hybridisation 111 genes were found to be upregulated and 36 down-regulated more than 2-fold (Figure 1). Candidate genes involved in frost tolerance were found among these genes. Comparison of Hi-H and low frost tolerance hardened (Lo-H) genotypes showed 4 genes to be up-regulated and 3 down-regulated more than 2-fold (Opseth 2004).

Conclusions The results of the present study demonstrate that SSH in combination with a cDNA microarray is a successful approach for candidate gene isolation. All the candidate genes that were among the SSH-cDNA libraries were found to be differentially expressed upon cold-acclimation of high frost tolerant *Festuca* genotypes. The results will improve our understanding of the genetic regulation of cold-acclimation and frost-tolerance in forage grasses and cereals, and provide functional allele-specific markers that can be implemented in molecular breeding strategies. (http://www.grasp-euv.dk/default.asp)

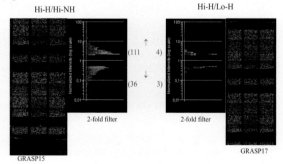

Figure 1 Composite images of two hybridisations and a statistical plot showing a 2-fold filtering of the expression data

Reference

Opseth, L (2004) Isolation of genes involved in cold-tolerance in *Festuca pratensis* Huds. by suppression subtractive hybridization and microarray. Cand. Scient. Thesis, Agricultural University of Norway

Isolation and characterization of a *CBF* gene from perennial ryegrass (*Lolium perenne* L.)

Y. Xiong and S. Fei

Department of Horticulture, Iowa State University, Ames, IA, USA, Email: sfei@iastate.edu

Keywords: *Lolium perenne* L., CBF, cold acclimation, freezing tolerance

Introduction Maximum freezing tolerance of many temperate plant species is achieved after exposure to a period of non-freezing low temperatures, a phenomenon called "cold acclimation". Multiple mechanisms appear to operate in conferring freezing tolerance in plants. The discovery of a class of transcription factor genes, *CBF* genes (C-repeat binding factor), in *Arabidopsis* demonstrated that *CBF* genes can serve as 'Master switches' to activate downstream cold-related (*COR*) genes during cold-acclimation (Liu *et al*., 1998). They act by binding to the core sequence (CCGAC) which is present in *COR* genes and thus activate the expression of *COR* genes and enhance freezing tolerance. The components of this cold acclimation pathway are conserved across a number of plant species including both eudicot and monocot species (Jaglo *et al*., 2001). Perennial ryegrass (*L. perenne* L.) is an important turf and forage grass but it lacks adequate winter hardiness which is a limiting factor for its adaptation to the northern regions of the US. Development of improved germplasm with enhanced winter hardiness would be highly desirable. The objective of this research was to isolate and characterize *CBF* gene(s) in perennial ryegrass and to determine the genomic location and the function of the *CBF* gene(s).

Materials and methods Total RNA was extracted using TRIZOL Reagent from a perennial ryegrass plant (cv. Caddieshack) that was exposed to 4°C for 72 hours. The RT-PCR approach was used to obtain a partial sequence of *CBF* gene from perennial ryegrass with degenerate primers that were designed based on conserved sequences from known *CBF* genes. 5' and 3' RACE were used to obtain the full length sequence of the *CBF* gene. Sequence alignment was performed using BLAST. Northern blot was performed on leaf, root and stem tissues that received no cold treatment and on leaf tissues treated with 4°C cold for 15 min, 30 min, 1h, 4h, 6h, 9h, 24h and 48h. MEGA 3 software was used to construct a dendrogram for the *CBF* genes from both perennial ryegrass and other species.

Results We have identified a *CBF*-like (designated as *LpCBF*) gene in perennial ryegrass. *LpCBF* gene is 942bp long without introns. *LpCBF* gene has all the conserved domains of known *CBF* genes. Dendrogram showed that *LpCBF* is clustered with *CBF* orthologs from other grass species, particularly with the rice *CBF3* gene (*OsDREB1C*). Ninety one percent of the *LpCBF3* DNA sequence is identical to the rice *CBF3* gene and more than half of the amino acids in *LpCBF* are identical to those in rice *CBF3*. Our results indicated that the expression of *LpCBF* was induced between 15min and 30 min of cold treatment with the highest expression level at 30min of cold treatment. No *LpCBF* expression was detected after 2 hour of cold treatment which indicates that the *CBF* gene we isolated from perennial ryegrass functions in the early steps of cold acclimation. Similar to the expression patterns of other known *CBF* genes, no *LpCBF* expression was detected in non treated leaves, stems, crowns and roots.

Conclusions The *LpCBF* gene shares significant sequence similarity with other known *CBF* genes and has similar expression patterns as other *CBF* genes. Additional homologs of the *LpCBF* gene are being isolated and their map locations will be determined. The potential functions of the *CBF* genes are currently being determined with a heterologous transformation system using *Arabidopsis*.

References

Jaglo, K.R., Kleff, S., Amundsen, K.L., Zhang, X., Haake, V., Zhang, J.Z., Deits, T., and Thomashow, M.F. (2001). Components of the *Arabidopsis* C-repeat/dehydration-responsive element binding factor cold-response pathway are conserved in *Brassica napus* and other plant species. Plant Physiol. 127: 910–917.

Liu, Q., Kasuge, M., Sakuma, Y., Abe, H., Miura, S., Yamaguchi-Shinozaki, K., & Shinozaki, K. (1998). Two transcription factors, *DREB1* and *DREB2*, with an EREBP/AP2 DNA binding domain separate two cellular signal transduction pathways in drought- and low-temperature-responsive gene expression, respectively, in Arabidopsis. Plant Cell 10: 1391–1406

Isolation and characterisation of genes encoding ice recrystallisation inhibition proteins (IRIPs) in the cryophilic antarctic hair-grass (*Deschampsia antarctica*) and the temperate perennial ryegrass (*Lolium perenne*)

U.P. John[1,2,3], R.M. Polotnianka[1,2,3], K.A. Sivakumaran[1,2], L. Mackin[1], M.J. Kuiper[4], J.P. Talbot[1,2], O. Chew[1,2], G.D. Nugent[1,3], N.O.I. Cogan[1], M.C. Drayton[1], J.W. Forster[1], G.E. Schrauf[5] and G.C. Spangenberg[1,2,3]

[1]*Primary Industries Research Victoria, Plant Biotechnology Centre, La Trobe University, Victoria 3086, Australia* [2]*Australian Centre for Plant Functional Genomics, Australia* [3]*Victorian Centre for Plant Functional Genomics, Australia* [4]*Victorian Partnership for Advanced Computing, Victoria 3053, Australia* [5]*Facultad de Agronomia, Universidad de Buenos Aires, Argentina Email: ulrik.john@dpi.vic.gov.au*

Keywords: antarctic hair-grass, perennial ryegrass, ice recrystallisation inhibition proteins (IRIPs), freezing tolerance

Introduction Antarctic hairgrass (*D. antarctica* Desv.), the only grass species indigenous to Antarctica, has a well developed tolerance of freezing, strongly induced by cold-acclimation. In response to low temperatures *D. antarctica* exhibits recrystallisation inhibition (RI) activity, localised to the apoplasm, that prevents further growth of ice crystals following freezing.

Materials and methods A gene family from *D. antarctica* encoding ice recrystallisation inhibition proteins (IRIPs) and orthologues from perennial ryegrass were isolated and comparatively characterised. IRIP genes are found in a range of Pooideae species but the products they encode vary widely in length due to apparent plasticity in the number of LRRs.

Results and conclusions Cold acclimation induces RI activity more than 64-fold and 8-fold in leaves and roots of *D. antarctica* respectively. Similarly, RI activity is induced by cold-acclimation in leaves and roots of perennial ryegrass (*L. perenne* L.) in excess of 16- and 4-fold respectively. Leaf apoplastic extracts from plants of both species grown at 22°C possess no RI activity while activity is induced in response to cold-acclimation by factors of at least 73-fold in *D. antarctica* and 1.7-fold in *L. perenne*. The genomes of *D. antarctica* and *L. perenne* both appear to harbour multiple IRIP-related sequences. IRIPs are apoplastically-targeted proteins with two potential ice-binding motifs: between 1 and 9 leucine rich repeats (LRRs), and 16 IRIP repeats. Three-dimensional structures of IRIPs from *D. antarctica* (*Da*IRIP) and *L. perenne* (*Lp*IRIP) were constructed by comparative homology modelling. *Da*IRIPs are predicted to adopt conformations with two ice-binding surfaces that are better matched to the prism face of ice than those of *Lp*IRIPs (Figure 1). An *Lp*IRIP was genetically mapped on perennial ryegrass LG1 using gene-associated single nucleotide polymorphisms. Levels of IRIP transcripts are greatly enhanced (c. 47-fold) in leaves of *D. antarctica* following cold-acclimation, but only moderately so (c. 4-fold) in roots of *L. perenne*. Moreover in extracts from *E. coli* expressing heterologous proteins RI activity is specifically conferred by *Da*IRIP. It is proposed that enhancement of IRIP-mediated RI activity relative to *L. perenne* and other temperate grasses and cereals has contributed to the cryo-tolerance of D. antarctica and thus its ability to colonise Antarctica.

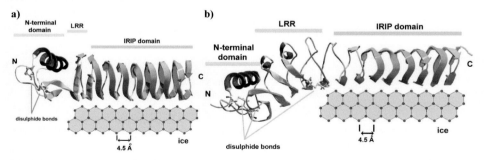

Figure 1 Structural models of IRIPs. a) Ribbon backbone diagram of theoretical structure of *Da*IRIP and b) *Lp*IRIP, aligned along the prism face of ice (parallel to the *a*-axis)

Development of genetic markers for drought tolerance in *Festuca-Lolium* complexes

J.P. Wang and S.S. Bughrara
Department of Crop and Soil Sciences, Michigan State University, East Lansing, MI 48824, USA,
Email: bughrara@msu.edu

Keywords: genetic markers, drought tolerance, *Festuca-Lolium* complexes

Introduction Drought stress is one of the most complex environmental constraints on turf. It is a major factor limiting the growth of cool-season turf grasses in a wide range of climatic regions. As water conservation becomes increasingly limiting, the development of drought tolerant lines becomes important. However, the progress in breeding turfgrass for drought resistance has been very slow, primarily because of the genetic complexity of drought stress responses and lack of screening procedures for rapid selection of germplasm with superior drought tolerance. Marker assisted selection (MAS) provides breeders with valuable tools to develop newer germplasm with improved drought tolerance (Quarrie *et al.*, 1999). Drought tolerance involves a cascade of events and is controlled genetically by multiple genes. To clarify the genetic network involved, key agronomic traits need to be clarified into individual components to reduce complex analysis (Tollenaar and Wu, 1999). After specific components of the genes corresponding to drought tolerance are isolated and cloned, they can be converted into PCR-based markers to assist the selection and allow us to rapidly identify genetic lines that had the desired allele and discard those without.

Materials and methods *Festuca mairei* (Fm) plants were selected from an Fm population collected from Morocco which was adapted to the hot and dry summers of Northwest Africa (Borrill *et al.*, 1971). Two genotypes of *Lolium perenne* (Lp) were obtained from the turfgrass cultivars 'Citation II' (Lp1) and 'Calypso' (Lp2) and also 16 *Festuca-Lolium* plants were grown.

Results Plants were deprived of water until they were severely stressed. Leaf (lamina) elongation, leaf water potential, leaf water content, soil water content, osmotic potential, root length and mass, and tiller survival rate were detected or evaluated during the stress period on both control and stressed plants. The differentially expressed fragments (cDNA fragment) identified from Fm during the drought stress were re-amplified from both the drought tolerant and susceptible complexes to correlate the differential expression pattern of the re-amplified bands and the drought tolerance performance.

References
Quarrie SA, Lazic-Jancic V, Kovacevic D, Steed A, Pekic S. (1999). Bulk segregant analysis with molecular markers and its use for improving drought resistance in maize. *Journal of Experimental Botany* 50: 1299–1306.
Tollenaar M, Wu J. (1999). Yield improvement in temperate maize is attributable to greater stress tolerance. *Crop Science* 39: 1597–1604.

Monitoring of gene expression profiles and identification of candidate genes involved in drought tolerance in *Festuca mairei* with cDNA-AFLP

J.P. Wang and S.S. Bughrara
Department of Crop and Soil Sciences, Michigan State University, East Lansing, MI 48824, USA,
Email: bughrara@msu.edu

Keywords: drought stress, cDNA-amplified restriction length polymorphism (cDNA-AFLP), transcript derived fragments (TDFs), differential expression

Introduction Drought stress is one of the most complex environmental constraints on plants. Response of plant to drought stress is manifested by various changes in physiological and metabolic processes, which are reflected at the molecular level. cDNA-amplified fragment length polymorphism (cDNA-AFLP) is a high-throughput transcript profiling technique for temporal and spatial gene expression analysis (Bachem *et al.*, 1996). To understand the molecular genetic basis of drought tolerance of grasses, we applied the cDNA-AFLP procedure to identify the genes responding to drought stress of *Festuca mairei* (Fm), which showed a xeriphytic adaptation (Marlatt *et al.*, 1997).

Materials and methods Five Fm plants were deprived of water until they were severely stressed and the other five plants as treatment control were watered daily throughout the drought stress period. During the drought stress treatment, leaf samples from both stressed and control Fm plants were detached for RNA isolation. 128 primer combinations for cDNA-AFLP were conducted on nine time-points of Fm during drought stress. Differentially expressed fragments were identified and recovered from polyacrylamide gel. Reverse northern hybridization was conducted to confirm the differential expression pattern.

Results cDNA-AFLP revealed 11,346 TDFs with fragment size distributed from 50 to 1000bp. 464 fragments were identified as differentially expressed across the nine time-points, and 434 (94%) were recovered from acrylamide gel. The expression patterns of these differentially expressed fragments included up-regulated (29.7%), down-regulated (54.3%), transient-expressed (12.1%) and up- then- down- regulated (3.7%). 406 TDFs (> 100 bp) were subjected to reverse northern analysis. 172 (42.4%) TDFs showed a consistent differential expression pattern with the cDNA-AFLP analysis. The sequences of these confirmed TDFs were compared to the Arabidopsis protein database with BLASTN to target the potential function of these gene fragments. About 10% of them were novel transcripts, while about 70% could be assigned putative function. The putative function of the differentially expressed genes could be classified as: 1) defence related or protective proteins, 2) transport facilitation, 3) amino acid and carbohydrate metabolism, and 4) signal transduction and transcriptional and translational control.

Conclusions The results of the present study demonstrate that cDNA-AFLP is an efficient high through-put transcript profiling technique for gene discovery. Fm responses to drought at a comprehensive molecular regulation level involving signal transduction, transport facilitation, metabolism regulation and protective defence.

References
Bachem, C.W.B., van der Hoeven, R.S., de Bruijn, S.M., Vreugdenhil, D., Zabeau, M., and Visser, R.G.F. (1996). Visualization of differential gene expression using a novel method of RNA fingerprinting based on AFLP: Analysis of gene expression during potato tuber development. *Plant J.* 9:745-753.
Marlatt, M. L., C.P. West, M.E. McConnell, D.A. Spleper, G.W. Buck, J. C. Correll, and S. Saidi. (1997). Investigations on xeriphytic *Festuca* spp. from Morocco and their associated endophytes. Neotyphodium/ Grass Interactions , Edited by Baxon and Hill. Plenum Press, New York.

Section 5

Use of molecular markers and bioinformatics in breeding

Towards a comparative map of white clover (*Trifolium repens*) and barrel medic (*Medicago truncatula*)

M. Febrer[1], G. Jenkins[2], M. Abberton[3] and D. Milbourne[1]

[1]*Teagasc, Crop Research Centre, Oak Park, Carlow, Ireland; [2]Department of Biological Sciences, University of Wales, Aberystwyth, Wales; [3]Institute of Grasslands and Environmental Research (IGER), Aberystwyth, Wales.*
Email: mfebrer@oakpark.teagasc.ie

Keywords: white clover, *Medicago truncatula*, molecular markers, synteny

Introduction Grassland is of pivotal importance to the Irish agricultural industry. This dependence of grass is reflected in the large proportion of land area under grass, approx. 80% of the total land acreage in Ireland. The presence of white clover (*Trifolium repens* L.) in grassland significantly improves the overall nutritional value of the forage by increasing the relative amounts of nitrogen present. Genetic improvement of white clover through breeding of varieties should increase the productivity of grasslands. Advances in plant biotechnology offer the possibility of developing tools that will radically enhance our ability to breed improved plant varieties. The objective of this study is (1) to construct a genetic map of white clover and (2) to assess the level of genome synteny of white clover and *M. truncatula* (the model for legume species) with the use of different molecular markers developed in *M. truncatula*.

Materials and methods Genomic DNA from a F1 hybrid white clover population was extracted using CTAB method. The population consisted of 94 progeny bred in IGER (Wales) from parents, S1S4 and R3R4, and was used as mapping population. In order to provide a basis for the comparison of the genomes of white clover and *M. truncatula*, we are in the process of developing a genetic map of clover, using AFLP markers, microsatellite markers and a set of PCR-based markers derived from genes that have previously been placed on the *M. truncatula* genetic map (Choi *et al.*, 2004).

Results To date, the segregation of 170 AFLP markers has been assayed in the mapping population (Figure 1). However, additional AFLP markers will be developed in order to obtain a comprehensive map. This map will then be tagged with a number of molecular markers of known chromosomal location from existing maps of white clover and *M. truncatula*. We have assessed the amplification in clover of 89 (EST and BAC end sequence derived) PCR-based markers mapped in *M. truncatula* by Choi *et al.* (2004). Sixty five (73%) successfully amplified in white clover, of which 22 were single copy amplicons (Table 1). The single copy products are being assessed for polymorphism in our mapping population and, if possible, be placed on the genetic map.

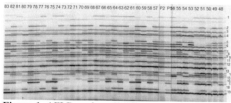

Figure 1 AFLP markers (1-19) representing Pac Maag primer combination on the white clover parents (P1, P2) and some of the progeny (48-83)

Table 1 Successful amplification of *M.truncatula* PCR-based primers applied to white clover

Total no. markers tested to date	95
No. successful in *M. truncatula*	89
No. successful in white clover	65
No. giving single bands in white clover	22
Proportion of amplification in both sp.	73%

Reference

Choi H-K., Kim D., Uhm T., Limpens E., Lim H., Mun J-H., Kalo P., Penmetsa R.V., Seres A., Kulikova O., Roe B.A., Bisseling T., Kiss G.B. and Cook D.R. (2004). A sequence-based map of *Medicago truncatula* and comparison of marker colinearity with *M. sativa*. Genetics, 166: 1463-1502.

Use of cross-species amplification markers for pollen-mediated gene flow determination in *Trifolium polymorphum* Poiret

M. Dalla Rizza[1], D. Real[2,3], R. Reyno[2] and K. Quesenberry[4]
[1]*Biotechnology Unit, National Institute of Agricultural Research, INIA Las Brujas, Ruta 48 km 10, Canelones, Uruguay* [2]*Forage Legume Department, National Institute of Agricultural Research, INIA Tacuarembó, Ruta 5 Km 386, Tacuarembó, Uruguay* [3]*Cooperative Research Center for Plant-Based management of Dryland Salinity, The University of Western Australia, University Field Station, 1 Underwood Avenue, Shenton Park, WA 6009, Australia* [4]*Department of Agronomy, University of Florida, Gainesville, FL 32611-0500, USA*
Email: mdallarizza@lb.inia.org.uy

Keywords: cross-species amplification, comparative genomics, population biology

Introduction The species *Trifolium polymorphum* Poiret is endemic to Uruguay and is widespread in native grasslands throughout the country. Preliminary observations suggested that the aerial flowers are chasmogamous (open at maturity for potential cross-pollination) while the basal flowers are cleistogamous. Several approaches have been practised to determine the reproductive system of forage legumes by the aid of co-dominant markers (Real *et al.*, 2004; Dalla Rizza *et al.*, 2004). The aim of this study is to explore cross-species amplification as a quick approach to obtain co-dominant markers to study the breeding system of *T. polymorphum*.

Materials and Methods Ten single patches in the field of approx. 1 m^2 were selected for this study. Inside each patch, 20 individual plants were physically mapped, and subsequently transplanted to pots in a naturally lit glasshouse. For cross-*T .polymorphum* amplification, 50 ng of genomic DNA extracted as reported by Dalla Rizza *et al.*, (2004) were used, following the amplification procedure described by Kölliker *et al.* (2001). Preliminary 4 perfect, 3 imperfect and 2 compound SSR classified markers were employed for the analysis using silver stain to reveal the amplicons in 8% non denaturing polyacrylamide gel electrophoresis.

Results From the 9 microsatellites primer pairs screened, 6 showed cross amplification success in *T. polymorphum* comparable in size to the product reported in *T. repens*. This is a highly unexpected value considering the differences in geographical origin of the two species and suggesting a probably remote common origin. As reported by Kölliker *et al.* (2001), this value observed for *T. polymorphum* is higher than *Trifolium pratense* L. and only surpassed by *Trifolium ambiguum* M.B. and *Trifolium nigrescens* Viv. At least 8 alleles were observed with the perfect microsatellite TRSSRA01H11 between distanced patches, more than observed in *T. repens* (Figure 1). Considering that no more than 2 alleles at a single locus would be expected in a diploid species probably this SSR has more than one locus, as also reported in *T. repens*. Between patches, high level of polymorphism was found meanwhile inside the patches the variability found among plants was low (lower than 20% in all the cases).

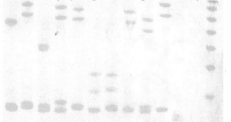

Figure 1 TRSSRA01H11 amplification on 10 *T. polymorphum* genotypes (first 10 lanes on left) and 1 genotype of *T. repens*

Acknowledgements The authors wish to thank Prof. G. Spangenberg for kindly sharing the *T. repens* microsatellites used in this study and also to thank D. Torres, M. Zarza, A. Viana and R. Mérola for their practical contributions.

Conclusions The high success rate of cross-species amplification obtained in *T. polymorphum* enables to extend the knowledge obtained in *T. repens* to *T. polymorphum* for plant breeding studies.

References
Kölliker, R., E.S. Jones, M.C. Drayton, M.P. Dupal. & J.W. Forster (2001). Development and characterisation of simple sequence repeat (SSR) markers for white clover (*Trifolium repens* L.). Theor Appl Genet 102: 416-424.
Real D., M. Dalla Rizza,, K.H. Quesenberry & M. Echenique (2004) Reproductive and molecular evidence for allogamy in Lotononis bainesii Baker. Crop Science, Vol. 44 (2), 394-400.
Dalla Rizza M., D. Real, K.H. Quesenberry & E. Albertini (2004) Plant reproductive system determination under field conditions based on codominant markers. (in press: Journal of Genetics and Breeding).

Clover ASTRA: a web-based resource for *Trifolium* EST analysis

G.C. Spangenberg[1,2,3], T. Sawbridge[1,2], E.K. Ong[1], C.G. Love[1,2], T.A. Erwin[1,2], E.G. Logan[1,2] and D. Edwards[1,2,3]
[1]*Primary Industries Research Victoria, Plant Biotechnology Centre, La Trobe University, Bundoora, Victoria 3086, Australia* [2]*Victorian Bioinformatics Consortium, Australia* [3]*Molecular Plant Breeding Cooperative Research Centre, Australia Email: german.spangenberg@dpi.vic.gov.au*

Keywords: white clover, expressed sequence tags (ESTs), gene annotation, bioinformatics ASTRA database, simple sequence repeats (SSRs), single nucleotide polymorphisms (SNPs)

Introduction White clover (*Trifolium repens* L.) is a major temperate forage legume.

Materials and methods A resource of 42,017 white clover expressed sequence tags (ESTs) from single pass DNA sequencing of randomly selected clones from 16 cDNA libraries representing a range of plant organs, developmental stages and environmental treatments was generated (Sawbridge *et al.*, 2003). Each of these sequences has been annotated by comparison to the GenBank and SwissProt public sequence databases. Automated intermediate Gene Ontology (GO) annotation has also been determined.

Results and conclusions Sequence annotation was determined for 74%, 58% and 54% of sequences using GenBank, SwissProt and GO annotation respectively, with only 20% of sequences determined to have no comparative annotation and thus appear to be unique to white clover (Fig 1). EST redundancy was resolved through assembly with CAP3, identifying at total of 15,989 unigenes within this dataset. All sequences and annotation are maintained within an ASTRA format MySQL database, with web based access for text searching, BLAST sequence comparison and Gene Ontology hierarchical tree browsing. Each white clover sequence was mapped onto an EnsEMBL genome viewer for comparison with the complete genome sequence of *Arabidopsis thaliana* and expressed sequences from related legumes.

In total 5,407 white clover simple sequence repeat (SSR) molecular markers were identified using SSRPrimer and specific PCR amplification primers were designed. 18,517 candidate single nucleotide polymorphism (SNP) molecular markers and 1,706 insertion deletions (InDels) were also identified across 1,409 loci using AutoSNP. The integration of different analysis tools and the resulting data within a central resource allow users to identify novel candidate genes and associated molecular genetic markers that can be applied to functional genomics and germplasm enhancement in this agronomically important forage crop.

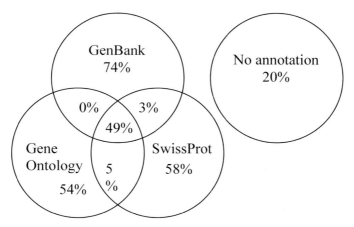

Figure 1 Annotation of white clover genes within the ASTRA database

Reference
Sawbridge, T., Ong, E.K., Binnion, C., Emmerling, M., Meath, K., Nunan, K., O'Neill, M., O'Toole, F., Simmonds, J., Wearne, K., Winkworth, A. and Spangenberg, G. (2003). Generation and analysis of expressed sequence tags in white clover (*Trifolium repens* L.). *Plant Science*, 165, 1077-1087.

SNP discovery and haplotypic variation in full-length herbage quality genes of perennial ryegrass (*Lolium perenne* L.)

R.C. Ponting[1,3], M.C. Drayton[1,3], N.O.I. Cogan[1,3], G.C. Spangenberg[1,3], K.F. Smith[2,3] and J.W. Forster[1,3]
[1]*Primary Industries Research Victoria, Plant Biotechnology Centre, La Trobe University, Bundoora, Victoria 3086, Australia* [2]*Primary Industries Research Victoria, Hamilton Centre, Hamilton, Victoria 3300, Australia* [3]*Molecular Plant Breeding Cooperative Research Centre, Australia Email: bec.ponting@dpi.vic.gov.au*

Keywords: SNP, lignin, fructans, haplotype

Introduction The development of forages with enhanced nutritive value through improvements of herbage quality (digestibility, carbohydrate content) is potentially capable of increasing both meat and milk production by up to 25%. However, the expense and time-consuming nature of the relevant biochemical and biophysical assays has limited breeding improvement for forage quality. The development of accurate high-throughput molecular marker-based selection systems such as single nucleotide polymorphisms (SNPs) permits evaluation of genetic variation and selection of favourable variants to accelerate the production of elite new varieties.

Materials and methods SNP marker discovery in perennial ryegrass has been based on PCR amplification and sequencing of multiple amplicons designed to scan all components of the transcriptional unit. Full-length genes with complete intron-exon structure and promoter information corresponding to well-defined biochemical functions are ideal for the determination of complete SNP haplotype data. The gene classes involved in this study were the lignin biosynthetic genes *Lp*CCR1 (cinnamoyl CoA-reductase, 12.1 kb, AY061889), *Lp*CAD2 (cinnamyl alcohol dehydrogenase, 7.2 kb, AF472592) and the fructan biosynthetic genes LpFT1 (= *Lp*6SFT =sucrose:fructose 6-fructosyltransferase, 9.98 kb, AF481763) and *Lp*1-SST (sucrose:sucrose 1-fructosyltransferase, 4.63 kb, AY245431).

Results Multiple SNPs located at regular intervals across the transcriptional unit were detected within and between the heterozygous parents of the F_1(NA$_6$ x AU$_6$) genetic mapping family, and were validated in the progeny set. The total numbers of putative SNPs identified were 54 (*Lp*CAD2), 212 (*Lp*CCR1), 265 (*Lp*FT1) and 240 (*Lp*1-SST). The *Lp*1-SST gene locus was assigned for the first time to a genetic map location on LG7 of perennial ryegrass through SNP mapping (Fig 1). Haplotype structures in the parental genotypes were defined and haplotypic variation was assessed in diverse germplasm. This data was used to determine the average rate of decay of linkage disequilibrium (LD) over distances comparable with gene length. This analysis will provide crucial information for the validation of strategies to correlate haplotypic variation and phenotypic variation.

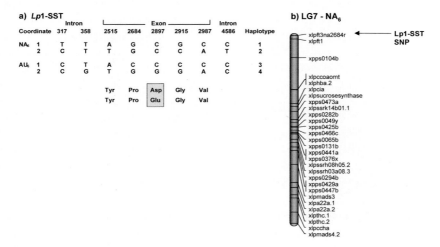

Figure 1 a) Partial SNP haplotype structures for the *Lp*1-SST gene and b) genetic map location of a validated *Lp*1-SST SNP locus on LG7 of the NA$_6$ parental genetic map derived from the perennial ryegrass F_1(NA$_6$ x AU$_6$) genetic mapping family

Development and use of a tool for automated alignments of genes in the rice BAC's GenBank card against other species

P. Barre, G. Darrieutort, J. Auzanneau and B. Julier

INRA Poitou-Charentes, Unité de Génétique et d'Amélioration des Plantes Fourragères, route de Saintes, 86600 Lusignan France Email: barre@lusignan.inra.fr

Keywords : sequence alignment, BLAST, Sequence Tagged Site, *Lolium perenne.*, homologous genes

Introduction In many cases, the analysis of the genetic bases of any trait requires molecular markers and if possible co-dominant PCR-based ones. In perennial fodder species, the number of publicly available markers (microsatellites and Sequence Tagged Site (STS)) is limited. Our goal is to use sequences from model grass species, i.e. rice, wheat, maize, barley, in *L. perenne* in order to develop STS markers in interesting regions such as under a QTL (Quantitative Trait Loci) or around a candidate gene,. As the genome sequence of rice is now available, the objective was to use the sequences of genes included in the BAC's GenBank card from rice. As there are almost no available sequences in *L. perenne*, we are designing consensus primers from an alignment of at least two different species. The problem is that for all the genes included in a BAC, just a few have their sequences known in at least two species. It is very laborious to check "by hand" if each gene has an homologous sequence known in another species.

Materials and methods We have developed a tool for automated sequence alignments of genes from the rice BAC's GenBank card against several species specific *Génoplante-info* data bases and screened the results. This tool includes two applications:

The first application consists of:
- a graphic interface for the user to enter BAC numbers and data base names on his local computer (PC);
- a procedure to transfer the files with BAC numbers and data base names from the local computer to the Génoplante-info computer;
- a procedure to perform sequence alignments between genes extracted from the BAC's GenBank Card and the *Génoplante-info* data bases selected by the user;
- a procedure to transfer the files with alignment results from the Génoplante-info computer to the local computer.

The second application consists of:
- a graphic interface for the user to enter his thresholds for parameters involved in the definition of homology between genes such as the e value, the bit score, the percent of homology and the length of the alignment;
- a procedure to screen all the results of various alignments using sorting parameters defined by the user ;
- a visualisation for each gene of species where homologous genes exist on an Microsoft Excel sheet.

Results In an analysis of the genetic basis of aerial morphogenesis in perennial ryegrass varieties, we are analysing linkage disequilibrium in the region of OsGAI gene (AY464568) in different varieties. For this purpose, we entered 13 rice BAC's Genbank card names from TIGR (including OsGAI and others around). The 232 genes present in the BACs were aligned against 6 species specific databases (1392 BLAST) in 1h30. We found 39 genes with homologous sequences in at least two species that were potentially interesting for STS development and STS marker primers were produced.

Conclusions i) Biologists who have used this tool find it convenient and time saving to select genes for which STS markers could be developed even if more automation of primer development is required ii) The fact that the program is split in two applications allows one to apply different thresholds on the alignment results and compare the screenings without again computing all the alignments. iii) This tool, developed for the Génoplante-info environment, could be easily adapted to other environments.

Screening genes for association with loci for nitrogen-use efficiency in perennial ryegrass by pyrosequencing[TM]

O. Dolstra, D. Dees, J.-D. Driesprong and E.N. van Loo
Plant Research International BV, P.O. Box 16, 6700 AA Wageningen, Netherlands. Email: oene.dolstra@wur.nl

Keywords: *Lolium perenne*, breeding, nitrogen-use efficiency, SNPs, QTL, DNA pooling, pyrosequencing[TM]

Introduction The application of marker-assisted selection to improve quantitative traits in perennial ryegrass (*Lolium perenne*) is cumbersome. It requires a priori knowledge on the association of markers and genes. The knowledge on the chromosomal location of major genes for quantitative traits as well as on gene sequences is rapidly growing. However, determination of the genetic constitution of parents prior to their use in breeding still is impractical. More realistic is to collect association data along with the testing activities needed for breeding new varieties. This study uses changes in allele frequency due to selection as a criterion for gene-trait association. Selection-dependent changes are detected with single nucleotide polymorphisms (SNPs) of candidate genes using DNA-pools of F2 plants differing in nitrogen-use efficiency (NUE). The procedure and its feasibility are outlined for one locus.

Materials and methods DNA pools were established by mixing equal quantities of DNA from two selections of about 45 F2 plants obtained by selfing #1331. The F2 selections, i.e. N+ and N-, differ in NUE (Dolstra *et al.*, 2003). A pyrosequencer, PSQ96 is used to determine allele frequencies (cf Neve *et al.*, 2002). Muylle (2003) reported the gene polymorphism, SNP3, which was used here to test the procedure. To this end PCR amplification with a biotinylated forward primer (5'-TGG GAA GAC AAC GCT TAA-3') and the reverse primer 5'-TTC CGA CAT CAA ACT CCT GC-3' was done to get a pyrosequencing template of 185bp. The sequence of the primer for the actual pyrosequencing of SNP3 was 5'-GGT CAC CCA TGC AA-3'. The incorporation order of nucleotides was set to be AC/GATGAAG.

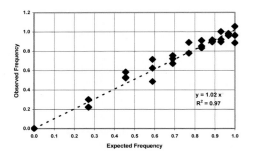

Figure 1 Relation between observed and expected

Results and conclusions Genomic DNA of the parents of #1331, both homozygous for a SNP3 allele, was mixed in different ratios to get variation in allele frequencies. Figure 1 shows the presence of a highly significant relationship between the expected and observed frequency of the C allele. The repeatability of the measurements is good as can be deduced from the data presented in Table 1. Small differences between expected and observed values are probably mainly due to incomplete background correction. The pools showed no significant change in allele frequencies, indicating the absence of any effect of selection for NUE (Table 1). There is probably no major NUE-QTL present in a chromosome area of 25 cM at each side of the SNP3 locus. This conclusion is supported by the finding that mapping showed SNP3 to be located on linkage group 2 where no NUE-QTLs were detected. A set of 25 markers like SNP3, evenly spread over the *Lolium* genome, would be sufficient to do a quick genome scan for the presence of major QTLs.

Table 1 Estimates of allele frequencies for the two contrasting F2 selections and their ancestors.

Materials	Allele Frequency (G/G+C)	
	Mean	SD
Parents		
P413	0.00	0.042
P510	1.01	0.059
F1		
#1331	0.56	0.009
F2 Selections		
N+	0.57	0.008
N-	0.41	0.045

References

Dolstra O., C. Denneboom, A.L.F. de Vos & E.N. van Loo (2003) Marker-assisted selection in improvement of quantitative traits of forage crops. In: International Workshop 'Marker Assisted Selection: a fast track to increase genetic gain in plant and animal breeding?', October 17-18, 2003 Villa Gualino, Torino, Italy, p1-5 (http://www.fao.org/biotech/docs/dolstra.pdf)

Muylle, H. (2003) Genetic analysis of crown rust resistance in ryegrasses (Lolium spp.) using molecular markers. Thesis Universiteit Gent, Gent, Belgium, p89

Neve B., P. Froguel, L. Corset, E. Vaillant, V. Vatin, P. Boutin (2002) Rapid SNP allele frequency determination in genomic DNA pools by Pyrosequencing ™. BioTechniques 32: 1138-1142

Gene-associated single nucleotide polymorphism (SNP) discovery in perennial ryegrass (*Lolium perenne* L.)

J.W. Forster[1,3], N.O.I. Cogan[1,3], A.C. Vecchies[1,3], R.C. Ponting[1,3], M.C. Drayton[1,3], J. George[1,3], J.L. Dumsday[1,3], G.C. Spangenberg[1,3] and K.F. Smith[2,3]

[1]*Primary Industries Research Victoria, Plant Biotechnology Centre, La Trobe University, Bundoora, Victoria 3086, Australia* [2]*Primary Industries Research Victoria, Hamilton Centre, Hamilton, Victoria 3300, Australia* [3]*Molecular Plant Breeding Cooperative Research Centre, Australia Email: john.forster@dpi.vic.gov.au*

Keywords: allele, SNP, abiotic stress tolerance

Introduction Perennial ryegrass (*Lolium perenne* L.) is the most important grass species for temperate pasture systems world-wide. Varietal improvement programs for this obligate outbreeding species are based on polycrossing of multiple parents to produce heterogeneous synthetic populations. The complexity of breeding systems creates challenges and opportunities for molecular marker technology development and implementation. Previous research has led to: the generation of a comprehensive suite of simple sequence repeat (SSR) markers, reference genetic map construction, comparative genetic studies, QTL identification, and population structure analysis. Emphasis has now shifted from the use of anonymous genetic markers linked to trait-specific genes to the development of functionally-associated genetic markers based on candidate genes. The successful implementation of this approach will allow effective selection of parental plants in germplasm collections based on superior allele content.

Materials and methods A hierarchical system for candidate gene identification has been devised, and targets in a range of functional categories have been defined, on the basis of association with key output traits. At present, over 150 perennial ryegrass genes have been introduced into an *in vitro* single nucleotide polymorphism (SNP) discovery process based on amplicon cloning from the heterozygous parents of the second generation $F_1(NA_6 \times AU_6)$ two-way pseudo-testcross genetic mapping family. This method has the benefits of permitting discrimination of paralogous sequences and direct determination of haplotype structure. SNP variation within and between the parental genotypes is based on amplicon sequencing and alignment, followed by validation using the single nucleotide primer extension (SNuPe) assay.

Results and conclusions Proof-of-concept for SNP detection, validation and subsequent mapping has been obtained using the *LpASRa2* abiotic stress tolerance associated gene. Putative SNPs have been identified for 76 genes, over a total of c. 78 kb of resequenced DNA, and a high proportion of these SNPs have been validated. SNP frequency in perennial ryegrass is high compared to other Poaceae species, at c. 1/57 bp across all components of the transcriptional unit. SNP haplotype and haplogroup structures have been defined for *LpASRa2* and other genes, and variability has been surveyed across diverse genotypic sources. A strategy has been developed for validation of associations between candidate gene haplotype variation and phenotypic variation for correlated agronomic characters. Herbage quality and abiotic stress tolerance genes will provide 'proof-of-concept' for this procedure, involving measurements of transcriptional activity, metabolite levels and field character expression. This information, in concert with detailed understanding of within- and between-population genetic variation, will inform future strategies for marker implementation in pasture grass breeding.

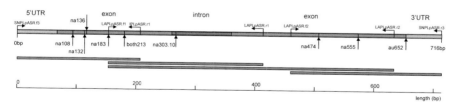

Figure 1 Location of SNP loci within the transcriptional unit of the *LpASRa2* gene, showing the locations of amplicon-specific primers and heterozygous SNPs relative to intron, exon and untranslated region (UTR) components. SNP loci detected as heterozygous within the NA_6 parental genotype of the $F_1(NA_6 \times AU_6)$ mapping family are indicated with the prefix na; the converse for the AU_6 parental genotype is indicated with the prefix au. A single SNP locus (at coordinate 213) was heterozygous between the two parents.

Development and testing of novel chloroplast markers for perennial ryegrass from *de novo* sequencing and *in silico* sequences

S. McGrath[1,2], T.R. Hodkinson[2] and S. Barth[1]

[1]*Teagasc Crops Research Centre, Oak Park, Carlow, Ireland;* [2] *Department of Botany, University of Dublin, Trinity College, Ireland, E-mail: smcgrath@oakpark.teagasc.ie*

Keywords perennial ryegrass, chloroplast SSR markers, cross species amplification

Introduction Chloroplast DNA is uniparentally inherited and non-recombinant in *Lolium perenne*. These properties make the chloroplast genome a useful tool for studying inter- and intra- specific relationships. Previous genetic studies on *L. perenne* have used chloroplast sequence data. However, the relative lack of variation in the chloroplast genome limits its usefulness for analysis at the single individual level within a species. However, chloroplast SSR markers have recently been shown to have high levels of polymorphism (Provan *et al.*, 2004). This is the first study to design and employ such markers for *L. perenne*. The objectives of this study are (1) to design and (2) optimise novel chloroplast SSR markers and (3) use them to analyse variation and diversity in *L. perenne* and related grass species.

Materials and methods DNA from *L. perenne*, *L. multiflorum*, *Festuca pratensis* and *Festuca arundinacea* was extracted using a modified CTAB method. These DNA samples were amplified and sequenced using the primers c and f for the *trn-L* intron and *trn-F* intergenic spacer region (Taberlet *et al.*, 1991), and the primers 1R and 2R for the *atpB-rbcL* intergenic spacer region (Samuel *et al.*, 1997), on an ABI 310 automated DNA sequencer. Further plastid DNA sequences from *L. perenne* and related species were also obtained from GenBank. All *de novo* and *in silico* sequences were analysed for microsatellite motifs using a modified version of the MISA perl script (Thiel, 2003). 27 pairs of primers were designed using Primer 3 software (Rozen & Skaletsky, 1998). PCR conditions were optimised to amplify loci for each primer pair. Of these, 19 were chosen for further analysis. 12 primer pairs were optimised as 5'=fluorescently labelled primers. These primers were used to amplify DNA from various populations of *L. perenne* and other grass species. PCR products were analysed using an ABI 3100 automated DNA sequencer and sized using GeneMapper™ v3.0 software.

Results The optimised primers showed little variation within *Lolium* species (Table1). The markers amplified across a broad range of grass species, *e.g.* from nine species for marker TeaCpSSR1 to 23 for marker TeaCpSSR11. Some of the alleles were shared between *L. perenne* and other species. Certain alleles were species specific, *e.g.* an allele of marker TeaCpSSR1 was specific to *Arundo donax*.

Table 1 Example for the allelic range of a set of four chloroplast SSR markers

Primer name	Gene region	*L. perenne*		Other grass species	
		# alleles	Size range in b.p.	# alleles	Size range in b.p.
TeaCpatpssr2	*atpB-rbcL*	4	217 - 229	3	227 - 230
TeaCpssr1	*23S-5S*	3	193 - 196	5	193 - 200
TeaCptrnssr2	*trnL-trnF*	5	178 - 195	4	176 - 200
TeaCpssr11	*trnV*	3	193 - 196	5	193 - 201

Conclusions As a next step, these optimised primers will be used to investigate the phylogeography of perennial ryegrass including mode and time-point of introduction of *L. perenne* into Ireland. They can also be used to test for maternal inheritance in festulolium hybrids. These primers are also a valuable tool to control cross pollination and selfings in breeding programs. Furthermore, their greatest potential lies in species differentiation, such as in seed testing or at the taxonomic level.

Acknowledgements We are grateful to the grass breeding group in Oak Park for their help and to Dr. Nicolas Salamin for his bioinformatics assistance. SMcG is supported by a Walsh Fellowship.

References

Provan, J., Biss,P.M., McMeel, D., and Mathews S. (2004) Universal primers for the amplification of chloroplast microsatellites in grasses (Poaceae). Molecular Ecology Notes 4: (2), 262-264

Rozen, S., Skaletsky, H.J. (1998) Primer3. http://www.genome.wi.mit.edu/genome_software/other/primer3.html.

Samuel, R., Pinsker, W., Kiehn, M. (1997) Phylogeny of some species of Cyrtandra (Cesneriaceae) inferred from the atpB/rbcL cpDNA intergene region. Botanica Acta 110(6) 503-510

Taberlet, P., Gielly, L., Pautou, G. and Bouvet, J. (1991) Universal primers for amplification of three non-coding regions of chloroplast DNA. Plant Molecular Biology 17: 1105-1109

Thiel, T. (2003) MISA - Microsatellite identification tool. http://pgrc.ipk-gatersleben.de/misa/

Ryegrass ASTRA: a web-based resource for *Lolium* EST analysis

G.C. Spangenberg[1,2,3], T. Sawbridge[1,2], E.K. Ong[1], C.G. Love[1,2], T.A. Erwin[1,2], E.G. Logan[1,2] and D. Edwards[1,2,3]
[1]*Primary Industries Research Victoria, Plant Biotechnology Centre, La Trobe University, Bundoora, Victoria 3086, Australia* [2]*Victorian Bioinformatics Consortium, Australia* [3]*Molecular Plant Breeding Cooperative Research Centre, Australia Email: german.spangenberg@dpi.vic.gov.au*

Keywords: perennial ryegrass, expressed sequence tags (ESTs), gene annotation, bioinformatics ASTRA database, simple sequence repeats (SSRs), single nucleotide polymorphisms (SNPs)

Introduction Perennial ryegrass *(Lolium perenne* L.) is a major grass species of temperate pastoral agriculture.

Materials and methods A resource of 44,524 perennial ryegrass expressed sequence tags (ESTs) from single pass DNA sequencing of randomly selected clones from 29 cDNA libraries representing a range of plant organs and developmental stages was generated (Sawbridge *et al.*, 2003). Each of these sequences has been annotated by comparison to the GenBank and SwissProt public sequence databases. Automated intermediate Gene Ontology (GO) annotation has also been determined.

Results and conclusions Sequence annotation was determined for 80%, 53% and 50% of sequences using GenBank, SwissProt and GO annotation respectively. Only 17% of sequences were determined to have no comparative annotation and thus appear to be unique to ryegrass. EST redundancy was resolved through assembly with CAP3, identifying at total of 12,170 unigenes within this dataset. All sequences and annotation are maintained within an ASTRA format MySQL database, with web based access for text searching, BLAST sequence comparison and Gene Ontology hierarchical tree browsing. Each ryegrass sequence was mapped onto an EnsEMBL genome viewer for comparison with the complete genome sequence of rice and expressed sequences from related species.

In total 3,214 ryegrass simple sequence repeat (SSR) molecular markers have been identified using SSRPrimer and specific PCR amplification primers were designed. A total of 2,716 candidate single nucleotide polymorphism (SNP) molecular markers and 345 insertion deletions (InDels) molecular markers were also identified in 493 loci using AutoSNP. The integration of different analysis tools and the resulting data within a central resource allow users to identify novel candidate genes and associated molecular genetic markers that can be applied to functional genomics and germplasm enhancement in this agronomically important forage and turf grass.

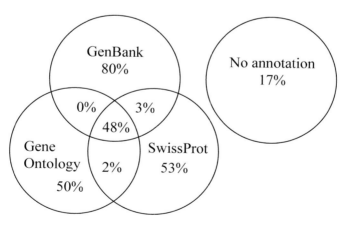

Figure 1 Annotation of ryegrass genes within the ASTRA database

Reference
Sawbridge, T., Ong, E.K., Binnion, C., Emmerling, M., McInnes, R., Meath, K., Nguyen, N., Nunan, K., O'Neill, M., O'Toole, F., Rhodes, C., Simmonds, J., Tian, P., Wearne, K., Webster, T., Winkworth, A. and Spangenberg, G. (2003). Generation and analysis of expressed sequence tags in perennial ryegrass (Lolium perenne L.). Pl*ant Science*, 165, 1089-1100.

Positive effect of increased AFLP diversity among parental plants on yield of polycross progenies in perennial ryegrass (*Lolium perenne* L.)

R. Kölliker, B. Boller and F. Widmer
Agroscope FAL Reckenholz, Swiss Federal Research Station for Agroecology and Agriculture, 8046 Zurich, Switzerland, Email:roland.koelliker@fal.admin.ch

Keywords: heterosis, molecular markers, polycross breeding, yield

Introduction In outbreeding forage crops such as perennial ryegrass (*Lolium perenne* L.), genetic diversity among parental plants may greatly influence the success of a cultivar through mechanisms such as heterosis and inbreeding depression. The aim of this study was to evaluate the use of amplified fragment length polymorphism (AFLP) markers to select polycross parents with contrasting levels of genetic diversity and to analyse genetic diversity and agronomic performance of first and second generation synthetic progenies (Syn1, Syn2).

Materials and methods Genetic diversity among 98 plants from advanced breeding germplasm including three maturity groups was analysed using 184 AFLP markers. Based on Euclidean squared distances, two polycrosses (PC) with contrasting levels of genetic diversity (PC narrow with low diversity, PC wide with large diversity), consisting of six parental plants each, were composed for three maturity groups. Genetic diversity within Syn1 progenies of all six PC was analysed using the same 184 AFLP markers scored in parental plants. Agronomic performance of Syn1 and Syn2 progenies was assessed by measuring dry matter yield of three cuts in the field and uniformity was assessed by comparing coefficients of variation for heading date.

Results Genetic diversity, expressed as average Euclidean squared distance, among parental plants selected to form narrow polycrosses was on average 36 % lower than the genetic diversity among plants selected for wide polycrosses. The differences between Syn1 progenies were less pronounced, with Syn1 progenies from narrow polycrosses showing 16 % lower genetic diversity when compared to Syn1 progenies from wide polycrosses. In general, diversity within Syn1 progenies derived from narrow polycrosses was comparable to the diversity among the respective parental plants while diversity within Syn1 progenies from wide polycrosses was considerably lower when compared to the respective parents. Syn1 and Syn2 progenies derived from wide polycrosses showed consistently higher dry matter yields when compared to progenies derived from narrow polycrosses. Averaged across both generations and all maturity groups, this difference reached 3.7 % and was highly significant (Figure 1 A). While there was no significant difference for the coefficients of variation for heading date between narrow and wide polycrosses of the early and intermediate maturity groups, Syn1 and Syn2 progenies from narrow polycrosses of the late maturity group showed significantly higher variation when compared to progenies from wide polycrosses (Figure 1 B). However, all progenies still showed coefficients of variation comparable to standard cultivars assessed in the same experiment.

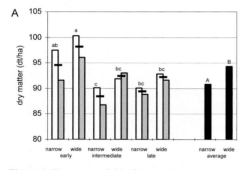 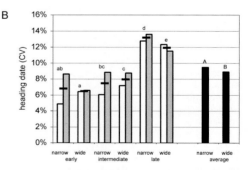

Figure 1 Dry matter yield of three subsequent cuts (A) and coefficients of variation for heading date (B) for first (Syn1; white columns) and second generation synthetic progenies (Syn2; grey columns) derived from polycrosses with contrasting levels of genetic diversity (narrow, wide) of three maturity groups (early, intermediate, late). Bars indicate average values of Syn1 and Syn2, black columns indicate average values across maturity groups. Different letters indicate significant differences (*P*<0.05).

Conclusions Our results provide evidence for an effective application of molecular markers to select genetically diverse parents in a polycross breeding program in order to increase agronomic performance of synthetic progenies without compromising phenotypic uniformity.

Genomic constitution of *Festulolium* varieties

D. Kopecky[1], V. Cernoch[2], R. Capka[2] and J. Dolezel[1]
[1]*Institute of Experimental Botany, Sokolovska 6, Olomouc, 77200, Czech Republic, Email: kopecky@ueb.cas.cz*
[2]*Plant Breeding Station, Fulnecka 95, Hladke Zivotice, 742 47, Czech Republic*

Keywords: chromosome translocations, *Festulolium*, genomic *in situ* hybridization, intergeneric hybrids

Introduction Hybrids between species of ryegrass (*Lolium*) and fescue (*Festuca*) combine useful agronomical characteristics such as rapid establishment from seed and fodder quality from ryegrass and tolerance against abiotic and biotic stressses from fescue. The superior potential of hybrids has stimulated breeding programs generating so called *Festulolium* varieties. While the varieties have been evaluated extensively for their agronomic characteristics, little information is publicly available on their genomic constitution. The aim of our study was to analyse genomic constitution of a representative set of commercially available European *Festulolium* cultivars. To do this, we have employed genomic *in situ* hybridization (GISH).

Materials and methods Four Czech cultivars ('Hykor', 'Felina', 'Korina' and 'Lesana', all 2n=6x=42) were of *L. multiflorum* x *F. arundinacea* origin backcrossed to *F. arundinacea*. Three other Czech cultivars 'Perun', 'Achilles' and 'Perseus', two Polish cultivars ('Felopa' and 'Rakopan'), one Lithuanian cultivar ('Punia') and one German cultivar ('Paulita') originated from crosses between *L. multiflorum* x *F. pratensis* and were all tetraploid (2n=4x=28). Preparation of mitotic metaphase plates and GISH was done according to Masoudi-Nejad *et al.* (2002). In all experiments, DNA of *L. multiflorum* was labelled with digoxigenin and used as a probe; genomic DNA of the *Festuca* species was used to block hybridization of common repetitive DNA sequences. The probe to block ratio was 1:150 with minor deviations. The detection of the hybridization signal was with the Anti-DIG-FITC conjugate; counterstaining was with propidium iodide (PI).

Results GISH is based on hybridization of species-specific DNA sequences in a probe labelled by fluorochrome with DNA of chromosomes fixed on microscopic slide. The method facilitates characterization of hybrids by the number of parental chromosomes, number of translocated chromosomes, number of translocations, number of break points, and total amount of chromatin of both parents. In this study, we screened 25 plants per cultivar, on average. In *L. multiflorum* x *F. arundinacea* hexaploid hybrids, a relatively stable number of parental chromosomes as well as some translocated chromosomes were observed. All *L. multiflorum* x *F. pratensis* hybrids showed large numbers of translocated chromosomes indicating frequent recombinations of both parental genomes. Proportions of parental chromatins varied from cultivar to cultivar.

Figure 1 Mitotic metaphase plates of *Festulolium* cultivars after GISH. Genomic DNA of *L. multiflorum* was used as a probe after labelling with FITC (light grey colour); genomic DNA of *F. pratensis* and *F. arundinacea* were used as blocks, respectively. Chromosomes were counterstained by PI (dark grey colour). [A] A representative example of mitotic chromosomes of 'Hykor' with a large number of introgressions of *L. multiflorum* into *F.*

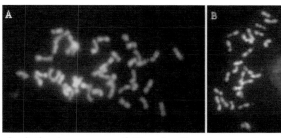

arundinacea. [B] A representative example of mitotic chromosomes of 'Perun' with a large number of translocations.

Conclusions The results of our study demonstrate the value of genomic *in situ* hybridization for rapid and relatively cheap screening of commercially available *Festulolium* hybrids for their genomic constitution and for the detection of chromosome translocations. The analyses revealed that none of the commercially available *Festulolium* cultivars contained equal proportions of ryegrass and fescue genomes. Individual cultivars were characterized by a specific ratio of parental chromatin and the number of chromosomal translocations.

Acknowledgement
This work was supported by a research grant no. S5038104 from the ASCR.

Reference
Masoudi-Nejad, A., S. Nasuda, R.A. McIntosh, and T.R. Endo (2002). Transfer of rye chromosome segments to wheat by a gametocidal system. Chromosome Res. 10:349-357.

Genetic changes over breeding generations of *Festulolium*

M. Ghesquière, P. Barre and L. Barrot
Institut National de la Recherche Agronomique, Unité de Génétique et Amélioration des Plantes Fourragères, 86600 Lusignan, France Email: Marc.Ghesquiere@lusignan.inra.fr

Keywords: tetraploid, gene evolution, genetic drift, *Lolium multiflorum*, *Festuca glaucescens,* SSR, STS

Introduction *Festulolium* hybrids are a valuable breeding source for tolerance to abiotic stress and to make grass more persistent under drought and in cold environments. In 2004, the *EU* Commission enlarged the definition of *Festulolium* which may now include all hybrids between *Lolium* sp. and *Festuca* sp. and not only those between *L. multiflorum* and *F. pratensis*. We here report allele frequencies at two unlinked PCR-based marker loci in populations derived from tetraploid (2n=4x=28) *L. multiflorum* x *F. glaucescens* hybrids where breeding history enables us to test the effects of selection vs that of genetic drift.

Material and methods The two markers used to monitor allele frequency were an STS-marker, OSRB (Lem & Lallemand, 2003), and an SSR-marker, B1B6 (Lauvergeat *et al*., 2005). The locus OSRB was mapped on the corresponding *Triticeae* linkage group 5 (Jones *et al*., 2002) and the B1B6 locus, on group 1 (Ghesquière, unpublished results). Two of the populations used were derived from an original population of 50 tetraploid primary hybrids between *L. multiflorum* and *F. glaucescens*. Another three populations were obtained following one generation of backcrossing (BC1) onto *L. multiflorum*. From 162 to 177 individuals per population were genotyped at both loci. In addition 22 individuals of *F. glaucescens*, 26 individuals of *L. multiflorum* and a sample of 26 primary hybrids were genotyped at the OSRB locus where allele species-specificity was not known. No assumption was made about allele dosage in banding patterns so that allele frequency was estimated directly from band frequency. Pedigree and polycross size at each generation of the breeding process enabled us to estimate the effective size (Ne) of all populations, assuming tetrasomic inheritance (Gallais, 2003), and to compute confidence interval of allele frequencies.

Results Seven and six segregating alleles were found at locus OSRB and B1B6 respectively. Allele differentiation between parent species was much lower at OSRB than at B1B6 where only one allele was found in *F. glaucescens* and none in *L. multiflorum*. Due to a low sampling error, all pairs of populations differed significantly in the frequency of at least one allele at one or other locus. However, in no instance was it possible to reject the null hypothesis that gene changes only resulted from genetic drift based on the confidence interval for Ne. In particular, it was striking to find no significant discrepancy in the BC1 populations although *Festuca* chromosomes had no homologous counterpart and, consequently, the frequency of *Festuca* alleles was expected to decrease.

Conclusions *Festulolium* populations are subject to many antagonistic evolutionary forces after primary hybridization. These include chromosome structural rearrangements, modification of gene expression from genome conflicts as well as indirect selection for life-history traits or fertility. In a breeding programme, there are genome effects on gene frequency as well as the effects of selection by the breeder for desirable traits and genetic drift due to a narrowing of effective population size at each generation. However, the allele frequencies presented here are consistent with changes only through genetic drift. It appears that either or both loci are essentially neutral with respect to selection, or that breeding for *Festuca* traits (such as persistency) may have almost balanced exactly any counter-selection of the *Festuca* genome due to lower fertility or a lower transmission rate of chromosomes. Thus, *Festulolium* populations provide interesting material for surveying gene changes in recent interspecific hybrid populations and a better understanding of polyploidization that so frequently occurred in the evolution of grasses. However, we conclude that to detect the effects of selection will require populations of large Ne and/or extensive genome coverage by markers or use of the GISH procedure.

References

Jones, E.S., N.L. Mahoney, M.D. Hayward, I.A. Armstead, J.G. Jones, M.O. Humphreys, I.P. King, T. Kishida, T. Yamada, F. Balfourier, G. Charmet & J. Forster (2002). An enhanced molecular marker based genetic map of perennial ryegrass (*Lolium perenne*) reveals comparative relationships with other *Poaceae* genomes. *Genome*, 45, 282-295.

Lauvergeat, V., P. Barre, M. Bonnet & M. Ghesquière (2005). Sixty simple sequence repeats (SSR) markers for use in the *Festuca/Lolium* complex of grasses. *Molecular Ecology Notes*, in the press.

Lem, P. & J. Lallemand (2003). Grass consensus STS markers: an efficient approach for detecting polymorphism in *Lolium. Theoretical and Applied Genetics*, 107, 1113-1122.

Gallais, A. (2003). Quantitative genetics and breeding methods in autopolyploid plants. Institut National de la Recherche Agronomique, Paris, 515 pp.

Phenotypic variation within local populations of meadow fescue shows significant associations with allele frequencies at AFLP loci

S. Fjellheim[1], Å.B. Blomlie[2], P. Marum[3] and O.A. Rognli[4]
[1]Department of Chemistry, Biotechnology and Food Science, Agricultural University of Norway, P.O. Box 5003, N-1432 Ås, Norway; [2]The Norwegian Crop Research Institute, Løken Research Centre, Volbu, N-2940 Heggenes, Norway; [3]Graminor AS, Bjørke Forsøksgård, Hommelstadvegen 60, N-2344 Ilseng, Norway; [4]Department of Plant and Environmental Sciences, Agricultural University of Norway, P.O. Box 5003, N-1432 Ås, Norway E-mail:odd-arne.rognli@nlh.no

Keywords: germplasm, meadow fescue, molecular diversity, AFLP, association mapping

Introduction To identify markers useful for Marker Assisted Selection (MAS), mapping families are usually constructed and used for Quantitative Trait Loci (QTL) mapping. Association mapping offers an alternative strategy for marker development using already characterized germplasm, preferably from natural populations. Simultaneous phenotypic and molecular screening of gene bank accessions can reveal associations between molecular marker alleles and phenotypic traits, and lead to a more targeted construction of mapping families for fine-mapping. In this investigation, we combine molecular (AFLP) and phenotypic data of 15 Norwegian local populations and 5 Nordic cultivars in order to identify markers associated with phenotypic traits of interest.

Materials and methods Fifteen local populations and 5 Nordic cultivars (Løken, Kalevi, Fure, Norild, and Svalöfs Sena) of meadow fescue were scored for 19 morphological and phenological traits (DUS characters), and genotyped for 74 AFLP markers (Fjellheim, 2004). Local populations covered geographic variation both for longitude, latitude, and altitude. Forty and 20 genotypes/population were scored for phenotypic characters and for AFLP markers, respectively. AFLP marker data were converted to allele frequencies (Zhivotovsky, 1999), and stepwise multiple regression analyses conducted using mean values of phenotypic traits as responses (Y) and AFLP allele frequencies from each population as predictors (X).

Results and discussion Models with the first 4 markers explained more than 90% of the variation with high associated R^2 (predicted) values (above 80) for 7 of the characters (early growth, leaf width, herbage yield 2. cut, inflorescence forming tendency, growth habit, inflorescence emergency, and width of flag leaf). Table 1 and Figure 1 present the model for leaf width.

Table 1 Stepwise multiple regression model for leaf width based on AFLP marker allele frequencies of 15 local populations (VIF=variance inflation factor)

Marker	R^2	R^2 (pred)	P	VIF
P77M66-62	73.9	65.0	0.000	2.8
P77M72-234	87.7	79.5	0.003	2.4
P77M72-139	94.4	88.2	0.004	1.3
P77M66-202	98.1	91.7	0.001	2.1

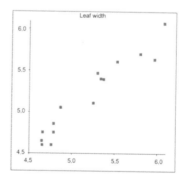

AFLPs that map close to QTL for heading date, flag leaf length, and flag leaf width were present among markers included in the regression models for these traits, and indicate that the marker-trait associations detected are real. The results from the present analysis demonstrate that association mapping is a promising method for screening germplasm.

Figure 1 Predicted (Y) vs. true (X) values for the regression model of leaf width using the 4 best AFLP markers of the stepwise regression model

Reference

Fjellheim S (2004) Molecular and phenotypic characterisation of Nordic meadow fescue (*Festuca pratensis* Huds.) with a view on phylogeographic history. Doctor Scientiarum Thesis 2004:27, Agricultural University of Norway, ISBN 82-575-0612-5

Zhivotovsky LA (1999) Estimating population structure in diploids with multilocus dominant DNA markers. Mol Ecol 8: 907-913

Marker-assisted selection for fibre concentration in smooth bromegrass

C. Stendal, M.D. Casler and G. Jung
University of Wisconsin-Madison and USDA-ARS, Madison, WI 53706.. Email:mdcasler@wisc.edu

Keywords: RAPD markers, NDF, breeding, *Bromus inermis*

Introduction The concentration of neutral detergent fibre is the best single laboratory predictor of voluntary intake potential in forage crops. However, the assay of thousands of plant samples for NDF selection in a breeding program requires a large amount of labour and time, potentially increasing cycle time and reducing the rate of progress. A previous study (Diaby and Casler, 2005) identified 16 random amplified polymorphic DNA (RAPD) markers that were strongly associated with NDF concentration in one or more of four smooth bromegrass (*Bromus inermis* Leyss) populations. The objective of this study was to validate these associations by implementing marker-assisted selection for these 16 RAPD markers.

Materials and methods A total of 244 smooth bromegrass clones representing four populations were established as spaced plants in two replicates at Arlington, WI, USA. Leaf tissue samples were harvested eight times over 2 years and analysed for NDF. Each plant was scored for all 16 RAPD markers. Plants were sorted according to presence or absence of each marker and a contrast was performed to test the difference in mean NDF between those clones with the marker vs. those clones without the marker. Marker indices were generated as combinations of marker scores, weighted by the percentage of phenotypic variation explained by each marker. Ties for marker index scores were broken by selection on the basis of pedigree and prior NDF data.

Results There were significant marking scoring differences between this study and that of Diaby and Casler (2005), which were scored by different people. This resulted in a distribution of selection effects with an equal number of significant positive and negative effects. As a result, marker indices failed to validate, with low selection differentials and little variation explained (Table 1). Using only those markers with significant positive effects in both studies, new marker indices had large selection differentials, ranging from 36 to 58% of the selection differentials for phenotypic selection (control) and highly significant effects (Table 2).

Table 1 Marker index validation statistics[#]

Population	SD	% of control	R^2	P-value
Alpha	2	3	0	0.5096
WB19e	-11	-20	2	0.0007
Lincoln	8	13	1	0.0062
WB88S	7	12	1	0.0134

Table 2 New marker index statistics[#]

Population	SD	% of control	R^2	P-value
Alpha	43	58	23	<0.0001
WB19e	31	58	18	<0.0001
Lincoln	22	36	7	<0.0001
WB88S	24	42	10	<0.0001

[#] SD = selection differential. Control = selection differential for phenotypic selection for NDF concentration.

Conclusions Marker selection indices with large and significant selection differentials can be developed from RAPD markers, providing evidence that marker-assisted selection may be used to potentially improve the efficiency of selection for low NDF. The only marker indices that showed potential for use in marker-assisted selection were those based on results of both the current marker-validation study and the previous marker-discovery study of Diaby and Casler (2005). Reproducibility problems between different personnel scoring RAPD marker bands will limit the use of RAPD-based selection indices to the tenure of the person scoring bands, creating a potential need for redevelopment of marker selection indices as program personnel change. Conversion to a more reproducible marker system that is insensitive to personnel would provide a better long-term solution than RAPD markers. These results suggest that one cycle of selection for low NDF should be successful based on these marker selection indices. Polycross populations from selections based on phenotype vs. marker indices will be employed to test this hypothesis.

References
Diaby, M. and M.D. Casler (2005). RAPD Marker variation among divergent selections for fiber concentration in smooth bromegrass. *Crop Science*, 44,27-35.

Endophyte ASTRA: a web-based resource for *Neotyphodium* and *Epichloë* EST analysis

K. Shields[1,2], M. Ramsperger[1,2], S.A. Felitti[1,2], C.G. Love[1,3], T.A. Erwin[1,3], D. Singh[1,3], E.G. Logan[1,3], D. Edwards[1,2,3] and G.C. Spangenberg[1,2,3]

[1]*Primary Industries Research Victoria, Plant Biotechnology Centre, La Trobe University, Bundoora, Victoria 3086, Australia* [2]*Molecular Plant Breeding Cooperative Research Centre, Australia* [3]*Victorian Bioinformatics Consortium, Australia Email: kate.shields@dpi.vic.gov.au*

Keywords: grass endophytes, expressed sequence tags (ESTs), gene annotation, bioinformatics ASTRA database, simple sequence repeats (SSRs), single nucleotide polymorphisms (SNPs)

Introduction Large-scale gene discovery has led to the production of 13,964 expressed sequence tags (ESTs) collectively from the grass endophytes *Neotyphodium coenophialum, N. lolii* and *Epichloë festucae.*

Materials and methods A total of 9,783 ESTs were generated from fungal tissue grown *in vitro,* while 4,181 ESTs were generated from tissue grown *in planta.* Each of these sequences was annotated by comparison to the GenBank and SwissProt public sequence databases. Automated intermediate Gene Ontology (GO) annotation was also determined.

Results and conclusions Sequence annotation was determined for 38%, 45% and 43% of sequences using GenBank, SwissProt and GO annotation respectively, with 43% of sequences were determined to have no comparative annotation and thus appear to be unique to grass endophytes (Fig 1). EST redundancy was resolved through assembly with CAP3, identifying at total of 7,585 unigenes within this dataset, comprised of 5,846 singletons and 1,739 contigs. All sequences and annotation are maintained within an ASTRA format MySQL database, with web based access for text searching, BLAST sequence comparison and Gene Ontology hierarchical tree browsing.

In total, 1,047 endophyte simple sequence repeat molecular markers (SSRs) have been identified using SSRPrimer and specific PCR amplification primers were designed. A total of 1,636 candidate single nucleotide polymorphism (SNP) molecular markers and 326 insertion deletions (InDels) were also identified across 300 loci using AutoSNP. The integration of different analysis tools and the resulting data within a central resource allow users to identify novel candidate genes for functional genomics and design of novel endophyte-grass associations.

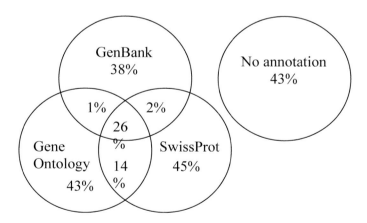

Figure 1 Annotation of endophyte genes within the ASTRA database

Section 6

Genetics and breeding for symbiosis

Genetic variation in the perennial ryegrass fungal endophyte *Neotyphodium lolii*

E. van Zijll de Jong[1,3], N.R. Bannan[2,3], A.V. Stewart[4], K.F.M. Reed[2], M.P. Dobrowolski[2,3], K.F. Smith[2,3], G.C. Spangenberg[1,3] and J.W. Forster[1,3]

[1]*Primary Industries Research Victoria, Plant Biotechnology Centre, La Trobe University, Bundoora, Victoria 3086, Australia.* [2]*Primary Industries Research Victoria, Hamilton Centre, Hamilton, Victoria 3300, Australia.* [3]*Molecular Plant Breeding Cooperative Research Centre, Australia.* [4]*Pyne Gould Guinness Seeds, Ceres Research Station, P.O. Box 3100, Christchurch, New Zealand. Email: eline.vanzijlldejong@dpi.vic.gov.au*

Keywords: genetic diversity, SSR, UPGMA, migration

Introduction The common fungal endophytes (*Neotyphodium* species) of temperate pasture grasses are associated with improved tolerance to water and nutrient stress and resistance to insect pests, but are also the causal agents of animal toxicoses. Considerable variation exists among grass-endophyte associations for these beneficial and detrimental agronomic traits. The extent to which this variation may be attributed to the endophyte genotype, the host genotype or environmental interactions is currently unknown. The development of molecular genetic markers for endophytes based on simple sequence repeat (SSR) loci and the demonstration of the specific detection of endophytes *in planta* with these markers (van Zijll de Jong *et al.*, 2005) allows efficient assessment of endophyte diversity in grass populations.

Materials and methods To investigate the contribution of endophyte genotype to variation in endophyte-related traits in the perennial ryegrass-*N. lolii* association, 18 *in planta*-validated endophyte SSR markers were screened across 37 accessions of wild and cultivated perennial ryegrass from around the world. The incidence of endophyte in these accessions was variable, ranging from 2-100%.

Results Assessment of the genetic diversity of endophytes in 281 endophyte-positive genotypes detected low levels of genetic variation, with a single endophyte predominant in most accessions. The more divergent endophytes detected in some accessions probably belong to other taxonomic groupings such as *N. occultans*. The geographical pattern of diversity in endophytes was non-random, showing similarities with both ancient and more recent routes of dispersion of perennial ryegrass.

Conclusions The low level of genetic variation within *N. lolii* suggests that variation in endophyte-related traits may be more associated with genotypic differences in perennial ryegrass. Further assessment (van Zijll de Jong *et al.*, 2005) of the genetic diversity of endophytes in perennial ryegrass, in plants collected from farms with variable incidence of endophyte-related livestock toxicosis as well as varieties containing endophyte strains with reduced toxicity effects, support the role of both endophyte and host genotype in the expression of endophyte-related traits. This study may be extended by assessment of variation in symbiosis-related genes through the development of functionally-associated single nucleotide polymorphism (SNP) markers.

Reference

Van Zijll de Jong, E., Smith, K.F., Spangenberg, G.C., Forster, J.W. (2005) Molecular genetic marker-based analysis of the forage grass host – *Neotyphodium* endophyte interaction. In: Roberts, C.A., West, C.P., Spiers, D.E. (eds.) *Neotyphodium* in Cool Season Grasses: Current Research and Applications, pp.123-133.

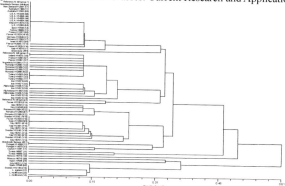

Figure 1 A phenogram of the genetic diversity of endophytes in *Lolium* accessions from around the world based on measurements of average taxonomic distance for 18 SSR loci and cluster analysis with unweighted pair group method of arithmetic averages (UPGMA).

Molecular breeding for the genetic improvement of forage crops and turf

Isolation and characterisation of novel BTB domain protein encoding genes from fungal grass endophytes

M. Ramsperger[1,2], S.A. Felitti[1,2], D. Edwards[1,2,3] and G.C. Spangenberg[1,2,3]

[1]Primary Industries Research Victoria, Plant Biotechnology Centre, La Trobe University, Bundoora, Victoria 3086, Australia [2]Molecular Plant Breeding Cooperative Research Centre, Australia [3]Victorian Bioinformatics Consortium, Australia Email: marc.ramsperger@dpi.vic.gov.au

Keywords: grass endophytes, BTB domain proteins (BDP), bacterial artificial chromosomes (BACs), large insert genomic libraries, gene and promoter discovery

Introduction Pasture grasses belonging to the Pooideae sub-family of the Poaceae family frequently host symbiotic fungal endophytes. These include the sexual *Epichloë* species and the anamorphic asexual *Neotyphodium* species, which are thought to have evolved from *Epichloë* species either by the direct loss of sexual reproduction or by interspecific hybridisation. The two key temperate pasture grasses, tall fescue (*Festuca arundinacea* Schreb) and perennial ryegrass (*Lolium perenne* L.) interact with the fungal endophytes *N. coenophialum* and *N. lolii*, respectively. Large insert genomic DNA libraries are valuable resources for the discovery and isolation of genes and their regulatory sequences, for physical mapping, map-based cloning of target genes as well as for whole genome sequencing. BTB (Bric-a-brac, tram-track, broad complex) domains are highly conserved motifs of 120 amino acids in length. The domains are rich in hydrophobic amino acids, and mediate protein-protein interaction that lead to homomeric dimerisation and in some cases heteromeric dimerisation of a large number of functionally diverse proteins. The presence of BTB domains defines a large family of genes involved in various biological processes, such as the regulation of transcription, DNA binding activity and structural organisation of macromolecular structures. Genes encoding BTB domain proteins (BDP) have previously been described in viruses, yeasts, plants, nematodes, insects, fish and mammals. However, BDP genes have not as yet been described for filamentous fungi.

Materials and methods Four related sequences encoding BDPs that are transcribed at high levels were isolated and characterised from cDNA libraries of the grass endophytes *N. lolii*, *N. coenophialum* and *E. festucae*.

Results and conclusions DNA sequence and Southern hybridisation analyses demonstrated that two distinct BDP gene variants are present in *N. coenophialum*, while a single gene is present in each of *N. lolii* and *E. festucae*. To assist in gene and promoter discovery in these fungal endophytes of grasses, large-insert genomic DNA libraries of *E. festucae*, *N. coenophialum* and *N. lolii* were generated using phage lambda, as well as a bacterial artificial chromosome (BAC) library was constructed for *N. lolii* with a 120 kb average insert size and 15-fold genome coverage. The genomic library from *N. lolii* was screened using the relevant cDNA (*Nl*BDP) and the resulting genomic sequence (6,982 bp) was shown to contain a BDP open reading frame (858 bp) lacking introns (Figure 1). The transcriptional activity of the *N. lolii* BDP gene was determined in cultured endophytes by northern hybridisation analysis and *in planta* by real time (RT) PCR.

Genes related to the grass fungal endophyte BDP genes were also identified *in silico* within genomic and cDNA sequences from other fungal species. This analysis has defined a grass endophyte-derived BTB domain that has not been previously characterised in filamentous fungi, and consequently a discrete class of BDP genes.

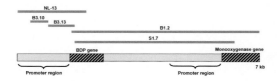

Figure 1 Characterisation of an *N. lolii* genomic clone containing a BTB domain protein (BDP) gene and a monooxygenase gene. Subcloned and sequenced fragments are shown. Upstream regulatory promoter (1,504 bp) and coding (858 bp) regions of the NlBDP gene were isolated and used for gene functional analysis in the grass-endophyte association

Genetic analysis of the interaction between perennial ryegrass and the fungal endophyte *Neotyphodium lolii*

E. van Zijll de Jong[1,3], A.C. Vecchies[1,3], M.P. Dobrowolski[2,3], N.O.I. Cogan[1,3], K.F. Smith[2,3], G.C. Spangenberg[1,3] and J.W. Forster[1,3]

[1]*Primary Industries Research Victoria, Plant Biotechnology Centre, La Trobe University, Bundoora, Victoria 3086, Australia.* [2]*Primary Industries Research Victoria, Hamilton Centre, Hamilton, Victoria 3300, Australia.* [3]*Molecular Plant Breeding Cooperative Research Centre, Australia. Email: john.forster@dpi.vic.gov.au*

Keywords: QTL, ELISA, symbiosis, candidate gene

Introduction The fungal endophyte *Neotyphodium lolii* is widely distributed in perennial ryegrass pastures, especially in Australia and New Zealand. The presence of the endophyte is associated with improved tolerance to water and nutrient stress and resistance to insect pests, but is accompanied by reduced herbivore feeding. The molecular mechanisms responsible for these endophyte-related traits are in general poorly understood. Comparisons of different grass-endophyte associations show that endophyte-related traits are affected by both endophyte and host genotype, and environmental interactions.

Materials and methods To investigate the role of host genetic factors, preliminary research has focused on quantitative trait loci (QTL) analysis of a full-sib mapping family with a single resident endophyte. This F_1 (Northern African$_6$ x Aurora$_6$) population is based on a two-way pseudo-testcross structure for which two genetic parental maps have been generated, populated by genomic DNA-derived simple sequence repeat (SSR), expressed sequence tag-derived SSRs (EST-SSRs) and EST-RFLPs. The $F_1(NA_6xAU_6)$ family also provides the resource for *in vitro* gene-associated single nucleotide polymorphism (SNP) discovery. A semi-quantitative enzyme-linked immunosorbent assay (ELISA) test was used to detect endophyte in the progeny of this family and ELISA scores in the field were examined using simple interval mapping (SIM) and composite interval mapping (CIM). The development of a quantitative real-time (RT) PCR assay to measure endophyte content *in planta* in the $F_1(NA_6xAU_6)$ family is described.

Results SIM and CIM mapping of the variation in ELISA scores assessed in the field for endophyte-positive progeny has lead to the identification of several genomic regions on the maps of both parents that significantly affect the trait. The ELISA score may be potentially correlated with endophyte incidence, efficiency of colonisation and associated metabolic traits. Candidate symbiosis-associated genes in perennial ryegrass have been identified based on sequence annotation and gene expression data, and mapping of these candidate genes will allow evaluation of co-location with endophyte-specific trait QTLs. Strategies for the identification of sequence variation in key symbiosis-related genes of the endophyte, such as the *dmaW* gene that plays a determinate role in ergovaline biosynthesis, will be described.

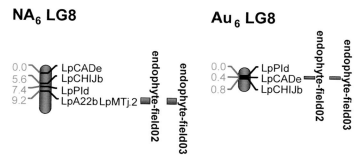

Figure 1 Identification of an endophyte-associated QTLs based on semi-quantitative ELISA scores derived from field-grown progeny of the $F_1(NA_6xAU_6)$ genetic mapping family. Measurements were taken in two successive seasons (summer 2002 and summer 2003).

Microarray-based transcriptome analysis of the interaction between perennial ryegrass (*Lolium perenne*) and the fungal endophyte *Neotyphodium lolii*

S.A. Felitti[1,2], P. Tian[1,2], T. Webster[1,3], D. Edwards[1,2,4] and G.C. Spangenberg[1,2,3,4]

[1]*Primary Industries Research Victoria, Plant Biotechnology Centre, La Trobe University, Bundoora, Victoria 3086, Australia* [2]*Molecular Plant Breeding Cooperative Research Centre, Australia* [3]*Victorian Microarray Technology Consortium, Australia* [4]*Victorian Bioinformatics Consortium, Australia*
Email: silvina.felitti@dpi.vic.gov.au

Keywords: fungal endophytes, perennial ryegrass, expressed sequence tags (ESTs), microarrays, transcriptome

Introduction *Neotyphodium lolii*, *Neotyphodium coenophialum* and *Epichloë festucae* are common symbiotic fungal endophytes of the temperate pasture grasses perennial ryegrass (*Lolium perenne*), tall fescue (*Festuca arundinacea*) and red fescue (*Festuca rubra*), respectively. A genomic resource of 13,964 expressed sequence tags (ESTs), representing 7,585 unique endophyte genes, has been established for *Neotyphodium* and *Epichloë* fungal endophytes.

Materials and methods The endophyte genomic resource established has enabled the design and fabrication of two endophyte-specific cDNA microarrays (Nchip[TM] microarray and EndoChip[TM] microarray). The Nchip[TM] and EndoChip[TM] microarrays have been applied to comparative transcriptome analyses of different asexual (*N. coenophialum* and *N. lolii*) and sexual (*E. festucae*) endophyte taxa under various saprophytic growth conditions, leading to the identification of differentially expressed genes.

Results and conclusions The Nchip[TM] and EndoChip[TM] microarrays permit the interrogation of 3,806 *Neotyphodium* genes (Nchip[TM] microarray), and 4,195 *Neotyphodium* and 920 *Epichloë* genes (EndoChip[TM] microarray), respectively. They represent tools for high-throughput transcriptome analysis, including genome-specific gene expression studies, profiling of novel endophyte genes, and investigation of the host grass-fungal symbiont interaction. Microarray-based transcriptome analysis was also undertaken in transgenic *Neotyphodium* endophyte expressing a chimeric *gusA* reporter gene [strain FM13 (Lp1/pNOM-101 and pAN7-1) (Murray *et al.*, 1992) kindly provided by B. Scott, Institute of Molecular BioSciences, Massey University, Palmerston North, New Zealand] compared to untransformed control, confirming the power of these tools for applications in characterising genetically modified endophytes. A recent extension of this proof-of-concept research has included the transcriptome analysis of transgene-expressing endophytes to assess gene expression responses to specific genetic modifications (e.g. gain-of-function, knock-outs and knock-downs). Within the context of a comprehensive spatial and temporal systems biology approach, a transcriptomics study of the mutualistic interaction between perennial ryegrass (*L. perenne* L.) and its fungal endophyte (*N. lolii*) was undertaken. In combination with a 15K ryegrass unigene microarray, the EndoChip[TM] microarray was applied to the detailed analysis of in planta gene expression in different ryegrass organs using endophyte-infected and endophyte-free ryegrass plants in an isogenic host genetic background. Data derived from endophyte microarray analysis has been validated using both northern hybridisation and RT-PCR analyses.

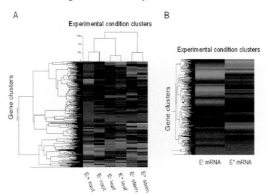

Figure 1 *Neotyphodium in planta* gene expression microarray-based transcriptome analysis. A) Hierarchical clustering of mean signal values (Euclidean distance) from infected and uninfected ryegrass organs on the Endophyte unigene microarray. B) Endophyte-infected stem prevalent gene expression

Reference
Murray, F.R., Latch, G.C.M. and Scott, D.B. (1992). Surrogate transformation of perennial ryegrass, *Lolium perenne*, using genetically modified *Acremonium* endophyte. *Mol. Gen. Genet.*, 233, 1-9.

A high-throughput gene silencing approach for studying the interaction between perennial ryegrass (*Lolium perenne*) and the fungal endophyte *Neotyphodium lolii*

S.A. Felitti[1,2], P. Tian[1,2], D. Edwards[1,2] and G.C. Spangenberg[1,2]
[1]*Primary Industries Research Victoria, Plant Biotechnology Centre, La Trobe University, Bundoora, Victoria 3086, Australia* [2]*Molecular Plant Breeding Cooperative Research Centre, Australia*
Email: silvina.felitti@dpi.vic.gov.au

Keywords: fungal endophytes, perennial ryegrass, expressed sequence tags (ESTs), gene silencing, siRNAs

Introduction Perennial ryegrass (*Lolium perenne* L.) and its fungal endophyte (*Neotyphodium lolii*) are known to establish a mutualistic association that impacts on the agronomic productivity of endophyte-infected ryegrass pastures. To study this interaction at the molecular level, a genomic resource consisting of 13,964 endophyte ESTs has been generated. However, the functions of a large proportion of these genes remain to be elucidated. Recent work has demonstrated the potential for RNA-mediated gene silencing to suppress gene expression in a sequence specific manner thus allowing for the subsequent analysis of gene function.

Materials and methods Vectors for RNA-mediated gene silencing in grass endophytes based on the Multisite Gateway[TM] recombination system were developed. These vectors can be used to readily create large numbers of genetically modified endophytes, for high-throughput gene characterisation. In order to validate this system, RNA silencing was performed in transgenic *Neotyphodium* endophyte expressing a chimeric *gusA* reporter gene [strain FM13 (Lp1/pNOM-101 and pAN7-1) (Murray *et al.*, 1992) kindly provided by B. Scott, Institute of Molecular BioSciences, Massey University, Palmerston North, New Zealand]. Vectors expressing sense and hairpin RNAs were produced and introduced into the *gusA*-expressing transgenic endophyte (Figure 1).

Results and conclusions Efficient gusA gene silencing was achieved through the expressed *gusA* hairpin RNA. Furthermore, small interfering RNAs (siRNAs) designed from the *gusA* sequence and used in endophyte transfection, were effective in silencing the expression of the *gusA* transgene. This high-throughput gene silencing technology has now been applied to characterise the function of a large number of endogenous endophyte genes corresponding to functionally annotated and unannotated sequences represented in the EST resource established. Overall, these results indicate that RNA silencing effectively operates in grass endophytes and this technology provides a new tool for the high-throughput functional characterisation of endophyte genes.

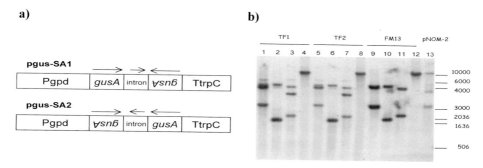

Figure 1 a) Maps of hpRNA *gusA* gene silencing vectors. Arrows indicate the orientation of the *gusA* gene. All the constructs were under control of the *Aspergillus nidulans gpd* promoter (Pgpd) and included *trpC* terminator (TtrpC). b) Southern blot analysis of endophytes transformed with hairpin RNA *gusA* gene silencing vectors pgus-SA1 or pgus-SA2. Genomic DNA was digested with *Hind*III (lanes 1, 5 and 9), *Nco*I (lanes 2, 6 and 10) and *Xba*I (lanes 3, 7 and 11) and was probed with a 250 bp *gusA* gene fragment. Lanes 4, 8 and 12 show undigested genomic DNA

Reference

Murray, F.R., Latch, G.C.M. and Scott, D.B. (1992). Surrogate transformation of perennial ryegrass, *Lolium perenne*, using genetically modified *Acremonium* endophyte. *Mol. Gen. Genet.*, 233, 1-9.

Metabolome analysis of the interaction between perennial ryegrass (*Lolium perenne*) and the fungal endophyte *Neotyphodium lolii*

P. Tian[1,2], S.A. Felitti[1,2], M.P. Dobrowolski[2,3], K.F. Smith[2,3], D. Edwards[1,2], R. Hall[4], J. Kopka[5] and G.C. Spangenberg[1,2,6]

[1]*Primary Industries Research Victoria, Plant Biotechnology Centre, La Trobe University, Bundoora, Victoria 3086, Australia.* [2]*Molecular Plant Breeding Cooperative Research Centre, Australia.* [3]*Primary Industries Research Victoria, Hamilton Centre, Hamilton, Victoria 3300, Australia.* [4]*Plant Research International, 6700 Wageningen, The Netherlands.* [5]*MPI of Molecular Plant Physiology, 14476 Golm, Germany.* [6]*Victorian Centre for Plant Functional Genomics, Australia. Email: pei.tian@dpi.vic.gov.au*

Keywords: fungal endophytes, perennial ryegrass, metabolomics, metabolic fingerprinting/profiling

Introduction Perennial ryegrass (*Lolium perenne* L.) and tall fescue (*Festuca arundinacea* Schreb.) frequently contain endophytic fungi (*Neotyphodium lolii* in perennial ryegrass and *N. coenophialum* in tall fescue). The presence of the endophyte has been shown to improve seedling vigour, persistence and drought tolerance in marginal environments as well as provide protection against some insect pests. Endophyte-infected grasses also produce a wide range of metabolites, including ergopeptine alkaloids, indole-isoprenoid lolitrems, pyrrolizidine alkaloids, and pyrrolopyrazine alkaloids. In contrast to information on alkaloids and animal toxicosis, the beneficial physiological aspects of the endophyte/grass interactions have not been well characterised. The physiological mechanisms which lead to increased plant vigour and enhanced tolerance to abiotic stresses unrelated to the reduction in pest damage to endophyte-infected grasses are unknown. Recent technological advances in metabolomics enable dynamic changes in the metabolome of an organism under varying experimental conditions to be studied. This provides opportunities for the investigation and validation of each and every detected metabolite, investigation of known metabolic pathways through searching of databases of known metabolites, molecular formula determination of unknown metabolites and creation of pathways from novel metabolites.

Materials and methods Endophyte-infected (E+) and endophyte-free (E-) ryegrass plants in an isogenic host genetic background were generated, characterised by SSR markers and ELISA and used for metabolic fingerprinting and metabolic profiling.

Results Within the context of a spatial and temporal systems biology approach, a metabolomics study of the mutualistic interaction between perennial ryegrass and its fungal endophyte was undertaken. Polar and lipophilic compounds were extracted from endophyte, E+ ryegrass and E- ryegrass standardised samples, and analysed (fractionation and mass determination) by gas chromatography (GC)/mass spectrometry (MS) and high performance liquid chromatography (HPLC) with photodiode array detection (PDA)/mass spectrometry using Q-TOF II MS with CapLC, Q-TOF Ultima API MS with Alliance HT and PDA, and GC-TOF Leco/Pegasus III MS with CTC PAL sampler and Agilent GC injector. Mass profile data were processed using MetAlign™ software and subjected to PCA. Mass peaks showing significant changes in comparative assessments between E+ ryegrass and E- ryegrass samples allowed the identification of sample-specific and differentially accumulating compounds (Figure 1), indicating specific biochemical pathways that may lead to enhanced tolerance to abiotic stress in the endophyte-grass host interaction.

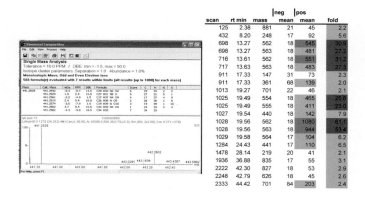

scan	rt min	mass	neg mean	pos mean	fold
125	2.38	881	21	45	2.2
432	8.20	248	17	92	5.6
698	13.27	562	18	545	30.9
698	13.27	563	18	481	27.2
716	13.61	562	18	551	31.2
717	13.63	563	18	483	27.3
911	17.33	147	31	73	2.3
911	17.33	361	68	139	2.0
1013	19.27	701	22	46	2.1
1025	19.49	554	18	465	26.6
1025	19.49	555	18	411	23.0
1027	19.54	440	18	142	7.9
1028	19.56	562	18	1080	61.1
1028	19.56	563	18	944	53.4
1029	19.58	564	17	104	6.2
1284	24.43	441	17	110	6.5
1478	28.14	219	20	41	2.1
1936	36.88	835	17	55	3.1
2222	42.30	827	18	53	2.9
2248	42.79	626	18	45	2.6
2333	44.42	701	84	203	2.4

Figure 1 Analysis of mass profile data from metabolic fingerprinting/profiling of the association between perennial ryegrass and *Neotyphodium lolii*

Endophyte effects on antioxidants and membrane leakage in tall fescue during drought

C.P. West[1], R.D. Carson[1], C.A. Guerber[1] and B. de los Reyes[2]
[1]University of Arkansas, 1366 W. Altheimer Dr., Fayetteville, AR, 72704 USA. Email: cwest@uark.edu
[2]University of Maine, Orono, ME, 04469, USA

Keywords: antioxidants, endophyte, drought, superoxide dismutase, tall fescue

Introduction Tall fescue [*Festuca arundinacea* (Schreb.)=*Lolium arundinaceum* (Schreb.) S.J. Darbyshire] infected (E+) by its fungal endophyte [(*Neotyphodium coenphialum* Morgan-Jones & Gams.) Glenn, Bacon & Hanlin] often shows greater persistence during summer drought than endophyte-free (E-) plants (Malinowski *et al.*, 2005). Survival of the apical meristem and growing zone of vegetative tillers likely involves biochemical adaptations whose benefits to the host are enhanced by endophyte presence. Antioxidant enzymes may scavenge free radicals during heat and drought, and thereby reduce membrane damage. Their roles in endophyte-mediated drought tolerance in tall fescue have not been tested. Our objective was to determine the endophyte influence on antioxidant enzyme activities, membrane leakage, and tiller survival in tall fescue during water deficit.

Materials and methods Two contrasting genotypes of tall fescue, 60 and 330, each with its respective wild-type endophyte genotype, were grown separately in glasshouse trials. In Trial 1, E+ and E- ramets were grown together in replicated plastic dishpans containing sandy-loam soil, and subjected to well-watered or drought-stress treatments. Vegetative tiller bases (2 cm) were sampled at six stages of progressive water deficit (early leaf rolling to complete sheath desiccation) and analysed for activities of ascorbate peroxidase, peroxidase, glutathione reductase, and superoxide dismutase (SOD) activities. In Trial 2, E+ and E- plants were grown in 60-cm deep PVC tubes. Water was withheld, and tiller bases were assayed for membrane leakage, expressed as membrane damage coefficient (Howarth *et al.*, 1997), at three stress stages. After complete desiccation, tubes were rewatered, and surviving tillers were counted after one week to quantify percentage tiller survival.

Results There were no effects of water deficit or endophyte infection status on the activities of ascorbate peroxidase, peroxidase, or glutathione reductase. In contrast, SOD activity in both genotypes increased faster in E+ than in E- plants in the stress treatment as water deficit intensified (Figure 1). As expected, membrane leakage was enhanced by withholding water (data not shown). In genotype 60, membrane damage was greater in E- than in E+ stressed plants at the intermediate stress stage (leaf blades severely rolled); however, there was no significant endophyte effect in genotype 330. Tiller survival rates for E+ and E- plants were 32% and 16% ($P<0.05$), respectively, for genotype 60, and 42% and 20% ($P<0.05$), respectively, for genotype 330.

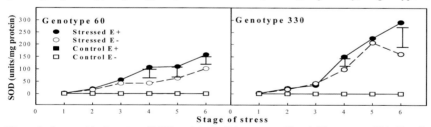

Figure 1 Superoxide dismutase (SOD) activity in (A) genotype 60 and (B) genotype 330. Bars indicate $LSD_{0.05}$. Symbols for control E+ data are under those of the control E- data.

Conclusions Stimulation of SOD activity by endophyte infection was associated with endophyte-enhanced tiller survival across host genotypes, supporting the hypothesis that SOD enhancement serves as a mechanism for endophyte-mediated drought tolerance in tall fescue. Some protection against membrane leakage was observed in one genotype, suggesting involvement of membrane protection in some endophyte-grass associations.

Acknowledgments We are grateful for the technical assistance of A. Haverly and S. Rajguru, and for funds provided by USDA-ARS Specific Cooperative Agreement 58-6227-3-014 through the Dale Bumpers Small Farm Research Center, Booneville, Arkansas, USA.

References

Howarth, C.J., C.J. Pollock, and J.M. Peacock. (1997). Development of laboratory-based methods for assessing seedling thermotolerance in pearl millet. *New Phytologist* 137, 129-139.
Malinowski, D.P., D.P. Belesky, & G.C. Lewis. (2005). Abiotic stresses in endophytic grasses. In: C.A. Roberts, C.P. West, & D. Spiers (eds.) *Neotyphodium in Cool-Season Grasses*. Blackwell Publ., Ames, Iowa, 187-199.

Section 7

Transgenics for research and breeding including risk assessment

Role of the BANYULS(*BAN*) gene from *Arabidopsis thaliana* in transgenic Alfalfa expression of anthocyanins and proanthocyanidins

S.M. Hesamzadeh Hejazi[1], S. Arcioni[2] and F. Paolocci[2]

[1]*Research institute of forests and rangelands(gene bank),Tehran-IRAN*
[2]*Istituto di Recerche sul Miglicramento, Genetico della piante Foraggere, Consiglio Nazionale della Ricerche,Perugia, Italy, Email: smhessamzadeh@rifr-ac.ir*

Keywords: *Agrobacterium tumefaciens, BAN* gene, condensed tannins, genetic transformation, *Medicago sativa*

Introduction Condensed tannins (CTs) are flavonoid oligomers, many of which have beneficial effects on animal (bloat safe) and human health. The *BAN* gene encodes anthocyanidin reductase (ANR), an enzyme proposed to convert anthocyanidins to their corresponding 2,3-cis-flavan-3-ols (Xie *et al.*, 2003). Ectopic expression of *BAN* in Alfalfa transgenic foliage results in accumulation of CTs. Thus, it has been assumed that the *BAN* gene also acts in starter units for the condensation of tannins in Alfalfa.

Material and methods Transformation was performed in two genotypes (Rgsy27,P1) of *M. sativa*. The binary vector pBI121.BAN was constructed by removing the *gus* gene and inserting the *BAN* gene into the BamHI and SacI sites of pBI121.1 to put *BAN* expression under control of the CaMV35S promoter and NOS 3' terminator. pBI121.BAN was transformed into *Agrobacterium tumefaciens* EHA105 and LBA4404 by the triparental mating method. To confirm chromosomal insertion of the *BAN* transgene, genomic DNA was extracted from leaves of putative transgenic Rgsy27 and P1(cv.Adriana) plants (Sambrook *et al.*, 1989). The NPTII selectable marker gene in the binary vector was used as a probe after labeling with ^{32}P-dCTP using the Ready –TO-GO DNA labeling beads (dCTP)Kit. For RT-PCR analysis of *BAN* transcript levels, total RNA isolated from leaf tissue using Nucleo Spin RNA Kit. CTs in the leaves of transgenic plants were visualized by staining tissues in a solution of methanol:6M HCl (1:1) containing 2%(w/v) DMACA for 3 min (Li *et al.*, 1996), then washing 3 times with MilliQ water. For quantitative analysis of CTs in transgenic plants, we used DMACA-HCl and Butanol-HCl assays at A643, A550 and A725.

Results We found a simple and efficient method for regeneration-transformation of Medicago including the perennial *M. sativa* var.Rgsy27 (2n=4x=32) and *M. sativa* var.Adriana (P1)(2n=4x=32). Development of this method (using 4 new media M1,M2,M3,M4 with low levels of 2,4-D and BAP) produced transgenic plants on all explants within 1.5-2 months which remained at the same ploidy level and maintained fertility.
The presence and structure of the transgene in the plants were analysed by genomic southern blots in the Rgsy27 and P1 genotypes. In each case both an internal fragment of the construct as well as one of the T-DNA borders were probed. The expected size of the internal fragments characteristic for each construct was detected in the plants analysed except for a few transgenic plants where the *BAN* fragment was not present. The number of T-DNA borders detected in these plants, representing the number of T-DNA copies integrated in the plant genome, varied from one to probably three or more. In several of the transformants we found aberrant sized *BAN* bands. This was probably due to the loss of one of the T-DNA BamHI sites, indicating that truncation of some of the T-DNA copies had occurred in these plants. The qualitative expression of the *BAN* gene was studied by RT-PCR and was compared with the expression of LAR and EF-1 genes. The strain LBA4404 was much better than EHA105 for transformation of both genotypes.

The *BAN* gene encodes anthocyanidin reductase (ANR), an enzyme proposed to convert anthocyanidins to their corresponding 2,3-cis-flavan-3ols. Ectopic expression of *BAN* in Alfalfa transgenic foliage results in accumulation of CTs. The percent of tannin in the dry matter of control and transgenic plants were: for P1(control) = 0.418, Rgsy27(control) = 0.259 , P1+*BAN* = 0.600 and for Rgsy27+*BAN* = 0.474.

Conclusions Preliminary results of our work indicated that in forage legumes BANYULS(*BAN)* could not autonomously induce tannin. However, when tissues are committed for tannin accumulation, the *BAN* gene can strongly increase CT levels.

References
Xie, D.Y., S.B.,Sharma,N.L.,Paiva, D.Ferreria, R. A., Dixon (2003). Role of anthocyanidin reductase encoded by *BANYULS* in plant flaonoid biosynthesis. SCIENCE, 299: 396-399.
Sambrook, J., Ef., Fritsch, T., Maniatis,(1989). Molecular cloning : alaboratory manual- cold Spring Harbor Laboratory Press, Cold Spring Harbor, NY, ed. 2nd.
Li, Y.-G. Tanner, G.J., Larkin, P.J., (1996). The DMACA-HCL protocol and the threshold proantho cyanidin content for bloat safety in forage legumes. J. Sci. Food Agric. 70: 89-101.

Development of alfalfa (*Medicago sativa* L.) transgenic plants expressing a *Bacillus thuringiensis* endotoxin and their evaluation against alfalfa caterpillar (*Colias lesbia*)

F. Ardila[1], M.C. Gómez[1], M.J. Diéguez[1], E.M. Pagano[1], M. Turica[2], R. Lecuona[2], V. Arolfo[3], D. Basigalup[3], C. Vázquez Rovere[4], E. Hopp[4], P. Franzone[1] and R.D. Rios[1]

[1]*Instituto de Genética "Ewald A. Favret"*, [2]*Instituto de Microbiología y Zoología Agrícola*, [3]*EEA INTA Manfredi, Ruta 9 Km 636 (5988), Manfredi, Argentina*, [4]*nstituto de Biotecnología. 1,2 and 4: CICVyA, INTA, Castelar, cc 25 (1712), Argentina. Email: fardila@cnia.inta.gov.ar*

Keywords: alfalfa, *Bacillus thuringiensis*, *Colias lesbia*

Introduction Alfalfa (*Medicago sativa* L.) is the most important forage crop in Argentina, with ca. 6 million cultivated hectares. The production of this crop is limited by the alfalfa caterpillar (*Colias lesbia*) which causes a loss equivalent to at least 10% of the biomass per year. No natural tolerance against this lepidoptera was found in alfalfa germplasm, hampering the development of tolerant cultivars by conventional breeding. This pest is usually controlled by using chemical insecticides but this has adverse effects on beneficial insects and the environment. Alternatively, low doses of commercial Bt insecticides (40 to 70 g/ha) also proved to efficiently limit the pest. This observation leads to us consider that the development of alfalfa transgenic plants expressing a suitable member of the *B. thuringiensis cry* gene family could be a useful tool for overcoming this alfalfa yield constraint. The aim of this work was to produce alfalfa transgenic plants expressing a Bt protein and to assess its biological activity against *C. lesbia* under laboratory conditions.

Materials and methods Petioles from alfalfa semi-dormant clones with high *in vitro* regeneration capacity (Moltrasio *et al.*, 2004) were submitted to a customized *A. tumefaciens* based transformation protocol. The *npt* II gene, under Nos promotor and Ocs terminator control, was employed as selectable marker. As a source of δ-entomotoxin, an unmodified 2,086 bp, 5′-terminal fragment of the *cry*IA(b) gene from *B. thuringiensis* var. *kurstakii* HD-1 (Geiser *et al.*, 1986), with high range of lepidoptera specificity, and controlled by 2 X 35S promoter, AMV enhancer and T7 terminator, was used. Kanamycin was employed as selection agent through all the *in vitro* culture steps. The regenerated plants were analyzed by PCR to evaluate the presence of the transgenes. Expression of the *cry* gene at mRNA level was evaluated by RT-PCR. The presence of the CryIA(b) protein in crude protein extracts from transgenic alfalfa leaves was determined by DAS-ELISA (adgia). A bioassay to test the entomocidal capacity of alfalfa transgenic plants was set up. For that purpose, first instar larvae of *C. lesbia,* obtained from field gathered eggs, were challenged by feeding them with transgenic ELISA positive or control plant leaves.

Results Fifty-one independent alfalfa transgenic plants were obtained and established in the greenhouse. All the plants proved to be PCR positive for both genes. Similarly, transcriptional activity from *cry*IA(b) gene was demonstrated in all the evaluated plants. Duplicated ELISA analyses showed that some plants expressed the Cry protein clearly above the control values. Some of the ELISA positive plants showed an insecticidal activity strongly enough to limit the development of alfalfa caterpillar larvae under laboratory conditions.

Conclusions The results of the present study provides proof of concept of the feasibility of limiting the development of alfalfa caterpillar through the employment of alfalfa transgenic plants expressing a truncated CryIA(b) protein. It is noteworthy to mention that the plants involved in this project were established in the greenhouse 8 years before the evaluations were performed. This underlines the remarkable structural and expression stabilities of these materials. The biological strategy described here paves the way to develop, in the short term, an effective tool for the field control of alfalfa caterpillar.

References

Geiser, M., S. Schweitzer & C. Grimm (1986). The hypervariable region in the genes coding for entomopathogenic crystal proteins of *Bacillus thuringiensis*: nucleotide sequence of the *kurhd*1 gene of subsp. kurstaki HD1, Gene 48 (1), 109-118.

Moltrasio R., C.G. Robredo, M.C. Gómez, A.H. Díaz Paleo, D.G. Díaz, R.D. Rios & P.M. Franzone (2004). Alfalfa (*Medicago sativa* L.) somatic embryogenesis: genetic control and introduction of favourable alleles into elite Argentinean germplasm". Plant, Cell, Tissue and Organ Culture 77 (2): 119-124.

Increased cuticular wax accumulation and enhanced drought tolerance in transgenic alfalfa by overexpression of a transcription factor gene

Z.-Y. Wang[1], J.-Y. Zhang[1], C. Broeckling[2], E. Blancaflor[2], M. Sledge[1] and L. Sumner[2]
[1]Forage Improvement Division, [2]Plant Biology Division, The Samuel Roberts Noble Foundation, 2510 Sam Noble Parkway, Ardmore, Oklahoma 73401, USA. Email: zywang@noble.org

Keywords: alfalfa, drought tolerance, transgenic plant, wax

Introduction Plant cuticular waxes play an important role in protecting aerial organs from damage caused by multiple environmental stresses such as drought, cold, UV radiation, pathogen infection and insect attack. The identification of leaf wax genes involved in stress tolerance is expected to have great potential for crop improvement. Cuticular waxes are complex mixtures of very long chain fatty acids, alkanes, primary and/or secondary alcohols, aldehydes, ketones, esters, triterpenes, sterols and flavonoids. Mutant analysis in *Arabidopsis* has contributed to the identification of the components and genes involved in wax deposition. However, no information is available on the effects of overexpression of these genes in crops of agronomic importance. Alfalfa (*Medicago sativa*) is the most important forage legume species in the world and a close relative of *Medicago truncatula*.

Materials and methods A novel AP2 domain-containing transcription factor gene, *WXP1*, was identified and cloned from the model legume *M. truncatula*. Chimeric transgene construct was made placing *WXP1* under the control of CaMV35S promoter. Large numbers of transgenic alfalfa plants were produced by *Agrobacterium*-mediated transformation. Detailed molecular and biochemical analyses were carried out for the transgenic alfalfa plants.

Results The predicted protein of *WXP1* has 371 aa; it is one of the longest peptides of all the single AP2 domain proteins in *M. truncatula*. *WXP1* is distinctly different from the most studied genes in the AP2/ERF transcription factor family, such as *AP2s*, *CBFs*, *DREBs*, *WIN1* and *GL15*. Transcript level of *WXP1* is inducible by cold, ABA and drought treatment in shoot tissues. Overexpression of *WXP1* under the control of CaMV35S promoter led to a significant increase in cuticular wax loading on leaves of transgenic alfalfa (Figure 1). The most striking phenotypic change in the *WXP1* overexpressed alfalfa plants was the more glaucous appearance in the leaves. Scanning electron microscopy revealed earlier accumulation of wax crystals on the adaxial surface of newly expanded leaves and higher

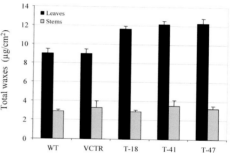

Figure 1 Cuticular wax accumulation in leaf and stem of transgenic and control alfalfa plants. WT, wild type; VCTR, empty vector control; T, transgenic lines overexpressing *WXP1*.

densities of wax crystalline structures on both the adaxial and abaxial surfaces of mature leaves. Gas chromatography – mass spectrometry (GC–MS) analysis revealed that total leaf wax accumulation per surface area increased 29.6 – 37.7% in the transgenic lines, and the increase was mainly contributed by C30 primary alcohol. Comparison of wax accumulation in leaves and stems of the transgenics indicates that *WXP1* is mainly involved in the acyl-reduction pathway for wax biosynthesis. *WXP1* overexpression induced a number of wax-related genes. Transgenic leaves showed reduced water loss and chlorophyll leaching. Transgenic alfalfa plants with increased cuticular waxes showed enhanced drought tolerance demonstrated by delayed wilting after watering was ceased and quicker and better recovery when the dehydrated plants were re-watered.

Conclusions We identified and characterized a novel AP2 domain-containing transcription factor gene (*WXP1*) by using an overexpression approach. Our study using isogenic lines (wild type control, empty vector control and transgenic plants) clearly demonstrates the positive effects of cuticular waxes on drought tolerance. Despite the fact that both drought tolerance and wax accumulation are complicated traits that are under the control of multiple genes, our results demonstrated for the first time that overexpression of a single transcription factor gene, *WXP1*, could turn on wax production and improve plant drought tolerance in agronomically important species.

Molecular breeding of white clover for transgenic resistance to *Alfalfa mosaic virus* and natural resistance to *Clover yellow vein virus*

P. Chu[1], G. Zhao[2,3] and G.C. Spangenberg[2]

[1]*CSIRO Plant Industry, GPO Box 1600, Canberra, ACT 2601, Australia;* [2]*Primary Industries Research Victoria, Plant Biotechnology Centre, La Trobe University, Bundoora, Victoria 3086, Australia;* [3]*Department of Grassland Science, Gansu Agricultural University, Lanzhou 730070, China. Email: paul.chu@csiro.au*

Keywords: virus resistance, white clover, alfalfa mosaic virus, clover yellow vein virus

Introduction *Trifolium repens* L. (white clover) is one of the most important pastoral plants in temperate Australia. Its productivity and persistence is being reduced significantly by *Alfalfa mosaic virus* (AMV), *Clover yellow vein virus* (ClYVV) and *White clover mosaic virus* (WClMV). These viruses are also widespread in other legumes and are inflicting large economic losses to farmers throughout the world (Campbell, 1984). To reduce the economic impact of these viruses, white clover plants resistant to both ClYVV and AMV are being developed for future commercial release. Since introducing viral transgenes from two or more viruses into a transgenic plant has the potential threat of viral recombination, we have decided to develop white clover with transgenic resistance to AMV and natural resistance to ClYVV.

Materials and methods Transgenic white clover plants expressing the AMV coat protein (CP) gene and showing immunity to that virus have been produced. White clover plants with natural resistance to ClYVV were identified by screening various cultivars of the white clover against different isolates of the virus. To combine the transgenic AMV immunity and natural ClYVV resistance into a single transgenic white clover cultivar, crosses between singly AMV immune and ClYVV resistant white clover plants were performed. The offspring plants were subjected to molecular analysis and infectivity screens to identify plants carrying AMV CP and showing immunity to both viruses. The best of three AMV + ClYVV resistant genotypes was selected for crossing with 12 Sustain-type elite plants in order to produce a new white clover cultivar with double virus resistance.

Results Crosses between singly AMV immune and ClYVV resistant white clover plants produced offspring plants that were resistant to both viruses. After crossing with the Sustain-type parents, almost 100% of the offspring plants carrying the AMV CP showed immunity to that virus, irrespective of the Sustain-type parent used. In contrast, there were significant variations in the percentage of ClYVV resistant progenies produced by the different Sustain-type parents (Figure1). Diallele crossing between these first generation double virus resistant progenies (T1 plants) increased the proportion of ClYVV resistant plants among the T2 population from 34% to 49% (Figure 2). Furthermore, T2 population showed a higher proportion of plants with milder symptoms than plants in the T1 population.

Figure 1   **Figure 2**

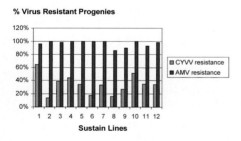

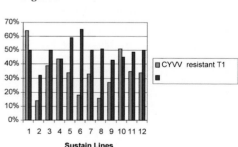

Conclusions The results showed that it is possible to produce white clover plants with transgenic resistance to AMV and natural resistance to ClYVV. The differential inheritance of ClYVV resistance from different Sustain genotypes suggests that multiple resistance genes may be involved in the natural ClYVV resistance.

References
Campbell C.L. and J.W. Moyer (1984). Yield responses of 6 white clover clones to virus infection under field condition. *Plant Disease*, 68, 1033-1035.

Molecular breeding of transgenic virus-immune white clover (*Trifolium repens*) cultivars

M. Emmerling[1], P. Chu[2], K.F. Smith[3], C. Binnion[1], M. Ponnampalam[1], P. Measham[2], Z.Y. Lin[2], N. Bannan[3], T. Wilkinson[3] and G.C. Spangenberg[1]

[1]*Primary Industries Research Victoria, Plant Biotechnology Centre, La Trobe University, Bundoora, Victoria 3086, Australia* [2]*CSIRO Plant Industry, Canberra, ACT 2601, Australia* [3]*Primary Industries Research Victoria, Hamilton Centre, Hamilton, Victoria 3300, Australia Email: michael.emmerling@dpi.vic.gov.au*

Keywords: transgenic white clover, alfalfa mosaic virus (AMV), clover yellow vein virus (CYVV), coat protein-mediated virus resistance

Introduction White clover (*T. repens* L.) is a major component of improved pastures throughout the temperate world. It is, however, highly susceptible to virus infection. Alfalfa mosaic virus (AMV), clover yellow vein virus (CYVV) and white clover mosaic virus (WCMV) all contribute to a significant reduction in dry matter yield and persistence of white clover. Sources of natural resistance to AMV in white clover or sexually compatible species are not available. Pathogen-derived resistance strategies, such as the expression of viral coat protein in transgenic plants, thus provides opportunities for the development of virus immune transgenic white clover.

Materials and methods Transgenic white clover plants expressing chimeric AMV coat protein (AMV-CP) genes were generated by *Agrobacterium*-mediated transformation. Selected transformation events were subjected to detailed molecular analysis and virus resistance phenotype. Virus immune events were chosen for transgenic germplasm development.

Results and conclusions The molecular characterisation of over 30 independent transformation events allowed the identification of transgenic genotypes carrying single T-DNA inserts, showing mitotic and meiotic stability of transgene expression (assessed at transcript and protein levels), and AMV immunity evaluated under containment glasshouse and field conditions. Two transformation events in white clover cultivar 'Irrigation' (H1 and H6), with field-immunity to aphid-mediated AMV infection were subjected to multi-site (Hamilton and Howlong, Australia), multi-year (over the 1998-2004 period) and multi-generation ($T_0 - T_4$) field evaluations and concurrently used for elite transgenic germplasm development. Following top crosses with elite parental breeding lines, diallel crosses of heterozygous offspring plants and identification of AMV-CP homozygous T_2 lines through quantitative PCR-based high-throughput zygocity screening, a breeding nursery with 1,300 transgenic white clover plants was established. A final selection of 37 *syn0* plants derived from both H1 and H6 transformation events was made based on transgene-mediated AMV immunity, non-transgenic CYVV resistance and agronomic characteristics such as plant height, stolon density, internode length, leaf length, flower number, summer growth and survival, autumn vigour and spring vigour. These *syn0* plants were poly-crossed to produce the world's first AMV-immune transgenic white clover cultivar.

In addition, transgenic white clover germplasm was developed in a different genetic background, leading to the selection of 600 T_1 elite white clover plants, following virus infectivity and transgene zygocity screenings. A corresponding breeding nursery was established in Hamilton, Australia in November 2004 for the selection of *syn0* parents for the development of a second transgenic white clover cultivar with transgenic AMV immunity and enhanced non-transgenic CYVV resistance. Furthermore, biosafety research to underpin the release of AMV immune transgenic white clover cultivars was undertaken, such as studies on transgene flow in white clover; development of protocols with highest sensitivity for AMV-CP transgene tracking and tracing in transgenic white clover including a range of downstream products; assessment of AMV-CP allergenicity; assessment of substantive equivalence of AMV CP-transgenic vs non-transgenic white clover at the transcriptome and metabolome level by microarray and FT-MS analysis, respectively.

Polyphenolic phenomena: transgenic analysis of some of the factors that regulate the cell-specific accumulation of condensed tannins (proanthocyanidins) in forage crops

M.P. Robbins, G. Allison, D. Bryant and P. Morris
Plant, Animal and Microbial Sciences Department, Institute of Grassland and Environmental Research, Plas Gogerddan, Aberystwyth, Ceredigion SY23 3EB, UK Email: mark.robbins@bbsrc.ac.uk

Keywords: condensed tannins, *Lotus*, transcription factors

Introduction Condensed tannins biosynthesised within crops are a well-established mechanism for protecting plant protein in the rumen of grazing livestock. Protein protection mediated by these polymeric flavonoid molecules has been characterised in *Lotus* spp. and offers an interesting contrast to the polyphenol oxidase (PPO) system that confers protein protection in red clover.

Materials and methods Clonal genotypes derived from *Lotus corniculatus* cv. Leo were supplied by Dr K.J. Webb and colleagues at IGER. Transgenic lines harbouring *R2R3-MYB* class transcription factors were produced at IGER while lines harbouring *Sn*, a maize *bHLH* gene originate from Dr F. Damiani at CNR, Perugia. Further details and initial characterisation of *Sn* constructs in three recipient genotypes are outlined in Robbins *et al*. (2003).

Results Transgenic plants were grown under containment conditions and leaves were scored (Table 1) for the presence of cells containing condensed tannins using dimethylaminocinnamaldehyde (Li *et al*., 1996).

Table 1 Presence of CT-containing cells in leaves of transgenic and recipient genotypes of *Lotus corniculatus*

	Low CT genotype S33, S50	High CT genotype S41	*Sn* lines S50	*MybPh2* lines S50
Vascular mesophyll	+	+	+	+++
Pallisade mesophyll	-	+	+	-
Spongy mesophyll	-	+	+	-

+, presence of CT cells; -, CT cells not detected

Other phenotypes resulting from the introduction and expression of *Sn* included enhancement of anthocyanin accumulation in selected cell types; ie. subepidermal cell layers of leaf midrib, leaf base and petiole tissues. Increases in trichome numbers were noted in genotype S33 in selected transgenic lines and this correlated with high level expression of the *Sn* transgene.

Conclusions Results of this study indicate that the ectopic expression of plant transcription factors in *Lotus corniculatus* can modulate the biosynthesis of natural products. Data from *Sn* plants is consistent with studies on *Arabidopsis* which demonstrate the interactive role of *bHLH* and *R2R3-MYB* class genes in the hierarchical control of plant development in higher plants (Zhang *et al*., 2003)

References
Li Y.G., Tanner G.J. & Larkin P.J. (1996). The DMACA-HCl protocol and the threshold proanthocyanidin content for bloat safety in forage legumes. *Journal of the Science of Food and Agriculture,* 70, 89-101.
Robbins M.P., Paolocci F., Hughes J-W., Turchetti V., Allison G., Arcioni, S., Morris, P. & Damiani F. (2003). *Sn*, a maize *bHLH* gene, modulates anthocyanin and condensed tannin pathways in *Lotus corniculatus*. *Journal of Experimental Botany*, 54, 239-248.
Zhang F., Gonzalez A., Zhao M., Payne C.T. & Lloyd A. (2003). A network of redundant bHLH proteins function in all *TTG1*-dependent pathways of *Arabidopsis*.

Minimising bloat through development of white clover (*T. repens*) with high levels of condensed tannins

M.T. O'Donoghue[1], C. Spillane[2] and E. Guiney[1]

[1]*Teagasc Crops Research Centre, Oak Park Carlow, Ireland.* [2]*Plant Molecular Genetics, Biochemistry Dept., University College Cork, Ireland. Email:todonoghue@oakpark.teagasc.ie*

Keywords: *BAN* gene, anthocyanidin reductase, condensed tannins, transformation, white clover

Introduction White clover constitutes a low percentage of the overall sward content in Irish pastureland despite EU directives limiting the use of nitrogenous fertilizers. This is mainly due to the tendency of large amounts of white clover to cause bloat. Bloat is a potentially fatal build up of proteinaceous foam in the guts of ruminants. Some lesser cultivated legumes such as *Lotus* species contain condensed tannins (CT) that decrease the incidence of bloated animals. The project's objective is to reduce the risk of bloat by generating white clover cultivars with high CT content. We are investigating whether expression of the *ANTHOCYANIN REDUCTASE* gene (*BAN*) in transgenic white clover and *Medicago truncatula* (model) plants leads to increased CT levels (Xie *et al.*, 2003).

Materials and methods The *BAN* gene was isolated via PCR from *Arabidopsis thaliana* genomic DNA and is being introduced into pCAMBIA2300, under the expression control of the constitutive *UBIQUITIN3* promoter (ATUBP3), to form the plasmid vector pMTOD001. A number of white clover cultivars from the Oak Park germplasm collection were tested for their ability to regenerate into whole plants from cotyledons. The cotyledons were excised from imbibed seeds and transferred to shoot inducing media. Regenerated shoots were transferred to root inducing media to promote root formation (Ding *et al.*, 2003).

Results Optimal amplification of the *BAN* gene (1.5kb) was achieved at a $MgCl_2$ concentration of 4mM (Fig1). The regenerating results of the white clover cultivars from Oak Park's germplasm were found to vary from 0% to 47.5%. The best regenerating cultivars have been identified as Tara and Susi (Table 1). This choice was based on plants that had regenerated both roots and shoots.

Table 1 The regenerating efficiencies of a selection of legume cultivars

Species	Cultivar	Percentage regeneration
T. repens	Tara	47.50%
T. repens	Susi	37.50%
T. repens	Huia	32.50%
T. repens	Haku	0%
M. truncatula	Jemalong	12.5%

Figure 1 The amplification of the ban gene (1.5kb) with four different $MgCl_2$: 2.5mM, 3mM, 3.5mM, 4mM

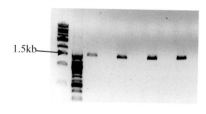

1.5kb

Conclusions The next step is to sequence and clone the gene before transforming the plasmid into white clover and *Medicago truncatula* via *Agrobacterium tumefaciens* mediated transformation. As the efficiency of transformation is relatively low (0.3%-6%) for white clover (Ding *et al.*, 2003) and can be a lengthy process, it is essential not to limit transformation further by using a poorly regenerating cultivar. There exists a wide variation of successful regeneration within the white clover cultivars at Teagasc (Table 1). Only cultivars with a high level of regeneration will be used for the transformation process.

References

Xie, D, S.B. Sharma, N.L. Paiva, D. Ferreira, and R.A. Dixon (2003). Role of Anthocyanidin Reductase, encoded by *BANYULS* in plant flavonoid biosynthesis. *Science*, 299, 396-399.

Ding Y., G. Aldao-Humble, E. Ludlow, M. Drayton, Y. Lin, J. Nagel, M. Dupal, G. Zhao, C. Pallaghy, R. Kalla, M. Emmerling, G. Spangenberg (2003). Efficient plant regeneration and *Agrobacterium*-mediated transformation in *Medicago* and *Trifolium* species. *Plant Science*, 165, 1419-1427.

Production and analysis of transgenic white clover (*Trifolium repens*) plants over-expressing organic acid biosynthetic genes

C.M. Labandera[1,2], S. Panter[1,2], A. Winkworth[1,2], J. Simmonds[1,2], A. Mouradov[1,2], U. John[1], P.W. Sale[3] and G.C. Spangenberg[1,2]

[1]*Primary Industries Research Victoria, Plant Biotechnology Centre, La Trobe University, Bundoora, Victoria 3086, Australia* [2]*Molecular Plant Breeding Cooperative Research Centre, Australia* [3]*Department of Agriculture, La Trobe University, Bundoora, Victoria 3086, Australia Email: Marcel.Labandera@dpi.vic.gov.au*

Keywords: white clover, organic acids, malate, aluminium tolerance, P-use efficiency

Introduction Aluminium (Al) toxicity is a major environmental limitation for plant production in acid soils, which represent more than one third of the world's agricultural land. Al-induced secretion in roots of organic acids (OA), such as malate and citrate, chelates the toxic Al cation excluding it from the root. This mechanism of Al-tolerance appears also to be associated with enhanced P-use efficiency. The development of transgenic plants for enhanced synthesis and secretion of OA from roots is a promising approach to confer Al-tolerance and enhanced P-acquisition efficiency. In order to understand the association between OA biosynthesis and secretion from roots in white clover (*Trifolium repens* L.), the physiological consequences of over-expressing 3 key white clover OA biosynthetic genes, individually and in combination, were assessed in transgenic plants.

Materials and methods White clover cDNAs encoding nodule-enhanced malate dehydrogenase (*TrneMDH*) and phosphoenolpyruvate carboxylase (*TrPEPC*) were isolated and sequenced. Transgenic white clover plants were generated by *Agrobacterium*-mediated transformation using binary vectors carrying chimeric OA biosynthetic genes from white clover under control of constitutive (CaMV35S) and/or root-prevalent (white clover phosphate transporter *TrPT1*) promoters.

Results and conclusions Nucleotide sequence analysis of *TrneMDH* revealed an open reading frame (ORF) of 1227 bp encoding for a protein of 408 amino acids. *TrPEPC* showed a 2904 bp ORF encoding a protein of 967 amino acids. Both *TrneMDH* and *TrPEPC* shared 93% sequence similarity with the respective *Medicago sativa* orthologues (*MsneMDH* and *MsPEPC*). A white clover mitochondrial citrate synthase (CS) cDNA (*TrCS*) was isolated and sequenced. Sequence analysis of *TrCS* revealed an ORF of 1419 bp encoding a protein of 472 amino acids. *TrCS* shared 74% identity to other dicotyledonous plant CS cDNAs. Molecular analysis of independent transgenic white clover plants confirmed the stable integration of the chimeric 35S::TrMDH, PT::TrMDH, 35S::TrPEPC, 35S::TrCS and PT::TrMDH-35S::TrPEPC transgenes. Transgenic white clover plants stably expressing the transgenes were identified by RT-PCR. Representative transformation events overexpressing *TrneMDH*, *TrPEPC* and *TrCS*, individually or in combination, were screened for OA synthesis and secretion. Selected transgenic lines were subjected to growth performance analysis under different aluminium and phosphorus levels, to allow for the identification of Al-tolerant transformation events with minimal disruption of carbon balance, and hence unaltered growth potential.

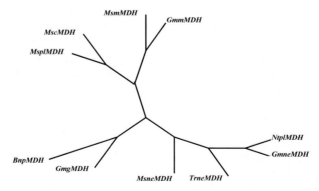

Figure 1 Unrooted phylogenetic dendrogram for deduced amino acid sequences of plant MDHs including white clover nodule-enhanced MDH (*TrneMDH*)

LXR ™ white clover: development of transgenic white clover (*Trifolium repens*) with delayed leaf senescence

Y.H. Lin[1], J. Chalmers[1], E. Ludlow[1], C. Pallaghy[2], G. Schrauf[3], Pablo Rush[3], A.M. García[4], A. Mouradov[1] and G.C. Spangenberg[1,5]

[1]Primary Industries Research Victoria, Plant Biotechnology Centre, La Trobe University, Bundoora, Victoria 3086, Australia [2]Department of Botany, La Trobe University, Bundoora, Victoria 3086, Australia [3] Universidad of Buenos Aires, Catedra de Genetica, Facultad de Agronomia, Buenos Aires, Argentina [4]CIGEN-CONICET [5]Phytogene Pty. Ltd., Plant Biotechnology Centre, La Trobe University, Bundoora, Victoria 3086, Australia
E-mail: german.spangenberg@dpi.vic.gov.au

Keywords: white clover, senescence, transgenic plants, cytokinins, isopentenyl transferase

Introduction Leaf senescence is a type of programmed cell death characterized by loss of chlorophyll, lipids, protein, and RNA. Cytokinins are a class of plant hormones that play roles in many aspects of plant growth and development, including leaf senescence, apical dominance, the formation and activity of shoot meristems, nutrient mobilization, seed germination, and pathogen responses. They also appear to mediate a number of light-regulated processes, such as de-etiolation and chloroplast differentiation. It is known that the concentrations of endogenous cytokinins decline in plant tissues as senescence progresses. This observation provides the opportunity to manipulate the senescence program in transgenic plants to enhance biomass and seed production, through the regulated expression of cytokinin biosynthesis genes.

Materials and methods Transgenic white clover plants carrying a chimeric isopentenyl transferase (IPT) gene (*ipt*) from *Agrobacterium tumefaciens* under control of the *Arabidopsis thaliana atmyb32* promoter (Figure 1a) and a chimeric neomycin phosphotransferase gene (*npt2*) as selectable marker were generated. The enzyme IPT catalyzes the rate-limiting step for *de novo* cytokinin biosynthesis i.e. the addition of isopentenyl pyrophosphate to the N6 of 5'-AMP to form isopentenyl AMP.

Results and conclusions Following a PCR screening using *npt2* and *ipt* primers (Figure 1b), the transgenic nature of the white clover plants was confirmed by Southern hybridisation analysis revealing the integration of 1 – 10 T-DNAs in the genome of independent *atmyb32::ipt* white clover plants (Figure 1c). Expression of the *atmyb32::ipt* gene in transgenic white clover plants was confirmed by RT-PCR. Selected *atmyb32::ipt* white clover plants showed a marked delay in leaf senescence but otherwise developed normally. The delay in senescence was revealed by an increase in chlorophyll content in *atmyb32::ipt* leaves relative to leaves of untransformed and *atmyb32::gusA* control white clover plants. Delayed senescence was also observed in detached leaves. Detached leaves from selected *atmyb32::ipt* white clover plants (LXR™ white clover) showed no visible signs of senescence over 20 days post-detachment. Assessment under containment conditions of growth characteristics of LXR™ white clover plants revealed significant (p<0.05) increases in number of leaves, stolon length, total leaf area, cumulative and relative leaf area appearance rates compared with untransformed negative control plants. The first small-scale field release of selected LXR™ white clover transformation events has been established in Junin, Argentina in 2004.

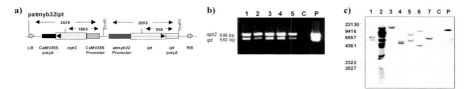

Figure 1 Molecular analysis of *atmyb32::ipt* white clover plants. a) Binary vector carrying chimeric *atmyb32::ipt* gene for the production of LXR™ white clover plants; b) PCR screening of LXR™ white clover plants using *npt2* and *ipt* specific primers; and c) Southern hybridisation analysis of LXR™ white clover plants using *ipt* hybridisation probe

Genetic transformation of rhodesgrass (*Chloris gayana* Kunth.) by particle bombardment

J. Matsumoto[1], S. Tsuruta[1], T. Gondo[2] and R. Akashi[1]
[1]*Faculty of Agriculture, University of Miyazaki, Miyazaki 889-2192, Japan.* [2] *National Agricultural Research Center for Hokkaido Region, Hitsujigaoka, Sapporo, 062-8555, Japan. Email: matumoto0654@hotmail.com*

Keywords: rhodesgrass, embryogenic calli, multiple shoots formed calli, particle inflow gun

Introduction Rhodesgrass (*Chloris gayana* Kunth) has been cultivated as one of the most important warm-season grasses in the world. One of the major limitations for cattle production on forage grasses, especially warm-season grasses is poor digestibility if compared to temperate grasses (Gondo *et al.*, 2003). It is believed that the low digestibility of warm-season grasses is due to high lignin contents (Akashi *et al.*, 2003). Recently, modification of the lignin content of plants appears to be feasible using genetic engineering strategies. We have established a methodology for high-frequency somatic embryogenesis and multiple shoot formation from seed-derived shoot apical meristems in rhodesgrass. Also, we have studied several factors involved in particle bombardment transformation.

Materials and methods Shoot apices as initial explants were isolated from aseptically germinated seedlings and were cultured in vitro. Embryogenic calli and the multiple shoot forming calli could be induced and maintained on MS basal medium with various combinations of 2,4-D and BAP. The induced embryogenic calli and multiple shoot forming calli were bombarded in a particle inflow gun with pAHC25, containing a modified bialaphos resistance gene (*bar*) and the GUS reporter gene. Following bombardment, cells were incubated at 27°C for 1 day and assayed for GUS activity. Transgene expression of regenerated plants was confirmed by PCR amplification analysis.

Results The best response of embryogenic calli and the multiple shoot forming calli was observed with 2.0 mg/L 2,4-D and 0.1 mg/L 2,4-D+2.0 mg/L BAP, respectively. Also, many bialaphos resistance cells were obtained from the bombarded calli. These regenerated plants displayed GUS activity and integration of the transgenes were confirmed by PCR amplification.

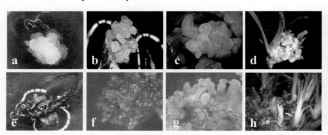 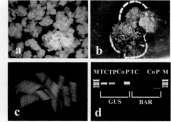

Figure 1 Differentiation of regeneration from somatic embryogenic calli and multiple shoot forming calli in rhodesgrass. a) Primary callus. b) Embryogenic callus. c) Somatic embryogenic callus. d) Germination of somatic embryo and plant regeneration. e) Primary callus. f) Multiple shoot forming callus. g) Multiple shoot formation under dark conditions. h) Multiple shoots under light conditions.

Conclusions In this report, we induced embryogenic calli and multiple shoot forming calli. These multiple shoot forming calli provided target tissue for genetic transformation using particle bombardment of rhodesgrass. The recovered transgenic plants were established in soil and acclimatised in the green house.

Figure 2 Transformation of rhodesgrass using particle bombardment. a) Transient GUS expression 24 hours after bombardment. b) Multiple shoot formation under bialaphos selection conditions. c) GUS expression in leaf tissue after transformation. d) Detection of *bar* and the GUS gene on transformed callus and plants by PCR. M: 100bp DNA ladder, TC: Transformed callus, TP: Transformed plant, Co: Non-transformed plant, P: Plasmid pAHC25 DNA.

References
Gondo, T., Y. Ishii, R. Akashi and O. Kawamura (2003). Efficient Embryogenic Callus Induction Derived from Mature Seeds and the Examination of the Genetic Transformation Condition by Particle Bombardment in Bahiagrass (P*aspalum notatum* Flugge). *Grassland Science* 49,33-37
Akashi, R., T. Gondo and O. Kawamura (2003). Fundamental Studies on the Improvement of Some Warm-season Grasses by Means of Plant Biotechnology. *Grassland Science* 49 (1): 79-87

Modulation of the gibberellin content in transgenic turf-type bahiagrass for improved turf characteristics and reduced mowing requirements

F. Altpeter, M. Agharkar and H. Zhang

University of Florida - IFAS, Agronomy Department, PMCB, Laboratory of Molecular Plant Physiology, 2191 McCarty Hall, P.O. Box 110300, Gainesville FL 32611-0300, USA. Email: faltpeter@mail.ifas.ufl.edu

Keywords: bahiagrass, *Paspalum notatum* (Flugge), genetic transformation, dwarf, gibberellin 2-oxidase

Introduction Bahiagrass is extensively used for utility turf along highways and for residential lawns in the southern USA and in the subtropical regions around the world. The objective of this experiment was to enhance turf quality of bahiagrass and reduce the mowing frequency by over-expression of a gibberellin catabolizing enzyme, Gibberellin 2-oxidase.

Materials and methods Gibberellin 2-oxidase8 cDNA was isolated from Arabidopsis using primers as suggested by Schomburg *et al.* (2003). Co-transfer of a constitutive *nptII* (Altpeter *et al.*, 2000) and GA-2 oxidase expression cassette, into seed derived callus cultures from turf-type bahiagrass (cv. 'Argentine') was followed by selection with paromomycin sulphate during callus subculture and, or regeneration (Altpeter and James, 2004). Transgenic plants were confirmed by NPTII ELISA (Agdia), PCR, RT-PCR and altered phenotype.

Results GA-2 oxidase cDNA was isolated from Arabidopsis, confirmed by sequencing and subcloned under the control of the constitutive 35S promoter. An efficient protocol for gene transfer to bahiagrass was established and supported the stable co-integration and constitutive expression of the selectable *nptII* gene and the Gibberellin 2-oxidase in bahiagrass. Transgenic plants over-expressing Gibberellin 2-oxidase showed an altered phenotype compared to wildtype bahiagrass (Figure 1 A,B).

Figure 1 A: Shorter internodes and darker green colour in transgenic bahiagrass line (left) compared to wild-type (right). B: Transgenic bahiagrass (right) with a dwarf phenotype compared to wild-type bahiagrass (left).

Conclusions Over-expression of Gibberellin 2-oxidase8 from Arabidopsis in bahiagrass resulted in dwarf phenotypes with darker green leaf colour. Transgenic plants produced normal roots and were established successfully in soil. Data correlating Gibberellin 2-oxidase over-expression in transgenic bahiagrass with physiological parameters will be collected.

References

Altpeter, F., J. Xu, & S. Ahmed. (2000). Generation of large numbers of independently transformed fertile perennial ryegrass (*Lolium perenne* L.) plants of forage- and turf-type cultivars. Mol. Breeding 6:519-528.

Altpeter, F & V. James (2004) Genetic transformation of turf-type bahiagrass (Paspalum notatum Flugge) by biolistic gene transfer. Intern. Turfgrass Soc. Res. J. (accepted for publication).

Schomburg, F.M., C. M. Bizzell, D. J. Lee, J. A. D. Zeevaart & R. M. Amasino (2003) Overexpression of a novel class of Gibberellin 2-Oxidases Decreases Gibberellin Levels and Creates Dwarf Plants. Plant Cell 15: 151–163.

Inducible over-expression of the CBF3 abiotic stress regulon in transgenic bahiagrass (*Paspalum notatum* Flugge)

V.A. James and F. Altpeter

University of Florida - IFAS, Agronomy Department, PMCB, Laboratory of Molecular Plant Physiology, 2191 McCarty Hall, P.O. Box 110300, Gainesville FL 32611-0300, USA. Email: faltpeter@mail.ifas.ufl.edu

Keywords: bahiagrass, *Paspalum notatum* (Flugge), genetic transformation, abiotic stress tolerance, CBF3

Introduction Bahiagrass is an important turf and forage grass in the Southern US and in the subtropical regions around the world. The objective of this experiment was to further enhance the productivity and persistence of bahiagrass during seasonal periods of drought and / or freezing and in salt affected regions by over-expression of the stress inducible transcription factor CBF3. Transcription factors like CBF3 are capable of activating the expression of multiple genes involved in protection against environmental stresses (Kasuga *et al.*, 1999).

Materials and methods The CBF3 gene, HVA1 or Dhn8 promoter candidates were isolated from genomic wild or cultivated barley DNA by PCR. Primers for isolation of target genes were designed according to the published cultivated barley sequences. Plant transformation vectors were constructed on basis of vector pJFnptII (Altpeter *et al.*, 2000). Biolistic gene transfer was carried out 6 weeks after initiation of callus cultures from mature seeds. Transgenic plants expressing the selectable *nptII* gene were regenerated on paromomycin containing medium and confirmed with NPT II-ELISA (Agdia) (Altpeter and James, 2004). Transgenic plants over-expressing CBF3 are currently identified by real time RT-PCR and will be subjected to cold stress (-5° C) in a completely randomized block design in a controlled environment chamber. Leaf tissue damage will be visually scored 1 day after cold stress and biomass production will be evaluated four weeks after recovery from cold stress.

Results A DREB1A transcription factor ortholog (CBF3), the dehydrin 8 (Dhn8) and HVA1 promoters were isolated from genomic, wild or cultivated barley. Plant transformation vectors placing CBF3 or GUS under control of the stress-inducible barley HVA1 (Figure 1B) or Dhn8 promoter were constructed and introduced into bahiagrass (Figure 1A). An efficient protocol for gene transfer to bahiagrass was established and supported the stable integration of transcription factor CBF3 and analysis of stress inducible promoter candidates (Figure 1C).

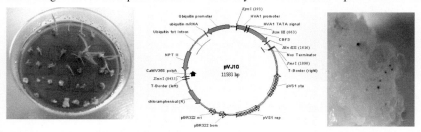

Figure 1 A: Regeneration of bahiagrass transformed with vector pJV10; B: plant transformation vector pVJ10 with CBF3 and selectable marker expression cassettes; C: Cold induced (4°C for 16 h) Gus reporter gene expression under regulatory control of the HVA1 promoter.

Conclusions An efficient protocol for tissue culture and generation of transgenic plants of the commercially important bahiagrass cultivar 'Argentine' has been developed. The CBF3 transcription factor and promoter candidates for cold, salt and, or drought inducible expression of transgenes were isolated by PCR and subcloned to drive CBF3 transcription factor or GUS reporter gene expression respectively. Dhn8 and HVA1 promoters were able to drive stress inducible reporter gene expression in bahiagrass (Figure 1C). Data correlating CBF3 over-expression in transgenic bahiagrass with freezing stress response is being obtained.

References

Altpeter, F., J. Xu, & S. Ahmed. (2000). Generation of large numbers of independently transformed fertile perennial ryegrass (*Lolium perenne* L.) plants of forage- and turf-type cultivars. Mol. Breeding 6:519-528.

Altpeter, F & V. James (2004) Genetic transformation of turf-type bahiagrass (Paspalum notatum Flugge) by biolistic gene transfer. Intern. Turfgrass Soc. Res. J. (accepted for publication).

Kasuga, M., Q. Liu, S. Miura, K. Yamaguchi-Shinozaki, and K. Shinozaki (1999). Improving plant drought, salt, and freezing tolerance by gene transfer of a single stress-inducible transcription factor. Nature Biotech. 29:287-291.

Genetic engineering for breeding for drought resistance and salt tolerance in Agropyron spp. (wheatgrass)

M. Fugui, Y. Jinfeng and H. Xiuwen
Inner Mongolia Agricultural University, 010018, Huhhot, China, Email: mfgui@ yahoo.com.cn

Keywords: genetic transformation, drought resistance, salt tolerance

Introduction Genetic engineering for breeding for drought resistance and salt tolerance in wheatgrass, lucerne and tall fescue is one of the main projects in a major national programs as part of the 10th five-year national plan: "Research of gene transfer in plants and its industrialisation". It is a large project that has financial support for work on forage crops in China and many research institutes and universities take part in it. The Inner Mongolia Agricultural University is in charge of the project on wheatgrass. The research was started in Nov. 2002. The general situation and the primary results are introduced and summarised in this paper.

Material and method Four species of wheatgrass (*Agropyron mongolicum, A .cristatum* cv. 'Fairway', *A. desertorum* cv. 'Nordan' and *A .cristatum* ×*A. mongolicum* cv. 'Hycrest-Mengnong') were used as the basic plant material. Based on an established regeneration system, the p5CS gene, which regulates the last step of proline synthesis, was transformed into these species, with phosphinothricin acetyltransferase (*bar*) conferring herbicide resistance as the selectable gene. The transformation was conducted through microprojectile bombardment of callus derived from immature inflorescence and the transgenic plants were examined by PCR Southern and RT-PCR analysis.

Results The results showed that callus initiation from immature inflorescence and plant regeneration could occur under induction in all four species. The medium for callus induction was the improved culture MS+2,4-D (2.0mg/L), with induction frequency up to 83.5%. The differentiation medium was hormone free MS+KT (0.2mg/L) and the differentiation ratio about 74.5%. One hundred percent of roots could be induced under the culture medium as 1/2 MS.

Transgenic plants were obtained by microprojectile bombardment of callus induced from immature inflorescence. The results of PCR and Southern analysis of transgenic plants indicated that the exogenous p5CS gene had integrated into the genome of the plants. RT-PCR assay showed that the p5CS transgene was expressed at a transcript level.

Conclusions Wheatgrass is one of the grasses which are suitable for genetic transformation. The culture system of callus and plant regeneration was optimized. Integration of the exogenous p5CS in the research material was demonstrated by molecular examinations. The transgenic frequency for the p5CS gene was about 0.1%. The transgenic plants obtained provide new genetic resources for further breeding improvement.

A novel genotype independent protocol for in vitro plant regeneration from mature seed derived callus of tall fescue (*Festuca arundinacea* Schreb.)

S. Chennareddy, R.V. Sairam and S.L. Goldman *Plant Science Research Centre, University of Toledo, Toledo, Ohio-43606, USA, Email: srudrab@utnet.utoledo.edu*

Keywords plant regeneration, callus, tall fescue

Introduction Tall fescues (*Festuca arundinacea* Schreb.) are cool season forage and turf grasses of significant agricultural importance in different grassland countries. Genetic improvement of tall fescues by conventional selection procedures is slow, since these are predominantly, cross-pollinated, hexaploid and generally infertile (Jauhar, 1993). Genetic Engineering approaches for incorporation of agronomically useful traits may contribute to the development of improved tall fescue cultivars (Spangenberg *et al.*, 1998). However for any genetic engineering studies, it is essential to develop a genotype-independent, reproducible and efficient *in vitro* plant regeneration protocol. In the present study, we analyzed the effects of different sterilization procedures for *in vitro* seed germination and studied the effects of different concentrations and combinations of 2,4-D and BAP on callus induction, growth and regeneration potential of two cultivars of tall fescue.

Materials and methods Seeds of two varieties of tall fescue, EnviroBlend (FESEVB) and EnviroSHADE (FESEVS) were surface sterilized in 70% ethanol for 1 min followed by various treatments with H_2SO_4 with Bleach and bleach alone or $HgCl_2$ and germinated in petriplates with 20 ml of MS medium (with 0.8% Agar) (Murashige and Skoog, 1962) supplemented with varying concentrations of 2,4-D (2 mg/l, 4mg/l and 6 mg/l). Ten seeds were placed on each petri dish, with 100 seeds per treatment for each cultivar. After one week of seed germination at 24± 2 °C, the emerging shoot and root were chopped to suppress germination and stimulate callus formation and sub cultured onto fresh medium and kept in dark. To evaluate the regeneration potential, the calli were transferred to MS medium with different combinations of hormones and incubated at 24 ± 2 °C under a 16/8-hour dark photoperiod provided by cool-white fluorescent lights at a quantum flux density of 30μmol $s^{-1}m^{-2}$.

Results Surface sterilization of seed with 0.1% $HgCl_2$ for 10 minutes was found to be optimum for seed germination. The treatment with 4.0-mg/l 2, 4-D was found to be the best for callus induction for both the varieties and further increase in 2,4- D concentration decreased the callus induction frequency. Highest frequency of callus induction was observed in FESEVS (88%) followed by FESEVB (86%) on MS medium supplemented with 4.0 mg/l 2, 4-D. Callus was maintained every 4 weeks on MS medium containing 4.0 mg/l 2,4-D and 0.1 mg BAP. Callus turned greenish and shoots appeared after 4 weeks on regeneration medium. Higher frequency of shoots were regenerated per callus clump on MS +0.5 mg/l BAP alone compared to other hormonal combinations in FESEVB (13.07) followed by FESEVS (11.13). When 2, 4, D was present (either 0.1 mg/l or 1.0 mg/l) in regeneration medium a drastic reduction in shoot regeneration frequency was observed. One hundred percent rooting was observed in both the varieties on MS media supplemented with 0.2 mg/l NAA in two weeks. In the combination treatments of NAA and GA_3 all the plants were rooted in FESEVS while only 80% of the plants rooted in FESEVB. When BAP (1.0 mg/l) was added to the rooting medium, either with NAA alone or in combination with NAA and GA_3, a drastic reduction in rooting frequency was observed. No significant differences were observed for both callus induction and shoot regeneration in both varieties. All rooted plants were transferred to the soil and acclimatized in the greenhouse.

Conclusions Previously, *in vitro* regeneration and the production of transgenic plants of tall fescues were done by using suspension cells, which is time consuming and often produces more somaclonal variation. For the first time, we developed a genotype independent, rapid and efficient *in vitro* plant regeneration system in four months from mature seed derived callus of tall fescue making this grass more amenable for genetic engineering studies.

References

Jauhar PP, (1993) Cytogenetics of the Festuca-Lolium Complex. Relevance to Breeding. In Frankel R., Grossman M, Linskens HF, Maliga P, Riley, R (eds) Monographs on Theoretical and applied genetics, Vol. 18, Springer, Berlin Heidelberg New York, 243 pp.

Murashige T and Skoog F. (1962) A revised medium for rapid growth and bioassays with tobacco cultures. Physiol Plant. 15: 473-497.

Spangenberg G, Wang Z Y, Potrykus I, (1998). Biotechnology in Forage and Turf grass improvement. In: Frankel R., Grossman M, Linskens HF, Maliga P, Riley, R (eds) Monographs on theoretical and applied genetics, Vol. 23, Springer, Berlin Heidelberg New York, 200 pp.

Efficient *in vitro* regeneration system from seed derived callus of perennial ryegrass (*Lolium perenne)* and annual ryegrass (*Lolium multiflorum*)

S. Chennareddy, R.V. Sairam and S.L. Goldman *Plant Science Research Center, University of Toledo, Toledo, Ohio-43606, USA, Email: srudrab@utnet.utoledo.edu*

Keywords callus, regeneration, ryegrass

Introduction The commercially important ryegrasses in cool temperate climates throughout the world are annual ryegrass (*Lolium multiflorum* L.) and perennial ryegrass (*Lolium perenne* L). Improvements through conventional breeding have been slow as they are usually heterozygous and highly self-infertile. Hence, there is a need to use modern biotechnological tools to the development of improved rye grass cultivars for incorporating value added traits. Successful transformation of rye grasses has been done using suspension cells, which is time consuming and laborious (Spangenberg *et al.*, 1995, 1998). We report here a rapid and highly efficient *in vitro* plant regeneration system from seed derived callus in annual and perennial rye grasses.

Materials and methods Mature seeds of one variety of annual ryegrass (RGANN) *and* one variety of perennial ryegrass (RGPER), were surface sterilized in 70% ethanol for 1 min followed by various treatments with H_2SO_4 and Bleach, bleach alone or $HgCl_2$ with intermittent shaking. Surface sterilized seeds were germinated on MS medium (Murashige and Skoog, 1962) solidified with agar (0.8%) and supplemented with varying concentrations of 2,4-D (2 mg/l, 4mg/l and 6 mg/l), and petri dishes were kept in dark at 24± 2 °C. Ten seeds were placed on each petri dish (100 x 15 mm) containing 20 ml of solidified MS medium and 100 seed per treatment for each genotype. After one week of seed germination, the emerging shoot and root were chopped to suppress germination and stimulate callus formation. After 6 weeks on fresh medium, the resulting callus was maintained by sub-culturing every 4 weeks. To evaluate the regeneration potential, the calli were transferred to MS augmented with different combinations of plant growth regulators. For regeneration, cultures were incubated at 24 ± 2 °C under a 16/8-hour dark photoperiod.

Results Among various surface sterilization procedures, the treatment with 0.1% $HgCl_2$ for 10 minutes was found to be optimum for seed germination. The treatment with 4.0 mg/l 2,4-D was found to be the best for callus induction for both RGANN and RGPER and with further increase in 2,4- D concentration, the callus induction frequency decreased. Highest frequency of callus induction was observed in RGANN (94%) followed by RGPER (72%) on MS medium supplemented with 4.0 mg/l 2,4-D. Callus initiation and growth took about 6 weeks. Callus was maintained every 4 weeks on MS medium supplemented with 4.0 mg/l 2,4 D and 0.1 mg BAP. Shoot regeneration was observed, after 4 weeks of transferring the callus to regeneration medium, containing various combinations of BAP and 2,4.D. Highest frequency of shoots was regenerated per callus clump on MS +0.5 mg/l BAP in RGPER (13.67). Whereas in RGANN, highest frequency of shoots (11.13) was observed in MS medium with 0.1mg/l BAP. In both the varieties, a decrease in shoot regeneration frequency was observed in the combination treatments of BAP (0.1 mg/l, 0.5 mg/l) and 2, 4-D (0.1 mg/l 1.0 mg/l). Hundred percent rooting was observed when regenerated plants were transferred to MS media supplemented with 0.2 mg/l NAA alone or in combination with 0.5 mg/l GA_3. Rooted plantlets were transferred to soil and acclimatized in the greenhouse.

Conclusions We report here an efficient and rapid (four months) tissue culture regeneration protocol for both annual and perennial ryegrasses.

References

Murashige T and Skoog F. (1962) A revised medium for rapid growth and bioassays with tobacco cultures. Physiol Plant. 15: 473-497.

Spangenberg G, Wang Z Y, Nagel, J., Iglesias, V.A., Potrykus, I. (1995). Transgenic tall fescue (*Festuca arundinaceae*) and red fescue (*Festuca rubra*) plants from microprojectile bombardment of embryogenic suspension cells. J. Plant Physiol. 145:693-701

Spangenberg G, Wang Z Y, Potrykus I, (1998). Biotechnology in Forage and Turf grass improvement. In: Frankel R., Grossman M, Linskens HF, Maliga P, Riley, R (eds) Monographs on theoretical and applied genetics, Vol. 23, Springer, Berlin Heidelberg New York, 200 pp.

Nylon mesh as an improved support for bombarded calli or cell suspensions

S.J. Dalton, P. Robson, M. Buanafina, A.J.E. Bettany, E. Timms, D. Wiffen and P. Morris
Institute of Grassland & Environmental Research, Aberystwyth, Wales, UK. SY23 3EB
Email: sue.dalton@bbsrc.ac.uk

Keywords: genetic transformation, *Poa, Lolium, Festuca*

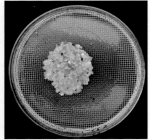

Introduction Using cell suspensions to transform some grass species by particle bombardment has a number of disadvantages including increased somoclonal variation in liquid cell culture and poor performance due to polysaccharide production. The use of calli avoids these problems, but the manipulation of calli through numerous media changes is laborious and time-consuming. We investigated a possible mechanism to facilitate the use of calli in transformation by immobilising calli on mesh.

Materials and methods A range of nylon meshes from 0.5-2mm (Cadisch) were tested to look at callus formation in pre-arranged targets. Embryogenic primary calli were transferred onto petri-dish sized mesh circles placed on callus medium and were arranged to form a target. After three to seven days growth the meshes were transferred to high osmolarity medium prior to bombardment in a particle inflow gun using parameters which varied depending on species and tissue. The following day the meshes were transferred back to callus induction medium and typically a week later the calli were transferred from the mesh to solid selection media. Cell suspension cultures were also tested and either plated to form pre-arranged targets on mesh as with calli and after bombardment were subsequently treated as callus, or were bombarded as usual, returned to cell suspension for initial selection and only plated onto mesh for regeneration. The results in Table 1 were collated over a large number of co-transformation experiments involving a number of genes of interest and two selection systems.

Results The calli of *Lolium multiflorum, Agrostis stolonifera, Festuca rubra* and *Poa pratensis* formed a thick mat of callus and grew through the mesh, leaving very little tissue behind when the mesh was moved. By contrast *Festuca arundinacea, Lolium perenne* and *Lolium temulentum* calli did not grow through the mesh and small pieces of callus were frequently left behind after transfer. Mesh of 1mm (Figure 1) proved best overall (data not shown). Enmeshed calli could be bombarded at higher pressures without disturbing the callus. The calli remained in the same orientation throughout recovery, whereas calli which are disturbed by bombardment may land with the bombarded surface face down in the medium and thus lose embryogenic potential. For those species where there is no alternative to callus, mesh was more convenient to use even if regeneration rates were not much improved (Table 1). In *F.arundinacea* and *L.multiflorum* which form good cell suspensions, better results were achieved with cell suspensions than with calli, whether plated on mesh to form a target or plated on mesh during selection. This allowed the suspension colonies to dry out slightly and encouraged embryogenesis.

Table 1 Regeneration of transgenic grass plants from bombarded callus and suspension cultures grown on mesh

Species and Tissue	Mesh	Successful bombardments	Percentage	No. of plants	No. plants per bombardment	Selection
F.arundinacea. Callus target	mesh	1/16	6	2	0.1	*CaMV-hpt*
Suspension as callus target	mesh	14/74	19	14	1.9	*CaMV-hpt*
Suspension plated during selection	control	45/56	80	102	1.8	*CaMV-hpt*
	mesh	16/16	100	38	2.4	*CaMV-hpt*
L.multiflorum Callus target	mesh	0/25	0	0	0	*CaMV-hpt*
Suspension as callus target	mesh	12/31	39	20	0.6	*CaMV-hpt*
Suspension plated during selection	control	21/63	33	34	0.5	*CaMV-hpt*
	mesh	6/18	33	8	0.4	*CaMV-hpt*
L.perenne callus target	control	3/17	18	3	0.2	*CaMV-hpt*
	mesh	7/45	15	8	0.2	*CaMV-hpt*
P.pratensis callus target	control	1/24	4	1	<0.1	*Ubi-nptII*
	mesh	13/78	17	30	0.4	*Ubi-nptII*

Conclusions The use of mesh to immobilise calli improves transformation by allowing large numbers of targets to be prepared, and by simplifying manipulations it allows treatments to be timed more precisely. Mesh also improves transformation in some cell suspensions. The technique was generally applicable but has proved particularly useful for the transformation of *Lolium.perenne, Poa pratensis, Festuca rubra* and *Agrostis stolonifera*.

A comparison of hygromycin and paromomycin selection strategies in the genetic transformation of seven *Lolium, Festuca, Poa,* and *Agrostis* species

S.J. Dalton, P. Robson, M. Buanafina, A.J.E. Bettany, E. Timms and P. Morris
Institute of Grassland & Environmental Research, Aberystwyth, Wales, UK. SY23 3EB
Email: sue.dalton@bbsrc.ac.uk

Keywords: genetic transformation, *Poa, Lolium, Festuca, Agrostis*

Introduction Hygromycin selection for the *hpt* gene, expressed from the CaMV-35S promoter, has been successful in transgenesis of a limited number of grass species. As an alternative to *hpt* selection Altpeter *et al.,* (2000) reported successful transformation using paromomycin selection for the *nptII* gene expressed by the maize ubiquitin promoter. We have tested the utility of a number of selection cassettes using previously sporadically transformable species which nevertheless had very good tissue culture and regeneration protocols.

Materials and methods Callus and cell suspension cultures were co-transformed with genes of interest and either the *hpt* gene encoding hygromycin resistance expressed from the CaMV-35S promoter, or the *nptII* gene encoding paromomycin resistance expressed from the maize ubiquitin promoter. Transformations were performed by bombardment in a particle inflow gun at various parameters depending on the tissue and species involved (Dalton, 1999). The results in Table 1 were collated from a large number of experiments and no account was taken of the effect of different genes of interest on regeneration except *L.multiflorum* and *F.arundinacea*, where large numbers allowed comparisons with co-transformations involving similar genes of interest.

Results There was genotype-associated variation in resistance to both paromomycin and hygromycin of cell cultures. However, paromomycin did not inhibit embryogenesis as much as hygromycin and many cultures in which regeneration was inhibited were still able to develop embryoids. Increasing the numbers of subcultures to fresh paromomycin-containing media reduced the numbers of escapes. Transformation frequency in terms of the number of bombardments in which plants were regenerated was greater in *L.perenne* when *hpt* was expressed from the CaMV-35S promoter and when *nptII* was expressed from the maize ubiquitin promoter. Selection with the *ubi-nptII* construct improved the recovery of transformants in all species.

Table 1 Transgenic plant regeneration under hygromycin (*CaMV-hpt*) and paromomycin (*ubi-nptII*) selection

Species	Tissue	Selection	Successful bombardments	Percentage	No. of plants	No. plants per bombardment
F.arundinacea	cell suspension	CaMV-hpt	60/108	56	112	1.0
F.arundinacea	cell suspension	ubi-nptII	20/44	45	63	1.4
L.multiflorum	cell suspension	CaMV-hpt	21/63	33	34	0.5
L.multiflorum	cell suspension	ubi-nptII	18/25	72	35	1.4
L.temulentum	cell suspension	CaMV-hpt	18/161	11	37	0.2
L.temulentum	cell suspension	ubi-nptII	36/91	40	58	0.6
L.perenne	callus	CaMV-hpt	10/67	15	12	0.2
L.perenne	callus	CaMV-nptII	0/15	0	0	0
L.perenne	callus	ubi-hpt	1/16	6	1	0.1
L.perenne	callus	ubi-nptII	7/20	35	7	0.4
P.pratensis	callus	CaMV-hpt	5/49	10	8	0.2
P.pratensis	callus	ubi-nptII	14/108	13	31	0.3
F.rubra	callus	CaMV-hpt	1/7	14	1	0.1
F.rubra	callus	ubi-nptII	5/9	55	6	0.7
A.stolonifera	callus	CaMV-hpt	5/53	9	9	0.2
A.stolonifera	callus	ubi-nptII	6/26	23	6	0.2

Conclusions Different commonly used constitutive promoters improve transformation frequencies when selecting for different marker genes, suggesting that relative differences in expression are important for different selection regimes. However, over the 7 grass species use of the *ubi-nptII* construct generally doubled the mean number of successful bombardments from 21% to 40% and the number of plants regenerated per bombardment from 0.34 to 0.71.

References

Altpeter *et al.* (2000). Generation of large numbers of independently transformed fertile ryegrass (*Lolium perenne* L.) plants of forage and turf-type cultivars *Mol Breeding.6:519-528*
Dalton *et al.* (1999) Co-transformed, diploid *Lolium perenne* (perennial ryegrass), *Lolium multiflorum* (Italian ryegrass) and *Lolium temulentum* (Darnel) plants produced by microprojectile bombardment. *Plant Cell Reports 18:721-726*

Agrobacterium tumefaciens-mediated transformation of perennial ryegrass (*Lolium perenne* L.)

H. Sato, M. Fujimori, Y. Mano, T. Kiyoshi and T. Takamizo
National Institute of Livestock and Grassland science, 768 Senbonmatsu, Nasushiobara, Tochigi, 329-2793, Japan. Email: s.hiroko@affrc.go.jp

Keywords: *Agrobacterium tumefaciens*, transformation, perennial ryegrass, GUS

Introduction An *Agrobacterium tumefaciens*-mediated transformation method has several advantages. However, this method has no example of success in perennial ryegrass (*Lolium perenne* L.). Since *Lolium* species are outcrossing, one cultivar consists of many genotypes. Each genotype can show a different ability for callus formation and plant regeneration (Takahashi *et al.*, 2004). Thus, it is important to select a good genotype for efficient and stable transformation. If the plant is maintained *in vitro*, we can perform transformation using calli induced from shoot tips of the same genotype at any time. Our objective is to confirm an *A. tumefaciens*-mediated transformation method for perennial ryegrass and to screen for suitable genotype.

Materials and methods Embryogenic calli were induced from shoot tips of each 50 genotypes of perennial ryegrass cultivars Saturn and Norlea. These calli were infected with *A. tumefaciens* strain EHA101 harboring a binary vector pIG121Hm and co-cultured in the presence of 100μ M acetosyringone for 4 days. The infected calli were cultured on MS selective medium containing 250mg^{-1} carbenicillin and 100mg^{-1} hygromycin. After 6-8 weeks, the hygromycin-resistant calli were transferred to MS regeneration medium containing 0.2mg^{-1} kinetin. Transformation was confirmed by histochemical GUS assay and PCR analysis of HPT and GUS gene.

Figure 1 Structure of the T-DNA region of pIG121Hm

RB and LB, right and left border; Pnos and Tnos, promoter and terminator of nopaline synthase gene; P35S, promoter of CaMV 35S RNA gene; NPT‖, neomycin phosphotransferase ‖ gene; GUS, β-glucuronidase gene; HPT, hygromycin phosphotransferase gene

Results Hygromycin-resistant calli and regenerated plants were obtained from several genotypes of Saturn and Norlea. Histochemical GUS assay showed that calli after 4 days co-cultivation and hygromycin-resistant calli were both GUS positive. Regenerated plants were subjected to PCR analysis and a single fragment of both 643bp for GUS and 375bp for HPT were detected at the equivalent sizes to the fragment from pIG121Hm. No amplification product was recognized from the non-transformed plant.

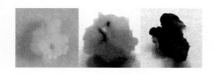

Figure 2 Expression of GUS gene (Saturn No.7)
A: non-transformed callus. B: callus after 4 days co-cultivation. C: hygromycin-resistant callus

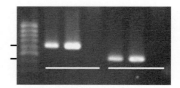

Figure 3 Detection of GUS and HPT gene by PCR (Norlea No.35)
M, 100bp DNA ladder; TP, transformed plant; P, pIG121Hm plasmid DNA; NP, non-transformed plant

Conclusions We established an *A. tumefaciens*-mediated transformation method of perennial ryegrass using selected genotypes with high transformation efficiency and plant regeneration.

References

Takahashi, W., T. Komatsu, M. Fujimori & T. Takamizo (2004). Screening of regenerable genotypes of Italian ryegrass (*Lolium multiflorum* Lam.). *Plant Production Science*, 7: 55-61.

Manipulating the phenolic acid content and digestibility of forage grasses by targeted expression of fungal cell wall degrading enzymes

M.M. de O. Buanafina, P. Morris, T. Langdon, S. Dalton, B. Hauck and H. Porter
Plant Animal and Microbial Science Department, Institute of Grassland & Environmental Research, Aberystwyth, SY23 3EB, UK. Email:marcia.buanafina@bbsrc.ac.uk

Keywords: forage grass, FAE, Xylanase, phenolics

Introduction Grass cell walls constitute 30-80% of forage dry matter, representing a major source of energy for ruminants. Ferulic acid (4-hydroxy-3-methoxy-cinnamic acid) and other hydroxycinnamic acids are ester linked to arabinosyl residues in arabinoxylans of grass cell walls and undergo oxidative coupling reactions resulting in the formation of a variety of dehydrodiferulate dimers which cross-link cell wall polymers. Although such cross-links have a number of important roles in the cell wall, they also hinder the rate and extent of cell wall degradation by ruminant microbial and fungal enzymes. We have shown previously the expression of a ferulic acid esterase gene from *Aspergillus niger* in *Festuca arundinacea* and the potential of the expressed FAE to break phenolic cross-links and release monomeric and dimeric ferulic acids on cell death in vacuole targeted FAE plants. This was enhanced several fold by the addition of exogenous recombinant xylanase (Buanafina *et al.*, 2002). We propose to decrease the level of phenolic cross-linking of cell wall carbohydrate by inducible expression of FAE to the apoplast, ER and golgi and by co-expressing FAE and endo-ß-1,4-xylanase from *Trichoderma reesei* to the apoplast and vacuole.

Material and methods Young *F.a.* suspension cultures were bombarded with plasmid DNA using a Particle Inflow gun (PIG) as in Dalton *et al.* (1999). FAE activity was determined as the amount of FA released under incubation with EF (ethyl 4-hydroxy-3-methoxycinnamate) as substrate for 24hr. Xylanase activity was determined measuring the absorbance at 590nm of plant extract after incubation with Azo-Xylan as substrate for 22 hrs. Ester bound compounds were extracted with NaOH under N_2, acidified and analysed by HPLC.

Results and discussion We found that targeting FAE to the apoplast and ER/golgi system resulted in a significant reduction in the levels of monomeric and dimeric cell wall phenolics in leaves of some plants when expressed constitutively. Apoplast targeting might be expected to directly affect cell wall composition by removing ferulic acids, whereas ER/golgi targeting may affect cell wall composition, indirectly, by reducing feruoylation of the arabinoxylans in the ER/golgi destined to the cell wall. We also show the potential of expressed FAE to break phenolic cross-links in vacuole targeted FAE plants, leading to increased initial rates of fermentation. We have now produced *F. a.* plants expressing endo-ß-1,4-xylanase from *Trichoderma reesei* co-transformed with FAE to determine whether co-expression of xylanase will increase further release of ferulates and increase digestibility.

References

Dalton S.J., Bettany, A.J.E., Timms, E., Morris,P. (1999) Co-transformed diploid *Lolium perenne*(perennial ryegrass), *Lolium multiflorum* (Italian ryegrass) and *Lolium temulentum* (darnel) plants produced by microprojectile bombardment. Plant Cell Reports 18:721-726.

Buanafina, M.M de O., Langdon, T., Hicks, H., Hauck, B., Dalton, S.J. and Morris, P. Targeted expression of a ferulic acid esterase from *Aspergillus niger* in leaves of forage grasses. In: Proc. 19th European Grassland Federation – General Meeting La Rochelle (France) 27030 May 2002. Pg. 66-67.

Buanafina, M.M de O., Langdon, T., Hicks, H., Hauck, B., Dalton, S.J. and Morris, P. Manipulating the phenolic composition of plant cell walls: targeted expression of an *Aspergillus niger* ferulic acid esterase in leaves of forage grasses. In: X Cell Wall Meeting Sorrento (Italy) August 29-September 3, 2004. Pg.18.

Improving forage quality of tall fescue (*Festuca arundinacea*) by genetic manipulation of lignin biosynthesis

Z.-Y. Wang, L. Chen, C.-K. Auh, A. Hopkins and P. Dowling
Forage Improvement Division, The Samuel Roberts Noble Foundation, 2510 Sam Noble Parkway, Ardmore, Oklahoma 73401, USA. Email: zywang@noble.org

Keywords: digestibility, grass, lignin, tall fescue, transgenic plant

Introduction Lignification of plant cell walls is a major factor limiting forage digestibility and concomitantly animal productivity. Improvement in forage grass cell wall digestibility has become an important goal of many plant-ruminant animal research programs. Lignins are complex phenolic heteropolymers associated with the polysaccharidic components of the wall in specific plant cells. Lignin in forage grasses comprises guaiacyl (G) units derived from coniferyl alcohol, syringyl (S) units derived from sinapyl alcohol, and *p*-hydroxyphenyl (H) units derived from *p*-coumaryl alcohol. Cinnamyl alcohol dehydrogenase (CAD) and caffeic acid *O*- methyltransferase (COMT) are key enzymes involved in lignin biosynthesis. Tall fescue is the predominant cool-season forage grass in the United States.

Materials and methods Lignification in stems of tall fescue (cv. Kentucky 31) were analyzed at three elongation and three reproductive stages. CAD and COMT cDNA sequences were cloned from tall fescue by screening a cDNA library. Transgenic tall fescue plants carrying either sense or antisense CAD and COMT gene constructs were obtained by microprojectile bombardment of single genotype-derived embryogenic suspension cells. Detailed molecular and biochemical analyses were carried out for the transgenic plants.

Results Anatomical comparisons of tall fescue stems at six different developmental stages revealed a gradient increase in lignification with progressive maturity. Digestibility is negatively correlated with lignin content, S lignin content, as well as S/G ratio. Relative *O*-methyltransferase activities increased during stem development, and in parallel with the lignification process of stem. The expression of COMT and CAD genes increased during the stem elongation stage and remained at high levels during the reproductive stages. The changes at anatomical, metabolic and molecular levels during plant development were closely associated with lignification and digestibility (Chen *et al.*, 2002).

Analysis of transgenic tall fescue plants by northern hybridization revealed that several plants had severely reduced mRNA levels. Enzyme activity analysis using different substrates showed that these transgenic plants had significantly decreased COMT or CAD enzymatic activities. These CAD and COMT down-regulated tall fescue plants had reduced total lignin content and altered lignin composition. No significant changes in cellulose, hemicellulose, neutral sugar composition, *p*-coumaric acid and ferulic acid levels were observed in the transgenic plants. In vitro dry matter digestibility increased by 7.2% to 10.5% in the transgenic lines, thus providing novel germplasm to be used for the development of grass cultivars with improved forage quality (Chen *et al.*, 2003, 2004).

Conclusions Consistent and closely related molecular and biochemical data demonstrated that the transgenic CAD and COMT lines were down-regulated in their lignin biosynthesis and had improved forage digestibility. Genetic manipulation of lignin biosynthesis is an effective approach to improve digestibility of grasses.

References

Chen, L., C. Auh, F. Chen, X.F. Cheng, H. Aljoe, R.A. Dixon & Z.-Y. Wang (2002). Lignin deposition and associated changes in anatomy, enzyme activity, gene expression and ruminal degradability in stems of tall fescue at different developmental stages. *Journal of Agricultural and Food Chemistry*, 50, 5558-5565.
Chen, L., C. Auh, P. Dowling, J. Bell, F. Chen, A. Hopkins, R.A. Dixon & Z.-Y. Wang (2003). Improved forage digestibility of tall fescue (*Festuca arundinacea*) by transgenic down-regulation of cinnamyl alcohol dehydrogenase. *Plant Biotechnology Journal*, 1, 437-449.
Chen, L., C. Auh, P. Dowling, J. Bell, D. Lehmann & Z.-Y. Wang (2004). Transgenic down-regulation of caffeic acid *O*-methyltransferase (COMT) led to improved digestibility in tall fescue (*Festuca arundinacea*). *Functional Plant Biology*, 31, 235-245.

Crown rust resistance in transgenic Italian ryegrass (*L. multiflorum*) expressing a rice chitinase gene and crosses with cytoplasmic male sterile hybrid ryegrass

W. Takahashi[1], M. Fujimori[2], Y. Miura[1], T. Komatsu[2], S. Sugita[2], A. Arakawa[2], Y. Nishizawa[3], H. Sato[2], Y. Mano[2], T. Hibi[4] and T. Takamizo[2*]

[1]Japan Grassland Farming and Forage Seed Association, Forage Crop Research Institute, 388-5 Higashiakada,Nishinasuno, Tochigi 329-2742, Japan
[2]National Institute Livestock and Grassland Science, Nishinasuno, Tochigi, 329-2793, Japan. Email: takamizo@affrc.go.jp
[3]National Institute of Agrobiological Sciences, 2-1-2 Kannondai, Tsukuba, Ibaraki 305-8602, Japan 4. Tamagawa University, Faculty of Agriculture, Machida, Tokyo, 194-8610, Japan

Keywords: disease resistance, *Puccinia coronata*, particle bombardment, cytoplasmic male sterility

Introduction. Italian ryegrass (*Lolium multiflorum* Lam.) is one of the most important forage grasses in the temperate region. In ryegrasses, crown rust (*Puccinia coronata*) is the most serious foliar fungal disease and brings about a reduction of herbage yield and loss of palatability to grass-eating domestic animals. In this study, we tried to increase tolerance to the pathogen by introducing a rice chitinase gene using particle bombardment.

Materials and methods Genetic co-transformation of Italian ryegrass (W3 genotype from cultivar Waseaoba) was performed with the *HPT* gene (pAcH1) and a rice chitinase gene (pAcC-RCC2) (Takahashi *et al.* in press). Hygromycin resistant transgenic plants were checked by PCR and Southern hybridisation was used to determine whether they possessed the rice chitinase gene. Northern analysis was also carried out. Transgenic plants carrying the rice chitinase gene were evaluated for their tolerance to crown rust by the inoculation of detached leaves with spores. Flowering transgenic plants carrying rice chitinase genes were crossed to a cytoplasmic male-sterile hybrid ryegrass in order to produce pollen-less transgenic disease resistant breeding material.

Results We checked the integration of foreign genes by PCR analysis in the 72 regenerants that formed green shoots. Of these, 58 plants were positive for the *HPT* gene, and 39 bore the *RCC2* gene. The different Southern hybridization patterns observed among the transgenic plants indicated that they resulted from independent transformation events. A stronger Northern hybridization band was observed in most transgenic plants compared to non-transgenic plants, but we could not distinguish the *RCC2* gene transcript from that of the endogenous chitinase gene (Figure1). In both non-transgenic and *RCC2* gene expressing transgenic plants, macroscopic regions of chlorosis were observed about 7 days after inoculation with *P. coronata* spores. However we were able to distinguish differences in the severity of disease symptoms between them by 10 days after inoculation. The lowest number of lesions was observed in plant 67 (Figure 2). However, the disease continued to gradually invade the inoculated leaves of the transgenic plants as the culture period was prolonged (>10 days), and the disease symptoms of the transgenic plants developed to the same degree as those of the non-transgenic plants about 13 days after inoculation. Transgenic plants expressing the *RCC2* gene were fertile and were crossed to a cytoplasmic male-sterile hybrid ryegrass (F_1 of Italian ryegrass carrying male-sterile cytoplasm x perennial ryegrass), whose progenies show stable male sterility. Inheritance of the *RCC2* gene in the progenies was confirmed by PCR. Tolerance to rust and male sterility is being checked further in the progenies.

C 4 8 23 24 51 55 58 67 98 99 102 105

Chitinase

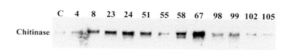

Conclusions Transgenic Italian ryegrass plants carrying the rice chitinase (*Cht-2*; *RCC2*) gene were obtained by particle bombardment and its expression was confirmed by molecular analysis. Bioassay of detached leaves indicated increased resistance to crown rust (*Puccinia coronata*) in the transgenic plants. Progenies between the transgenic plants and cytoplasmic male-sterile hybrid ryegrass were obtained.

References
Takahashi, W., M. Fujimori, Y. Miura, T. Komatsu, Y. Nishizawa, T. Hibi and T. Takamizo (2005) Increased resistance to crown rust disease in transgenic Italian ryegrass (*Lolium multiflorum* Lam.) expressing the rice chitinase gene. Plant Cell Reports, *In press.*

Discovery, isolation and characterisation of promoters from perennial ryegrass (*Lolium perenne*)

A. Lidgett[1,2], N. Petrovska[1,2], J. Chalmers[1,2], N. Cummings[1,2] and G.C. Spangenberg[1,2]

[1]*Primary Industries Research Victoria, Plant Biotechnology Centre, La Trobe University, Bundoora, Victoria 3086, Australia* [2]*Molecular Plant Breeding Cooperative Research Centre, Australia*
Email: german.spangenberg@dpi.vic.gov.au

Keywords: perennial ryegrass, bacterial artificial chromosomes (BACs), BAC library, promoters, reporter genes

Introduction The availability of a suite of promoters with a range of spatial, temporal and inducible expression patterns is of significant importance to enable targeted expression of genes of interest for molecular breeding of forage species. A range of resources and tools have been developed for promoter isolation and characterisation in perennial ryegrass (*Lolium perenne* L.) including genomic lambda and BAC libraries and a 15 K unigene microarray.

Materials and methods Discovery, isolation and characterisation of heterologous and endogenous promoters were undertaken in perennial ryegrass. Expression patterns of chimeric *gusA* reporter genes encoding bacterial β-glucuronidase (GUS) with two pollen-specific promoters from maize (Zm13, CDPK) were evaluated in transgenic ryegrass plants generated by biolistic transformation of embryogenic cells. The promoter of the gene encoding a Lolp2 pollen allergen from perennial ryegrass was isolated and its pollen-specificity was demonstrated in transgenic tobacco plants.

Results A perennial ryegrass BAC library consisting of 50,304 BAC clones with 113 kb average insert size, corresponding to 3.4 genome equivalents and 97% genome coverage was established. Upstream regulatory sequences of genes involved in lignin biosynthesis and fructan metabolism in perennial ryegrass were identified, isolated and characterised *in planta* as *gusA* transcriptional fusions. Chimeric *gusA* genes under control of the 5' regulatory sequences isolated from perennial ryegrass genes encoding caffeic acid *O*-methyltransferase (*LpOMT1*) and 4-coumarate-CoA ligase (*Lp4CL2*) revealed strong GUS activity in vascular tissue of transgenic tobacco plants. The promoter from the perennial ryegrass sucrose:sucrose 1-fructosyltransferase (*Lp1-SST*) conferred strong constitutive expression in leaves while the regulatory sequences from the perennial ryegrass fructosyltransferase homologue *LpFT1* showed weaker expression in stem bases. Conserved regulatory elements identified in genes showing inducible (e.g. by light, drought and sucrose) or cell-type specific (e.g. pollen cells) expression were identified in promoter sequences from perennial ryegrass genes (Figure 1).

Conclusions This research provides a toolbox of promoters with a range of specificities for targeted gene expression as part of a molecular breeding approach in perennial ryegrass deploying exclusively ryegrass genes and promoters for transgenic product development.

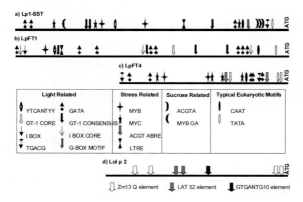

Figure 1 Analysis of perennial ryegrass regulatory sequences. a) 1.6 kb *Lp1-SST* promoter; b) 1.6 kb *LpFT1* promoter; c) 0.9 kb *LpFT4* promoter; and d) 0.95 kb *Lol p 2* promoter

Development and field evaluation of transgenic ryegrass (*Lolium* spp.) with down-regulation of main pollen allergens

N. Petrovska[1,2], A. Mouradov[1,2], Z.Y. Wang[1,3], K.F. Smith [2,4] and G.C. Spangenberg[1,2]

[1]*Primary Industries Research Victoria, Plant Biotechnology Centre, La Trobe University, Bundoora, Victoria 3086, Australia* [2]*Molecular Plant Breeding Cooperative Research Centre, Australia* [3]*The Samuel Roberts Noble Foundation, Ardmore, Oklahoma, USA* [4]*Primary Industries Research Victoria, Hamilton Centre, Hamilton, Victoria 3300, Australia Email: natasha.petrovska@dpi.vic.gov.au*

Keywords: ryegrass, pollen allergens, transgenic plants, hypo-allergenic pollen, field evaluation

Introduction Ryegrass (*Lolium* spp.) pollen is a widespread source of airborne allergens and is a major cause of hayfever and seasonal allergic asthma, which affect approximately 25% of the population in cool temperate climates. The main allergens of ryegrass pollen are the proteins Lol p 1 and Lol p 2. These proteins belong to two major classes of grass pollen allergens to which over 90% of pollen-allergic patients are sensitive. The functional role *in planta* of these pollen allergen proteins remains largely unknown. The generation, analysis and field evaluation of transgenic plants with reduced levels of the main ryegrass pollen allergens, Lol p 1 and Lol p 2 in the most important worldwide cultivated ryegrass species, perennial ryegrass (*L. perenne* L.) and Italian ryegrass (*L. multiflorum* Lam.) are described.

Materials and methods *Lol p 1* and *Lol p 2* cDNA and genomic clones were isolated. Transformation vectors were generated with *Lol p 1* and *Lol p 2* cDNA sequences in antisense orientation under the control of maize and ryegrass pollen-specific promoters.

Results and conclusions Embryogenic suspension cells of perennial and Italian ryegrass were subjected to biolistic transformation with *Lol p 1* and *Lol p 2* antisense vectors and transgenic plants were recovered. The transgenic nature of the perennial and Italian ryegrass plants was confirmed by Southern hybridisation analysis. Transgenic antisense *Lol p 1* and *Lol p 2* ryegrass plants showed a reduction in the levels of the respective pollen allergens assessed with antibodies raised against the recombinant allergenic proteins. Hypo-allergenicity of Lol p 1 down-regulated pollen was confirmed by immunoblots using IgE sera from Lol p 1 sensitised patients. Transgenic antisense *Lol p 1* ryegrass plants showed normal reproductive development and pollen viability. Selected antisense *Lol p 1* transformation events were evaluated in a small-scale field release carried out in Ardmore, Oklahoma, USA in 2004. Mitotic and meiotic stability of transgene integration and expression, as well as general morphology of the field-grown transgenic ryegrass plants were assessed. A more comprehensive assessment of pollen and gene flow of these transgenic ryegrass plants will be evaluated in a planned large-scale field release in USA in 2005. This will complement detailed gene and pollen flow studies undertaken in Hamilton, Australia using non-transgenic novel perennial ryegrass genotypes.

Figure 1 Western blot analysis of pollen protein extracts: a) using anti-Lol p 1 antibodies; untransformed control ryegrass plant (1) and Lol p1 down-regulated transgenic ryegrass plant; b) using anti-Lol p 2 antibodies; untransformed control ryegrass plant (1) and Lol p2 down-regulated transgenic ryegrass plant; c) using IgE antibodies from serum of grass pollen allergic patient; perennial ryegrass control plant (1), Italian ryegrass control plant (2) and Lol p1 down-regulated transgenic ryegrass plant (3)

Reference

Petrovska, N., Wu, X., Donato, R., Wang, Z.Y., Ong, E.K., Jones, E., Forster, J., Emmerling, M., Sidoli, A., O'Hehir, R. and Spangenberg, G. (2004) Transgenic ryegrasses (*Lolium* spp.) with down-regulation of main pollen allergens. *Molecular Breeding*, 14, 489-501.

Shutting the stable door after the horse has bolted? Risk assessment and regulation for transgenic forages

C.J. Pollock
Institute of Grassland and Environmental Research, Plas Gogerddan, Aberystwyth SY23 3EB, Wales UK.
Email: chris.pollock@bbsrc.ac.uk

Keywords: transgenic forages, regulation, costs, benefits, gene flow

Introduction The regulation of GM agriculture in Europe is a matter of considerable public interest. The development of a regulatory framework has imposed significant additional costs without generating broad public acceptance. However, the current risk assessment and management framework does provide a basis for considering how the broader issues surrounding the introduction of any novel agricultural technology might be approached.

The Status Quo The regulations are administered EU-wide on a case-by-case basis and are based upon an assessment of the likelihood of the release causing harm to human health and the environment. Consents to release into the food chain are time-limited, revocable and require a programme of post-market monitoring to validate the original risk assessment. Regulatory approval also involves an assessment of the wider impacts of changed agricultural practices associated with a GM crop release. The farm-scale trials of GMHT crops showed that use of broad-spectrum herbicides for weed control impacted on the non-agricultural food chain in a highly crop-dependent manner, with conventional crops being both the best and worst for sustaining wildlife. (Squire *et al.,* 2003 and references therein). The expectation is that future applicants would need to provide information that would allow such wider impacts to be assessed with a similar level of certainty. Given the significance of grassland as a land use system in Western parts of Europe, it is likely that such information would comprise a central element of any dossier on GM forages.

Future Challenges The current regulatory regime has two shortcomings. Firstly only harm is assessed, so it is not possible to argue that adverse impacts in one area are more than offset by advantages in another. Secondly, it only applies to one strand of modern intensive farming. Systems with equal or greater impact (e.g. mutation-bred HT crops) could be introduced without this regulatory process. In the UK, ACRE (the Advisory Committee for Releases into the Environment) has established a sub-group to examine some of these issues in terms of the forthcoming review of the regulatory system (http://www.defra.gov.uk/environment/acre/meetings/04/min040715.htm).

Implications for transgenic forages GMHT forage maize has, in principle, been approved for cultivation in the EU. It is, however, an annual with no wild relatives. A recent study on pollen movement in bent grass (Watrud *et al.,* 2004) has indicated how significant gene flow would be when there is a combination of obligate outbreeding and a large pool of compatible wild plants. A previous analysis of ryegrass and *Agrostis* populations using isozyme polymorphisms also established significant long-distance genetic interactions in ryegrass associated with breeding and cultivation that were absent in *Agrostis* (Warren *et al.,* 1998). Risk assessment for GM forages would, therefore, need to concentrate on the impact of the transgene in the natural environment, an interaction that is highly specific to the nature of the transgene. Development of transgenic forages in Europe is likely to be problematic without (a) a change in the regulations to allow consideration of benefit (b) the identification and development of traits that produce such a benefit and (c) the development of more effective genetic and management mechanisms to regulate gene flow.

References

Squire, G.R., D.R. Brooks, D.A. Bohan, G.T. Champion, R.E. Daniels, A.J. Haughton, C.Hawes, M.S. Heard, M.O. Hill, M.J. May, J.L. Osborne, J.N. Perry, D.B. Roy, I.P. Woiwod & L.G. Firbank (2003). On the rationale and interpretation of the Farm Scale Evaluations of genetically-modified herbicide-tolerant crops. *Philosophical Transactions of the Royal Society, Series B,* 358, 1779-1801.

Watrud, L.S., E.H. Lee, A. Fairbrother, C Burdick, J.R. Reichman, M. Bollman, M. Storm, G. King & P.K. Van de Water (2004) Evidence for landscape-level, pollen-mediated gene flow from genetically modified creeping bentgrass with *CP4 EPSPS* as a marker. Proceedings of the National Academy of Sciences (in press).

Warren,JM, A.F. Raybould, T. Ball, A.J. Gray & M.D. Hayward (1998) Genetic structure in the perennial grasses *Lolium perenne* and *Agrostis curtisii. Heredity* 81, 556-562.

Assessing the risk posed by transgenic virus-resistant *Trifolium repens* to native grasslands in Southeast Australia

R.C. Godfree, P.W.G Chu and A.G. Young
CSIRO Plant Industry, GPO Box 1600, Canberra, ACT 2601, Australia; Email: robert.godfree@csiro.au

Keywords: transgenic virus resistance, transgenic white clover, ecological risk assessment

Introduction In Australia, comprehensive environmental risk assessments must be performed on transgenic plants (GMOs) prior to their commercial release. A key element is the determination of whether the release of a particular GMO poses any weediness threat to the environment or other agricultural systems, which can occur by means of direct invasion or by introgression of transgenes into wild populations of the same or closely related species. For transgenic pasture plants this question could be of added importance because many of these species have been selected for traits encouraging long-term persistence and competitiveness in complex plant communities (Godfree *et al.*, 2004a). In situations where native vegetation is of high conservation value, such as Australia, the potential for transgenic pasture plants to invade native plant communities must therefore be quantified and analysed within a rigorous risk assessment framework. Over the past three years we have investigated the level of risk posed by transgenic virus-resistant (VR) *Trifolium repens* (white clover) to native grasslands and woodlands in the subalpine and montane regions of southeastern Australia. We have focused on identifying the viruses present in white clover populations in the subalpine zone, on determining the floristic composition of the communities that are most at risk, and on quantifying the likely selective advantage of VR *T. repens* in these environments.

Materials and methods Our assessment of the potential weediness of VR *T. repens* has involved three phases: 1) determination of the prevalence of *Alfalfa mosaic virus* (AMV), *Clover yellow vein virus* (ClYVV) and *White clover mosaic virus* (WClMV) in 31 wild populations of *T. repens* over a large part of the subalpine region of NSW, Australia, 2) use of surveys and plot-based coverage and biomass collection to identify communities and species that are of highest risk of further invasion by *T. repens*, and the functional significance of *T. repens* in these environments, and 3) initiation of a large-scale glasshouse and field trials comparing the performance of white clover plants both uninfected and infected by ClYVV, which is the dominant virus in the subalpine region. The field trial, which is ongoing, is being conducted at multiple spatial scales and in four distinct grassland and woodland communities.

Results Census results show that *Clover yellow vein virus* is the most abundant virus of white clover in the subalpine zone, being present in 80% of sites investigated in both grassland and woodland environments and infecting on average 18% of plants (range 0% to 59%). AMV and WClMV were present at only one site, and in lower frequencies (Godfree *et al.*, 2004b). These data suggested that selective pressure would likely be greatest on ClYVV-resistant *T. repens* in the subalpine environment. Ecological survey work conducted on wild *T. repens* populations indicated that moist, high-fertility *Poa* grasslands and *Eucalyptus-Poa* woodlands are the most heavily invaded communities, with *T. repens* comprising up to 25% cover and being more abundant in many plots than the nearest herbaceous native species (Godfree *et al.*, 2004c). Both of these communities contain a range of native species that appear to occupy a similar functional niche to *T. repens*. Finally, preliminary data suggest that the impact of ClYVV on growth, survival and fecundity of *T. repens* can be large in the glasshouse (up to 60% reduction in some parameters), but that these impacts, while detectable, are more transient and of smaller magnitude in native plant communities, primarily a result of the dominant climatic and topo-edaphic impacts on white clover in subalpine environments.

Conclusions Our work to date indicate that *T. repens* is a significant component of subalpine plant communities in south-eastern Australia, and that wild populations contain large amounts of *Clover yellow vein virus*. Given the presence of both host and pathogen in threatened plant communities in this region, development of an understanding of the importance of ClYVV in limiting the size of white clover populations in these environments is an important step in the ecological risk assessment of VR *T. repens* prior to commercial release. It is hoped that with further investigation, this will function as a model system and thereby provide insights into the functioning of host-pathogen relationships for a wider range of transgenic pasture species.

References

Godfree, R.C., A.G. Young, W.M. Lonsdale, M.J. Woods and J.J. Burdon (2004a). Ecological risk assessment of transgenic pasture plants: a community gradient modelling approach. *Ecology Letters*, 7, 1077-1089.

Godfree, R.C., P.W.G. Chu and M.J. Woods (2004b). White clover (*Trifolium repens*) and associated viruses in the subalpine region of south-eastern Australia: implications for GMO risk assessment. *Australian Journal of Botany*, 52, 321-331.

Godfree, R.C., B. Lepschi and D. Mallinson (2004c). Ecological filtering of exotic plants in an Australian sub-alpine environment. *Journal of Vegetation Science*, 158, 227-236.

Pollen-mediated gene flow from genetically modified herbicide resistant creeping bentgrass

L.S. Watrud[1], E.H. Lee[1], A. Fairbrother[1], C. Burdick[1], J.R. Reichman[1], M. Bollman[2], M. Storm[2], G. King[2] and P.K. van de Water[3]

[1]US Environmental Protection Agency ORD NHEERL, [2]Dynamac Corporation and [3]US Geological Service, 200 SW 35th Street Corvallis, OR 97333 USA. Email: watrud.lidia@epa.gov

Keywords: pollen, *Agrostis stolonifera* L., risk assessment

Introduction Approximately 162 ha of multiple experimental fields of creeping bentgrass (*Agrostis stolonifera* L.) genetically modified for resistance to Roundup ®herbicide, were planted in central Oregon in 2002. When the fields flowered for the first time in the summer of 2003, a unique opportunity was presented to evaluate methods to monitor potential pollen-mediated gene flow from the experimental GM crop fields to compatible sentinel and resident plants that were located in surrounding, primarily non-agronomic areas.

Materials and methods A sampling grid to study the spatial pattern of gene flow was based on assumptions of pollen viability of 3 hours and prevailing winds of 10 km per hour from the northwest at the anticipated time of pollen shed. Sentinel plants of *A. stolonifera* were placed in all map directions, but primarily to the south and southeast of the control district which contained the GM fields. Sampling grid intervals for the sentinel plants were smaller close to the control district and increased with distance from the perimeter of the control district. Greenhouse methods for detecting gene flow (**Figure 1**) were based on seedling survival after treatment with Roundup® herbicide in a track sprayer and positive TraitChek™ tests for the *CP4 EPSPS* marker. Confirmatory laboratory tests for the presence of the engineered *CP4 EPSPS* marker included PCR and DNA sequencing.

Figure 1 Greenhouse assays A: Seedlings; B: Track Sprayer; C: Resistant Plants; D: TraitChek™ test

Table 1. Incidence and prevalence of gene flow from GM crop to sentinel and resident *Agrostis* spp.

Species	% Plants with positive seedling progeny	Number of seedling progeny tested	Number positive seedling progeny	% Positive seedling progeny
Sentinel *Agrostis stolonifera*	54% (75/ 138)	32,000	625	2.00%
Resident *Agrostis stolonifera*	53% (16/30)	565,000	157	0.03%
Resident *Agrostis gigantean*	33% (13/39)	397,000	159	0.04%

Results. As shown in Table 1 the overall frequency of hybridisation was higher in sentinel plants (2%) than in resident *Agrostis* spp. (0.03-0.04%). While most hybridisations were observed within 2 km in the direction of prevailing winds from the perimeter of the control district, the maximal distances to which hybridisations were observed were 21 km in sentinel plants and 14 km in resident *Agrostis* spp.

Conclusions The methods we describe study the spatial pattern of gene flow to compatible non-crop plants can contribute to the environmental risk assessment of genetically modified crops by providing estimates of exposure to marker genes from GM crops on a landscape level that may include non-target non-crop and crop plants.

Reference
Watrud, L.S., E. H. Lee, A. Fairbrother, C. Burdick, J.R. Reichman, M. Bollman, M. Storm, G. King & P. K. Van de Water (2004). Evidence for landscape level pollen-mediated gene flow from genetically modified creeping bentgrass, Proc. National Acad. Sci. (US) 101, 1433-1438.

Use of cellular automata modelling approaches to understand potential impacts of GM grasses on grassland communities

R. Colasanti[1], R. Hunt[2] and L.S. Watrud[3]
[1]National Research Council1US [2]University of Exeter [3]USEPA NHEERL WED, Corvallis, OR US
Email: Colasanti.ricardo@epa.gov

Keywords: cellular automata, strategy theory, risk assessment

Introduction In order to predict the potential unintended ecological impacts of genetically modified (GM) grasses, we must understand how the engineered traits, in this case herbicide resistance, are expressed in an ecological context. It would be a daunting task to experimentally evaluate the full multiplicity of potential pair-wise interactions between GM plants and native plants under a broad variety of actual environmental conditions. We have employed the modelling methodology of cellular automata (CA), where a plant's distribution within a two-dimensional environmental grid is determined by rules relating to phenomena such as seed dispersal, clonal expansion and interactions with adjacent plants. We have used CA simulation to model interactions between GM grasses and the natural environment by describing the plants and the effect of the GM trait in terms of plant functional types. This approach takes the external factors which limit the amount of plant material present in any habitat and classifies them into two categories: (1) stress, defined with regard to the availability of nutrients and (2) disturbance, which refers to the destruction of plant material. The ecological characteristics of all the plants can be described based on three functional types C (competitor), S (stress-tolerator) and R (ruderal) as determined by their quantifiable physiological relationships to stress and disturbance. By ascribing the large number of plant ecological characteristics to a smaller number of functional types the problem, of describing how the engineered trait of herbicide resistance is expressed in an ecological context, becomes tractable.

Materials and methods All simulations involve two-dimensional cellular automata consisting of a 50 X 50 array of discrete cells. At any point in time each cell can contain any number of resource units but only one plant occupant. According to the instructions in the rule base each cell's state is updated individually in discrete time steps (iterations) and is determined by the previous state of the cell and that of its immediate neighbours.

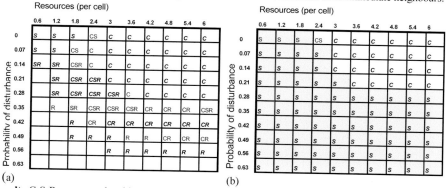

(a) (b)

Results C-S-R space; each grid cell contains the mean dominant plant type for a separate treatment of the 'all types all conditions' experiment. The disturbance axis of the matrix is analogous to the concentration of herbicide. Equal numbers of all plant types (35 of each) were planted into a 50 by 50 grid. Runs lasted for 1000 iterations; each point represents the mean of 20 replicate runs. Bold capitalised entries denote a plant monoculture. (a) C-S-R space for a community of all seven plant functional types. (b) C-S-R space for the seven plant types including a genetically modified plant of functional type S (indicated by a shaded grid cell).

Conclusions A conclusion that can be drawn from this is that the engineered trait of herbicide resistance preferentially benefits those plants that are suited to the opposing environmental driving factor. In this case the resistance is to the disturbance of herbicide, a factor that most affects slow growing stress tolerant plants. They would most benefit from a protection from disturbance rather than ruderal plant types that already have a strategy for 'overcoming' disturbance.

Reference
Grime J.P., J.G. Hodgson & R. Hunt. (1988). Comparative plant ecology: a functional approach to common British species. London: Unwin Hyman.

Section 8

Genetic diversity, genetic resources and breeding systems

The importance of exotic forage germplasm in feeding New Zealand's livestock

J. Lancashire

77, The Esplanade, Raumati South, Kapiti , 6010, New Zealand. Email: lancs@paradise.net.nz

Keywords: exotic forage germplasm , livestock feeding , biosecurity , economic value of plant breeding, New Zealand

Introduction A number of recent reports in New Zealand have expressed the view that restricted access to new plant genetic material from overseas is a major risk to the future growth of the primary sector (MAF, 2002; Douglas, 2003). The restrictions outlined in this paper are the result of the Hazardous Substances and New Organisms (HSNO) Act and regulations administered by MAF and the Environmental Risk Management Authority (ERMA) since July 1998. This paper reviews the historical role of exotic forage germplasm in plant improvement in New Zealand, and quantifies the current contribution of recently imported plant material to exports from the pastoral sector.

Discussion In the past 30 years there has been a large increase in the number of improved forage cultivars available to farmers (Lancashire,1985; Stewart & Charlton, 2003) . There are now 115 forage cultivars, 65% of which contain exotic germplasm, covering 26 species on the market, compared with 20 cultivars and 12 species in the early 1970's.

The current annual (2002) contribution of cultivars containing exotic forage germplasm to New Zealand pastoral exports of $14 billion is $735 million. This includes $74 million attributed to the perennial ryegrass endophyte (Bluett *et al.*, 2003) . This represents an almost 6 fold increase over the figure of $128 million in 1992, which is significantly faster than the 75% increase in pastoral exports over the same period.

There is potential loss of future opportunities which will result from the continuation of restricted access to novel plant material from overseas. These include responses to climate change (Campbell et.al., 1996), biosecurity breaches (Eerens *et al.*, 2001), developments in biotechnology (Bryan, 2001) and improvements in food type and quality (Lancashire, 2003).

Conclusions A return to the system of accredited institutions carrying out field evaluations under supervision is proposed .This procedure which operated successfully for most of the second half of the 20th century in New Zealand led to the recent very rapid increase in the successful utilisation of exotic germplasm .There is no record of serious biosecurity breaches as a result of this policy .It is suggested that in the case of exotic forage germplasm, the costs of mitigating or avoiding risk now far outweigh the consequences of avoiding risk (Sherwin, 2004).

References

Bluett, S. J., Thom, E.R. , Clark , D.A., Macdonald.,K.A and Minnee, E.M.K. 2003. Milk solids production from cows grazing perennial ryegrass containing AR1 or wild endophyte. Proceedings of the New Zealand Grassland Association. 65: 83-90

Bryan,G.T. 2001. Biotechnology in forage crops--capturing our potential .Proceedings of the New Zealand Grassland Association 63: 235-239 .

Campbell, B. D., Wardle, D.A., Woods, P.W., Field, T.R.O., Williamson, Derryn Y. and Barker, G.M. 1996. Ecology of subtropical grasses in temperate pastures :an overview .Proceedings of the New Zealand Grassland Association 57:189-197.

Douglas,J.A.2003. Plant importation . AgScience . Issue 11:13-15.

Eerens, J.P.J., Cooper, B. .,Willoughby, B.E..and Woodfield, D.R. 2001. Searching for clover root weevil (*Sitona lepidus*) resistance / tolerance-A progress report. Proceedings of the New Zealand Grassland Association 63: 177-181.

Lancashire, J.A. 1985 . Factors affecting the adoption of new herbage cultivars .In" Using Herbage Cultivars" Grassland Research and Practice Series No.3 pp 79-87. New Zealand Grassland Association (Inc). Private Bag 11008, Palmerston North.

Lancashire, J.A. 2003 .The potential for wealth creation in the New Zealand food industry .Future Times 2 :3- 5. The New Zealand Futures Trust (Inc) ,PO Box 12-008 , Wellington.

MAF.2002. In "Contribution of the Land Based Primary Industries to New Zealand's Economic Growth ".MAF, PO Box 2526, Wellington.

Sherwin , M. 2004 .Research mitigates risk in the primary industries .AgScience. Issue17:3.

Stewart , A.V .and Charlton , J.F.L.2003 In" Pasture and Forage Plants for New Zealand " .Grassland Research and Practice Series No 8. 2nd Edition. New Zealand Grassland Association (Inc). Private Bag 11008, Palmerston North.

Application of molecular diversity in a forage grass breeding program

A.A. Hopkins and M.C. Saha
Noble Foundation, Inc. 2510 Sam Noble Parkway, Ardmore, OK 73401 USA. Email: aahopkins@noble.org

Keywords: AFLP, SSR, genetic diversity

Introduction Little or no genotypic information is available for many forage grass populations. The degree of genetic similarity within and among populations greatly influences the choice of breeding strategies and germplasm for developing improved cultivars. Molecular markers have proven effective in classifying genetic diversity of a number of perennial grasses (e.g. Fu *et al.*, 2004; Kubik *et al.*, 2001). We present here an overview of our efforts to integrate molecular diversity data into our breeding program.

Materials and methods Amplified fragment length polymorphism (AFLP) markers were used to examine genetic variation among tall fescue (*Festuca arundinacea*) populations (Mian *et al.*, 2002), as well as among and within hardinggrass (*Phalaris aquatica*) populations (Zwonitzer *et al.*, 2003). Expressed sequence tag – simple sequence repeat (EST-SSR) markers originating from tall fescue were used to classify genetic variation among and within populations of Canada wildrye (*Elymus canadensis*) and Virginia wildrye (*Elymus virginicus*) (Saha *et al.*, 2004). Standard statistical analyses, such as Nei and Li (1979), were used to classify within and among population genetic variation.

Results Tall fescue populations from the southern Great Plains were found to differ genetically from several cultivars including KY-31 (Mian *et al.*, 2002), leading us to make further collections of tall fescue germplasm from this region. The pattern of genetic variation between hardinggrass populations agreed closely with geographical origins. We have used this information to construct a population based on accessions from Morocco, and a broad population based on accessions from a wide geographical area. We intend to determine in the future if these two populations represent potential heterotic groups. Genetic variation was minimal within wildrye populations, indicating that a given wildrye population can be handled as a pure line in our breeding program. Canada wildrye populations heavily infected with an endophytic fungus clustered together; further investigation revealed that presence of the endophyte did not bias interpretation of the plant genetic variation data. Sequencing of a specific locus indicated that single nucleotide polymorphisms (SNPs) were the source of variation among wildrye accessions.

Conclusions Genetic diversity data has proven to be valuable to our breeding program in directing acquisition of germplasm, constructing populations, and choice of breeding methods. In the future marker diversity data may also aid in identifying heterotic groups in forage grass species. The application of high throughput marker systems, such as SSR's developed from tall fescue, and the development of additional markers such as SNP's, will continue to enhance our ability to understand and utilize genotypic diversity in the forage grasses.

References

Fu, Y.B, A.T. Phan, B. Coulman, and K.W. Richards. (2004). Genetic diversity in natural populations and corresponding seed collections of little bluestem as revealed by AFLP markers. Crop Science 44, 2254-2260.

Kubik, C., M. Sawkins, W. A. Meyer, and B. S. Gaut. (2001). Genetic diversity in seven perennial ryegrass (*Lolium perenne* L.) cultivars based on SSR markers. Crop Science 41, 1565-1572.

Mian, M.A.R., A.A. Hopkins, and J.C. Zwonitzer. (2002). Determination of genetic diversity in tall fescue with AFLP markers. Crop Science 42, 944-950.

Nei, M. and W.H. Li. (1979). Mathematical model for studying genetic variation in terms of restriction endonucleases. Proceedings of the National Academies of Science USA. 76, 5269-5273.

Saha, M.C., M.A.R. Mian, K. Chekhovskiy, J.C. Zwonitzer, A.A. Hopkins. (2004). Genetic diversity among a collection of wildrye determined by EST-SSR markers. Agronomy Abstracts.

Zwonitzer, J.C., K. Chekhovskiy, M.C. Saha, A. Hopkins, and M.A.R. Mian. (2003). Assessing genetic diversity in hardinggrass (*Phalaris aquatica*) with SSR and AFLP markers. Third International Symposium on Molecular Breeding of Forage and Turf. p. 90.

Databases for managing genetic resources collections and mapping populations of forage and related species

I. Thomas, H. Ougham and D. Peltier

Institute of Grassland and Environmental Research, Plas Gogerddan, Aberystwyth, SY23 3EB, Wales, UK
Email:ian.thomas@bbsrc.ac.uk

Keywords: databases, genetic resources, mapping populations

Introduction Effective management of plant material used in crop improvement and underpinning research is greatly facilitated by a properly designed data structure accessible by all those working with the material. At IGER we have developed the Aberystwyth Genetic Resources Information System, AGRIS, for managing genetic resources acquired through collecting trips, seed exchange, breeding and transgenic programmes. Recently this has been complemented by MaPIS, a Mapping Populations Information System, which links with AGRIS and allows for storage and documentation of information about plant mapping populations, including pedigrees, status and physical locations of accessions and individual genotypes. IGER also maintains the European Central Crop Databases for *Lolium* species and *Trifolium repens*, and the UK National Inventory of all plant genetic resources conserved *ex situ* in the UK; by November 2004, the UKNI had contributed over 220000 accessions to the 900000 in the Europe-wide database EURISCO.

Methodology and databases AGRIS and MaPIS are currently based on the Microsoft Access relational database management system, with user-friendly forms and reports for easy and accurate data input and generation of plant labels, data summaries etc.(Figure 1) Data in AGRIS use an extended version of the FAO / International Plant Genetic Resources Institute's Multi-crop Passport Descriptors (Maggioni *et al.*, 1998). The UKNI and the *Lolium* and *T. repens* databases (http://www.igergru.bbsrc.ac.uk/Welcome/ECCDB/eccdb.htm) also currently use Access. However we propose to migrate all our existing Access databases to Microsoft SQL server in order to accommodate the rising level of usage and increasing data volumes now being experienced.

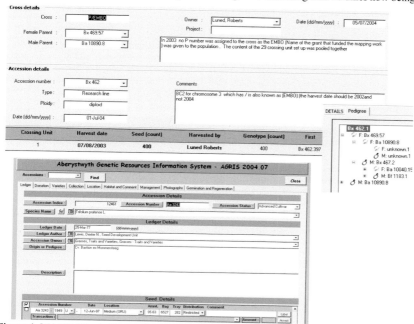

Figure 1 MAPIS (illustrating pedigree, cross and accession details) and AGRIS (showing ledger details)

Computer demonstrations of the databases described in this paper are taking place at the 4th International Symposium on the Molecular Breeding of Forage and Turf, Aberystwyth, 3-7 July 2005.

References

Maggioni, L., Marum, P., Sackville Hamilton, R., Thomas, I., Gass, T. & Lipman, E. (1998) Report of a Working Group on Forages - Sixth meeting, 6-8 March 1997, Beitostølen, Norway IPGRI, Rome, Italy.

Molecular breeding for the genetic improvement of forage crops and turf

The use of molecular markers in genetic variability analysis of a collection of *Dactylis glomerata* L.

R. Costa, G. Pereira, C. Vicente and M.M. Tavares de Sousa

Estação Nacional de Melhoramento de Plantas (National Station of Breeding Plants) Apartado 6 7350-951 Elvas, Portugal. E-mail: claudia_vicente@netcabo.pt

Keywords: *Dactylis glomerata*, genetic variability, RAPD markers, ISSR markers

Introduction *Dactylis glomerata* L. is one of the three most used perennial grasses in Europe (jointly with *Lolium perenne* and *Festuca arundinacea*). The main qualities of *Dactylis glomerata* are high productivity in pure cultures, high level of proteins and tolerance to drought, cold and shade (Mousset & Chosson, 1986). In the last ten years, techniques that allow direct discrimination at the DNA level have encouraged the study of the genetic variability within cultivated populations as well as identification of the diversity available in germplasm banks. DNA polymorphisms have been shown to be efficient in the identification of genetic variability in several groups of plants. They can be used as an auxiliary tool in breeding programs through obtaining genetic maps and in the identification of molecular markers useful for assisting selection. The aims of this work were: (1) to study the genetic variability of 100 genotypes in a collection of *Dactylis glomerata* from several areas of Portugal, grown in the experimental fields of ENMP-INIAP, using PCR based molecular markers obtained with RAPD and ISSR primers, together with morphological and agronomic characterization, (2) to select genotypes that may function as parents of new synthetic varieties.

Materials and methods DNA was extracted from fresh leaves of 100 genotypes of *Dactylis glomerata* using a modified CTAB method. DNA polymorphism was revealed using random amplified DNA (RAPD) based on oligonucleotides of 10 bases and a technique that requires the use of a primer composed by a repetitive sequence and an arbitrary sequence of 1 to 3 nucleotides in one of the extremities (ISSR). Markers were analysed with NTSYS-PC software (version 2.01b). A matrix of genetic variability was constructed with the data using the DICE coefficient (D). The grouping method UPGMA (Unweighted Pair Group Method with Arithmetic Average) was used to construct dendrograms.

Results and discussion A total of 89 RAPDs and 47 ISSRs were obtained and analysed using the grouping methods of UPGMA. The analysis of the RAPD dendogram showed that the genotypes could all be separated. However the similarity level was very high (94%) due to the fact that the majority of genotypes were from the same sub-species *Dactylis glomerata* ssp.*hispanica*. This high level of similarity is in accordance with data obtained from the morphological and agronomic characterisation of the genotypes. With regard to the ISSR dendrogram, the number of markers appeared not to be sufficient to separate all the genotypes. However those genotypes that were considered the best according to morphological and agronomic characterisations were all located in the same area of the dendogram and were relatively separated from other genotypes. In a joint analysis of the results, as expected the RAPD markers had more influence than the ISSRs. Genotypes were not aggregated according to geographic origin and it appears that genotypes that are not close geographically may be close genetically and vice-versa. The genotypes of most interest with regard to high yield, long growing season and the greatest spring and winter vigour are positioned in the lower section of the dendrogram (Figures 1 and 2). From this group of about 25 genotypes, parents were selected to form a new synthetic variety. The chosen genotypes were 2, 33, 40 and 41 and these are already established in the experimental field following the third year of crosses.

Reference

Mousset, C. & Chosson, J. F. (1986) - Caractéristigues physiologiques, morphologiques et agronomiques d'écotypes de dactyle de la Côte Nord de l'Espagne, de Galice et du Nord du Portugal. *C. R. de la réunion de Soc. Portugaise de Prodution Fourrage*, pp. 14.

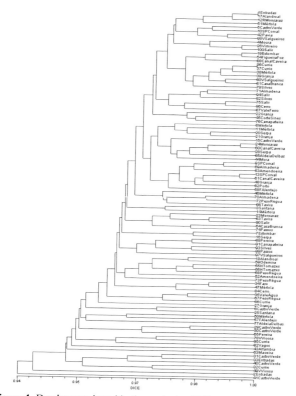

Figure 1 Dendrogram based in the analysis of 100 genotypes through RAPD´s + ISSR´s

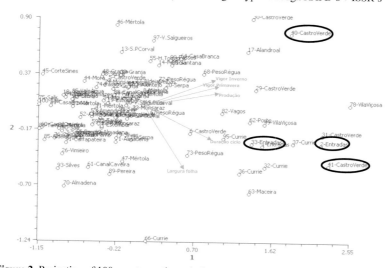

Figure 2 Projection of 100 genotypes through the agronomic characteristics

Genetic diversity in colonial bentgrass (*Agrostis capillaris* L.) revealed by *Eco*RI/*Mse*I and *Pst*I/*Mse*I AFLP markers

H. Zhao and S. Bughrara
Department of Crop and Soil Sciences, Michigan State University, East Lansing, MI 48824, USA
Email: bughrara@msu.edu

Keywords: colonial bentgrass, diversity, AFLP

Introduction Colonial bentgrass species (*Agrostis capillaris* L.) is a potential genetic resource for the improvement of other bentgrass species (*Agrostis* spp.) with regard to resistance to environmental stresses and diseases. Transferring resistance from colonial to other bentgrass species is a promising goal in turfgrass breeding programs (Belanger, 2003). Assessment of genetic diversity among accessions of colonial bentgrass species will contribute to eliminate undesirable duplications in the germplasm collection and increase the efficiency of research efforts. It will allow researcher to select diverse resistance genes from different sources to incorporate and pyramid these resistance genes into creeping or other bentgrass species cultivars. The objectives of this study were to investigate the genetic diversity of colonial bentgrass species consisting of 22 PI accessions from USDA collected from 11 countries, 14 accessions from north Spain and three commercial cultivars by using AFLP markers (*Eco*RI/*Mse*I and *Pst*I/*Mse*I enzyme combinations), and to compare the correlation between estimates of genetic diversity derived from these two enzyme combinations.

Materials and methods Thirty-six accessions were used in the study. The polymorphic bands produced from AFLP analysis (*Eco*RI/*Mse*I and *Pst*I/*Mse*I enzyme combinations) were scored. Genetic diversity analyses were conducted using Numerical Taxonomy and the Multivariate Analysis system (Exter Software Co., New York).

Results A total of 182 unequivocally recognizable polymorphic bands were obtained from *Eco*RI/*Mse*I AFLP analysis. The genetic similarity (Jaccard) coefficients (GS_j) ranged from 0.34 to 0.70 with a mean of 0.55. A dendrogram among 39 colonial accessions based on their cluster analysis of GS_j coefficients showed that no major 'ball cluster' was found (Figure 1). 120 polymorphic bands were scored from *Pst*I/*Mse*I AFLP analysis. Values of GS_j ranged from 0.99 to 0.45. With CPCA subroutine programs of NTSYS, a rotated PCA with the AFLP markers as observations was used to determine the number of groups based on Eigen values. Three groups were formed with an average GS_j=0.63 (Figure 2).

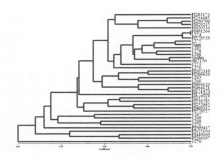

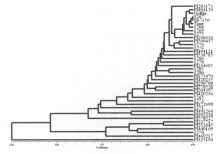

Figure 1 UPGMA dendrogram of 39 accessions revealed by *Eco*RI/*Mse*I AFLP markers

Figure 2 UPGMA dendrogram of 39 accessions revealed by *Pst*I/*Mse*I AFLP markers

Conclusions A high level of diversity in colonial bentgrass species was demonstrated with averages of 0.51 (*Eco*RI/*Mse*I) and 0.58 (*Pst*I/*Mse*I)(GS_j). Greater genetic diversity was detected by *Eco*RI/*Mse*I AFLP primer combinations because of genome region difference (hypermethylated vs. hypomethylated regions). A positive correlation (r=0.44, p=0.0099) between the two Jaccard similarity matrices was obtained by a Mantel test.

References
Belanger FC, Meagher TR, Day PR, Plumley K, Meyer WA (2003) Interspecific hybridization between *Agrostis stolonifera* and related *Agrostis* species under field conditions. Crop Sci. 43:240–246

Genetic diversity in zoysiagrass ecotypes based on morphological characteristics and SSR markers

M. Hashiguchi[1], S. Tsuruta[1], T. Matsuo[1], M. Ebina[2], M. Kobayashi[3], H. Akamine[4] and R. Akashi[1]
[1]Faculty of Agriculture, University of Miyazaki, Miyazaki 889-2192, Japan, [2]Okinawa Prefectural Livestock Experimental Station, Okinawa 905-0426, Japan, [3]National Institute of Livestock & Grassland Science, Tochigi 329-2793, Japan, [4]Faculty of Agriculture, Ryukyu University, Okinawa 903-0213, Japan
E-mail: mashiguchi@miyazaki-u.ac.jp

Keywords: zoysiagrass, genetic resource, genetic diversity, morphological characteristics, SSR markers

Introduction Zoysiagrass consists of a number of interfertile species, some of which are important grasses for turfgrass and grazing pasture in Japan. Recently, we developed simple sequence repeats (SSRs) markers from *Zoysia japonica* "Asagake" genomic DNA by enriched genomic library method (Yamamoto *et al.*, 2002). Here we identify genetic diversity in 38 ecotypes of zoysiagrass (*Z. matrella* and *Z. tenuifolia*) from a group of southwest islands of Japan based on morphological characteristics and SSR markers.

Materials and methods Thirty-eight zoysiagrass ecotypes and 3 cultivars were used in this study. These accessions were screened for 7 morphological characteristics and 13 SSR markers, which could produce 1 or 2 discrete amplified fragments in all ecotypes and cultivars.

Results Thirty-eight zoysiagrass ecotypes, except "Tanegashima 2", were classified into 2 groups based on 7 morphological characteristics. Cluster I consists of *Z. matrella* and 2 cultivars (Emerald and Tottori-kourai). On the other hand, cluster II contained *Z. tenuifolia* except "Minatogawa 2". Thirteen SSR markers were polymorphic in 38 ecotypes of zoysiagrass within 2 to 22 alleles per locus. 157 putative alleles were obtained, with an average of 12. Cluster analysis based on the 157 SSR bands revealed that the 38 ecotypes of zoysiagrass were classified into 6 groups.

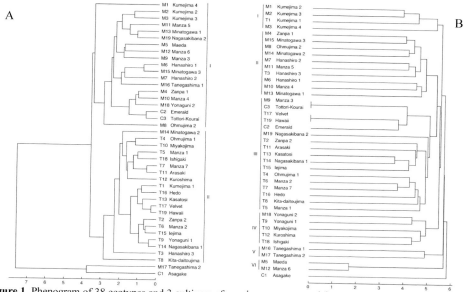

Figure 1 Phenogram of 38 ecotypes and 3 cultivars of zoysiagrass generated from morphological characteristics (A) and SSR markers (B) using UPGMA method. Scale on bottom indicates dissimilarity index

Conclusions Thirty-eight ecotypes of zoysiagrass were identified with morphological and SSR marker. Because of a high correlation between leaf width and other morphological characters, these accessions were classified into *Z. matrella* and *Z. tenuifolia* by 7 morphological characteristics. On the other hand, they could not be classified into two species using SSR markers. However, that classification tends to follow geographical origins. These SSR markers could be a useful tool to investigate genetic diversity of ecotypes of zoysiagrass as well as to identification and construction of genetic linkage map of agronomically and commercially important traits.

References Yamamoto *et al.* (2002) Mol. Eco. Notes, 2:14-16

Molecular breeding for the genetic improvement of forage crops and turf

Utilization of SSR to distinguish alfalfa cultivars

G.R. Bauchan, C. He and Z-L. Xia
U.S. Department of Agriculture, Agricultural Research Service, Building 006, Room 14, BARC-West, 10300 Baltimore Ave. Beltsville, Maryland 207050-2350 USA, Email: bauchang@ba.ars.usda.gov

Keywords: simple sequence repeat, microsatellites, germplasm, autumn dormancy, phylogenetics

Introduction Simple sequence repeat (SSR) or microsatellite markers are co-dominant, abundant and hyper-variable molecular markers from eukaryotic genomes that are being widely used in genetic mapping and phylogenetic studies. Currently, the number of available SSR markers is still very limited for use in alfalfa (*Medicago sativa*). Thus, this study was conducted to develop SSR from alfalfa genomic libraries and EST and BAC sequence data from *M. truncatula* for use in distinguishing the nine historically recognized U.S. germplasm sources and eleven fall dormancy check cultivars of alfalfa.

Materials and methods The nine historical cultivars of alfalfa included: African, Chilean, Falcata, Flemish, Indian, Ladak, Peruvian, Turkistan and Varia types as well as wild tetraploid M. falcata, two very non dormant *M. sativa* accessions and diploid *M. coerulea*, *M. sativa* ssp. *falcata* and *M. truncatula*. The SSR markers were also used for phylogenetic analysis for the 11 standard cultivars (Maverick, Vernal, 5246, Legend, Archer, ABI 700, Dona Ana, Pierce, CUF 101, UC-1887 and UC-1465) for autumn dormancy. The alfalfa genomic derived SSRs was developed from genomic DNA which was extracted from the alfalfa population W10, digested with Sma I, Alu I, Rsa I, Nac I, Hinc II and Xmn I, and fragments were cloned into pUC19. Colonies were probed with repeats of AC, AT, CT, CTT, GAT and GGT motifs. SSR derived from *M. trucatula* EST and BAC sequenced data were primer analyzed based on primers from several sources (Eujayl *et al*., 2004; Julier *et al*., 2003; Diwan *et al*., 2000). Primers were identified based on the nomenclature used in Diwan *et al*. (2000) or by the corresponding Genbank accession or tentative contig name (www.medicago.org). The PCR amplifications of SSR primers were based on the method of Diwan *et al*. (1997). All reactions utilized 30 ng of DNA in a final reaction volume of 10µl. Fluorescent labelling was used whereby a sequence-specific M13 forward primer is labelled with one of three fluorescent dye labels: FAM, HEX or NED. These fragments were then run on a 3100 DNA Analyzer (ABI) and scored using the Gene Scan and Genotyper software (ABI).

Results A dendrogram was constructed from these data, representing three main clusters: 1) diploid ssp. *falcata*; 2) *M. truncatula*; and 3) all remaining entries. *M. truncatula* (Jemalong), Ladak (Ladak), Arabian (UC-1465), Indian (Sirsa Type 9), Flemish (DuPuit), Peruvian (Hairy Peruvian), African 2 (Moapa) and Turkistan (Kayseri) were separated through multiple correspondence analysis. The remaining germplasms within the ssp. *sativa* species could not be separated due to a limited number of SSR markers used in this study. The results of phylogenetic analysis of the fall dormancy standards showed that the cultivars which are non-dormant tended to be clustered together separate from the cultivars which are dormant.

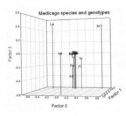

Figure 1. 3D scatter plot after multiple corres-pondence analysis of genetic distance among historically recognized alfalfa germplasm sources. Mt=*M. trucatula*, La=Ladak, Ar2=Arabian, In=Indian, Af2=African, Pe=Peruvian, and Tu=Turkistan

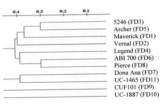

Figure 2 Dendogram of the 11 fall dormancy check cultivars of alfalfa.

Conclusions SSR can be utilized to discriminate a number of alfalfa cultivars, although additional SSR will need to be developed to clearly identify individual cultivars of alfalfa.

References

Diwan, N., J. H. Bouton, G. Kochert, P. B. Cregan, (2000). Mapping of simple sequence repeat (SSR) DNA markers in diploid and tetraploid alfalfa. Theor. Appl. Genet. 101:165-172

Eujayl, I., M. K. Sledge, L. Wang, G. D. May, K. Chekhovskiy, J. C. Zwonitzer, and M. A. R. Mian (2003). *M. truncatula* EST-SSRs reveal cross-species genetic markers for *Medicago* spp. Theor. Appl. Gen. 108:414-421.

Julier, B., S. Flajoulot, P. Barre, G. Cardinet, S. Santoni, T. Huguet, and C. Huyghe. (2003). Construction of two genetic linkage maps in cultivated tetraploid alfalfa (*M. sativa*) using microsatellite and AFLP markers. BMC Plant Biology. 3:9.

Genetic diversity among alfalfa cultivars using SSR markers

S. Flajoulot[1], J. Ronfort[2], P. Baudouin[3], T. Huguet[4], P. Barre[5], C. Huyghe[5] and B. Julier[5]
[1]Jouffray-Drillaud, 86600 Lusignan, France, [2]INRA, Station de Génétique et Amélioration des Plantes, 34130 Mauguio, France, [3]GIE GRASS, La Litière, 86600 Saint-Sauvant, France ; [4]UMR CNRS-INRA 2594/441, BP27, 31326 Castanet-Tolosan cedex, France, [5]INRA, Unité de Génétique et d'Amélioration des Plantes Fourragères, 86600 Lusignan, France. E-Mail: julier@lusignan.inra.fr

Keywords: F_{ST}, lucerne, microsatellite, tetraploid, panmixia

Introduction Alfalfa (*Medicago sativa*) is an autotetraploid, allogamous and heterozygous species. Cultivated varieties are synthetic cultivars, usually obtained through 3 or 4 generations of panmictic reproduction of a set of various numbers of parents. The parents can be clones, half-sib or full-sib families. The breeders apply selection pressure for some agronomic traits, to induce changes in the genetic background. The objective of this study was to investigate the differentiation level among seven cultivars originating from one breeding program, and between these cultivars and the breeding pool, with eight SSR markers.

Materials and methods We focused on seven varieties and the breeding pool of a French private breeder (GIE GRASS, formerly GIE Verneuil), registered between 1990 and nowadays. Each cultivar was represented by 20 plants, except one that was represented by 40 plants. The Breeding pool was represented by 49 plants. Eight SSR primer pairs originating from *M. truncatula* and mapped in alfalfa (Julier *et al.*, 2003) were selected to assay genetic variation in the cultivars. Genotypes were scored according to band intensity, since an allele can be present in more than one dose. The within population genetic diversity was estimated as the number of alleles (A) and the expected heterozygosities according to Hardy-Weinberg expectations (H_e) for each SSR locus. To test for a departure from Hardy-Weinberg expectations in the different studied populations, we used the software AUTOTET (Thrall and Young, 2000) to calculate the parameter F_{IS}. The level of differentiation between the cultivars was measured using the differentiation parameter F_{ST} computed in an ANOVA framework extended to autotetraploids, with the software Gene4X (Ronfort *et al.*, 1998).

Results The number of alleles per locus ranged from 3 to 24 depending on the locus, with an average of 14.9 alleles per locus. The number of alleles tended to be higher in the Breeding pool, but this tendency was not true after correction of the allelic richness for sample size. The mean number of alleles per plant (A) per locus ranged from 1.92 to 3.25 depending on the locus. For all SSR loci, the most frequent and infrequent alleles were the same among all the cultivars and the Breeding pool whatever the number of alleles per locus. One SSR locus, that showed a significant heterozygote deficiency related to the presence of null alleles, was removed from the data for the following analyses. Globally, all the studied cultivars and the Breeding pool showed no departure from Hardy-Weinberg equilibrium. The mean genetic diversity (H_e) for each cultivar and for the Breeding pool ranged from 0.665 to 0.717, indicating high within-population variability. H_e was not related to the year of registration nor to the number of parents of the cultivars. Over all loci and all populations, F_{ST} was highly significant ($P<0.001$), but the estimate were low ($F_{ST} = 0.0048$). The differentiation between each pair of cultivars was significant for 15 of the 21 pairs of cultivars, with F_{ST} ranging from 0 to 0.005.

Conclusions The SSR loci, with a scoring of the allelic doses, gave us the possibility to exploit the whole genotypic information. A large within-cultivar variation was observed, but the level of differentiation among cultivars was low, even if significant. Our results also show that the sampling effect due to the selection within the breeding pool of the female parental plants has a reduced impact on the differentiation between each cultivar and the breeding pool. Finally, this breeding procedure produces cultivars with small differences in allelic frequencies for neutral markers. Non neutral markers, related to breeding traits, would have probably given a stronger differentiation among cultivars. This type of description of genetic diversity in alfalfa populations is useful for the analysis of breeding programs or genetic resources management. The use of SSR markers to detect differences among cultivars (that usually is a critical step in variety registration due to low among-cultivar distinction for quantitative traits) requires further studies based on a wider range of variation.

References

Julier B, Flajoulot S, Barre P, Cardinet G, Santoni S, Huguet T, Huyghe C (2003) Construction of two genetic linkage maps in cultivated tetraploid alfalfa (*Medicago sativa*) using microsatellite and AFLP markers. BMC Plant Biol 3:9

Ronfort J, Jenczewski E, Bataillon T, Rousset F (1998) Analysis of population structure in autotetraploid species. Genetics 150:921-930

Thrall PH, Young A (2000) AUTOTET: A program for analysis of autotetraploid genotypic data. J Hered 91:348-349

Evaluation of genetic diversity in white clover (*Trifolium repens* L.) through measurement of simple sequence repeat (SSR) polymorphism

J. George[1,3], E. van Zijll de Jong[1,3], T.C. Wilkinson[2,3], M.P. Dobrowolksi[2,3], N.O.I. Cogan[1,3], K.F. Smith[2,3] and J.W. Forster[1,3]

[1]*Primary Industries Research Victoria, Plant Biotechnology Centre, La Trobe University, Bundoora, Victoria 3086, Australia* [2]*Primary Industries Research Victoria, Hamilton Centre, Hamilton, Victoria 3300, Australia* [3]*Molecular Plant Breeding CRC, Australia Email: julie.george@dpi.vic.gov.au*

Keywords: white clover, genetic diversity, UPGMA, SSR

Introduction White clover (*Trifolium repens* L.) is a key important temperate pasture legume. Due to the obligate outbreeding nature of white clover, individual genotypes within cultivars are highly genetically heterogeneous. Genetic diversity has been assessed within and between 16 elite cultivars derived from Europe, North and South America, New Zealand and Australia.

Materials and methods Each cultivar was represented by c.15 plants, to a total of 235 individuals. Sixteen simple sequence repeats (TRSSR) markers that had been previously used to construct a white clover reference genetic map were chosen on the basis of single, co-dominant locus status and polymorphic information content (PIC), based on previous studies. Genetic diversity among individuals was measured using the method of Tomiuk and Loeschcke (1991). Phenograms were constructed with the resulting genetic distance matrix using the unweighted pair group method of arithmetic averages (UPGMA). The single dose fragment (SDF) principle was used to score the data for analysis of molecular variance (AMOVA).

Results and discussion High levels of genetic variation were detected within and between white clover accessions. AMOVA showed variation within accessions. Most of the cultivars did not separate into distinct clades in the phenogram, but were rather distributed throughout the phenogram. There was no obvious genetic distinction between Mediterranean the cultivars AberHerald and AberVantage. Both AberHerald and AberVantage contain Swiss ecotypic material from similar localities and were each bred in the UK from a selection of 15 parents. These cultivars showed a similar pattern of distribution across the phenogram (Figure 1). Although the number of parental lines used for selection in the breeding program may be expected to influence the levels of genetic diversity in cultivars, no such correlation was observed. The overall level of intra population variation provides a greater understanding of the complexity of domesticated white clover germplasm. This knowledge is being exploited for improved parental selection for the construction of trait specific mapping families. In addition, the data will assist in the construction of larger scale germplasm collections for the design of population samples for linkage disequilibrium (LD)-based genetic analysis. The development of candidate gene-based marker systems for white clover will allow comparison of anonymous and functionally associated genetic diversity.

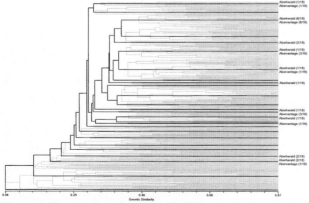

Figure 1 The overall white clover UPGMA genetic diversity dendrogram, demonstrating the influence of breeding history on genetic diversity, as exemplified by cultivars AberHerald and AberVantage. The identifiers of individual plants from these varieties are shown adjacent to their locations within the structure. Framework taxa are shown in bold lines, while the remaining accessions are indicated as dotted lines.

Genetic and phenotypic diversity of Swiss red clover landraces

D. Herrmann, B. Boller, F. Widmer and R. Kölliker
Agroscope FAL Reckenholz, Swiss Federal Research Station for Agroecology and Agriculture, 8046 Zurich, Switzerland, Email: doris.herrmann@fal.admin.ch

Keywords: Mattenklee landraces, red clover, diversity, ancestry

Introduction: Mattenklee landraces are persistent and locally adapted Swiss red clover populations. About 100 populations are preserved and may represent a significant yet poorly characterised genetic resource for temperate regions. Genetic characterisation is important in order to improve cultivars, manage genetic resources and to maintain or restore biodiversity. The objectives of this study were to analyse genetic diversity, investigate potential ancestry and to elucidate the congruence of phenotypic and genetic structure of Mattenklee landraces.

Materials and methods Eighty-nine Mattenklee landraces and 31 populations of five additional red clover groups (Swiss wild clover populations, Mattenklee cultivars, field clover cultivars, Dutch landraces, Dutch wild clover populations) were analysed using bulked AFLP (amplified fragment length polymorphism) analysis with twelve primer combinations. AFLP analysis was performed on two bulked samples per population consisting of twenty plants each. In addition, eight phenotypic characteristics were determined for a subset of 33 Mattenklee landraces and the Mattenklee cultivar Milvus (Table 1).

Results Genetic distances among red clover groups, expressed as co-ancestry coefficients derived from analysis of molecular variance, ranged from 0.06 to 0.25. Swiss wild clover populations revealed the largest genetic distance to any of the five groups investigated. In addition, principal coordinate analysis based on two bulked samples clearly separated Swiss wild clover populations from all other populations. Average Euclidean squared distances among the 33 Mattenklee landraces for which phenotypic characteristics were determined ranged from 12 to 52 with an average of 27.8 (Table 1). The 33 Mattenklee landraces investigated showed a broad variation for phenotypic characterisation as well as for AFLP marker diversity. Although average agronomic performance of phenotypic characteristics of Mattenklee landraces was lower compared to the cultivar Milvus, there was, for most of the traits, at least one Mattenklee landrace which showed a better performance (Table 1). Cluster analysis based on 212 polymorphic AFLP markers and phenotypic characteristics yielded dendrograms of largely different topologies. However, redundancy analysis using the eight phenotypic characteristics as explanatory variables revealed that the three characters "dry matter yield, first production year", "stem length" and "length of medial leaflet" had a significant influence on AFLP diversity ($P<0.05$).

Table 1 Phenotypic characteristics and AFLP marker diversity for 33 Mattenklee landraces and the Mattenklee cultivar Milvus

	Milvus	Mattenklee landraces		
		Mean	Highest value	Lowest value
Phenotypic characteristic				
Dry matter yield, first production year	155.66	141.63	158.54	118.92
Dry matter yield, second production year	110.27	87.18	108.55	74.50
Anthracnose disease score (1=healthy)	2.74	4.82	7.15	2.02
Downy mildew disease score (1=healthy)	2.31	4.55	5.97	3.62
Stem length (cm)	90.71	86.27	94.26	79.58
Time of flowering (days after 30th April)	29.66	27.80	30.71	25.27
Length of medial leaflet (mm)	49.54	47.89	51.13	43.54
Width of medial leaflet (mm)	28.56	28.60	30.47	26.86
AFLP marker diversity				
Euclidean squared distances	33.55	27.80	52.00	12.00

Conclusions Mattenklee landraces are a distinct genetic resource with considerable variation in AFLP marker diversity as well as in phenotypic characteristics. Swiss wild clover populations were clearly separated from Mattenklee landraces, which in turn grouped closely to Dutch germplasm. Multivariate analyses indicated a significant effect of three key characteristics on AFLP marker diversity. This study shows that Mattenklee landraces may serve as a valuable gene pool for red clover improvement and may contribute to the restoration of biodiversity in pastures and meadows. In addition, the ancestry of Mattenklee landraces is found more in introduced cultivars than in natural wild clover populations.

Improving the utilisation of germplasm of *Trifolium spumosum* L. by the development of a core collection using ecogeographical and molecular techniques

K. Ghamkhar, R. Snowball and S.J. Bennett
Centre for Legumes in Mediterranean Agriculture, University of Western Australia, 35 Stirling Highway, Crawley, WA 6009, Email: kioumars@cylenne.uwa.edu.au

Keywords: Amplified Fragment Length Polymorphisms, Arcview, MStrat, passport data, pasture

Introduction A core collection is a sub-set encompassing more than 70% of the variability of all accessions held in a collection (Brown, 1995). The development of one for *Trifolium spumosum* (bladder clover) could assist in future development of the cultivar within southern Australia. The aim of this work is to develop a core collection of *Trifolium spumosum* as a model for other pasture legume species using molecular and ecogeographical data.

Materials and methods Accessions with near complete ecogeographical data were selected from the Australian *ex situ* collection of *Trifolium spumosum*. This collection of 317 accessions was grouped into 5 geographical regions. MStrat Software (Gouesnard *et al.*, 2001) was used to select the preliminary core of 30% of the collection. Fluorescent Amplified Fragment Length Polymorphism (FAFLP) will be used to screen the diversity within the species. The primers producing the highest number of bands will be used to screen the preliminary core collection. Mstrat will be used to develop final core collections containing 30% of the preliminary core.

Results A preliminary core collection of 95 accessions was selected. In the randomly selected cores the scores (based on the Nei index) were different for each repeat, however, scores were constant for cores selected using the maximising strategy (OPT in Table 1). A final core of 32 accessions will be selected using AFLP and ecogeographical data. The AFLP markers with the green fluorescent labelled EcoR-I primer (TET) showed the greatest amount of data with the highest diversity (Figure 1). The genetic profiles of the preliminary core will be scored and recorded in a database with ecogeographical data.

Table 1 Active scores generated from the optimisation (OPT) and random (RAN) sampling methods using MStrat for core sizes of 76 and 109

Core size	Method	Final score#								
Repeat		1	2	3	4	5	6	7	8	9
76	OPT	101	101	101	101	101	101	101	101	101
	RAN	84	85	84	83	82	84	83	80	81
109	OPT	101	101	101	101	101	101	101	101	101
	RAN	89	91	92	82	88	90	91	92	86

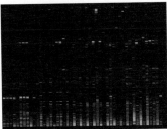

Figure 1 AFLP bands obtained from 48 samples in green fluorescent labelled EcoRI primers

Conclusions The present study aims to demonstrate that a combination of AFLP marker and ecogeographical data can be used to develop an effective core collection that maintains the majority of the genetic diversity. This model should be used to develop core collections of other pasture legume species that are too large for efficient utilisation.

References
Brown, A.H.D. (1995). The core collection at the crossroads. In: T. Hodgkin, A.H.D. Brown, Th. J.L. van Hintum, & E.A.V. Morales (eds.) Core collections of plant genetic resources John Wiley & Sons, 3-20.
Gouesnard, B., T.M. Bataillon, G. Decoux, C. Rozale, D.J. Schoen & J.L. David (2001). MSTRAT: an algorithm for building germplasm core collections by maximising allelic or phenotypic richness. *Journal of Heredity*, 92, 93-94.

Molecular characterization and tissue culture regeneration ability of the USA Arachis pintoi (Krap. and Greg.) germplasm collection

M.A. Carvalho, K.H. Quesenberry and M. Gallo-Meagher
University of Florida, Agronomy department, 304 Newell Hall, PO Box 110500, Gainesville, Florida 32611-0500 Email: clover@ifas.ufl.edu

Keywords: tropical legume, genetic resources, RAPD marker

Introduction *Arachis pintoi* Krap. and Greg. is a herbaceous, perennial legume, exclusively native to Brazil. It is considered a multiple use legume, being grown for forage; ground cover in fruits orchards, forest, and low tillage systems; erosion control; and ornamental purposes. Although several cultivars have been released in different countries, little is known about the genetic diversity of the germplasm stored in world genebanks. Our objective was to characterize and evaluate the genetic diversity of the germplasm of 35 accessions of *Arachis pintoi* at molecular level using RAPD markers. Concurrently, two tissue culture protocols were evaluated for their organogenesis ability. Further, variation in band profile was analyzed by comparing "Parent Plants" and tissue culture regenerated plants.

Materials and methods DNA was extracted from leaves of single Parent Plants using a modified CTAB protocol. Eighteen primers of ten nucleotides length from the Operon Technologies kit were used to amplify genomic DNA. Amplification products were separated by electrophoresis on 1.5% agarose gels and banding patterns were visualized by staining the gels in ethidium bromide solutions and viewing under UV radiation. Allele frequency, number of polymorphic loci, Nei's genetic distance, Nei's genetic diversity index, and Shannon-Weaver's genetic diversity index were the parameters calculated. Genetic distance (D= -ln I) was later used as a criterion for differentiation among accessions to prepare a cluster analysis. Two protocols were used to access the organogenic potential among germplasm accessions. Protocol 1 was proposed by Rey *et al.* (2000) and Protocol 2 proposed by Ngo and Quesenberry (2000). To compare these two protocols, callus rating and weight and number of regenerated plants were used. The RAPD band profiles of regenerated plants were compared to the Parent Plants to evaluate induction of somaclonal variability during the tissue culture process.

Results From the 18 primers tested, amplifications were obtained with only eight. Primers A4, B4, B5, C2, D4, D13, E4, and G5 amplified 100 different bands. The average number of amplified bands per primer was 12.5. The size of these 100 fragments ranged from 250 bp to 3500 bp. From the 100 bands amplified, 98 presented polymorphism. The average presence of bands per accession was 32, ranging from 20 to 44 bands. Ten bands were able to discriminate individual germplasm accessions. The proportion of polymorphic RAPD loci was 89%. Genetic diversity of the whole set of germplasm was estimated by Nei's gene diversity (h) and by Shannon-Weaver's diversity index (H). The average h was 0.29 ± 0.16, and average H was 0.45 ± 0.20. Average genetic distance was estimated as 0.36, and indicated that a great genetic diversity exits among the germplasm evaluated in this research. Genetic distances were used to prepare a dendogram for the 35 *A. pintoi* accessions, which separated them in four distinct groups. Callus induction was achieved on two different M.S. basal media protocols after 28 d of incubation. Analysis of variance demonstrated that Protocol 1 was superior to Protocol 2 for both variables related with callus growth. Great variability for these two variables was observed among the accessions. Shoot regeneration was achieved for several accessions on both media with no structures indicative of somatic embryogenesis being detected. Callus growth was not correlated with shoot regeneration. In Protocol 1 shoot regeneration was obtained from 15 accessions, whereas in Protocol 2, shoot regeneration was attained from 18 accessions. Root induction was very difficult to obtain, and invariably many shoots died during this process. At the end, 16 regenerated plants were recovered between the two protocols.

Conclusions Molecular characterization was achieved with RAPD molecular markers, which proved to be very informative and efficient to characterize the genetic diversity and relationships among germplasm accessions of this species. The variables used to assess the genetic diversity of the germplasm indicated that a large amount of genetic diversity exists among the germplasm evaluated in this research. Although differences in callus ratings and weight among protocols were observed we conclude that based on shoot development and plant regeneration both protocols were similar. RAPD band profiles of regenerated tissue culture plants were similar to their parent plants.

References

Ngo, H.L. and K.H. Quesenberry. 2000. Day length and media effects on *Arachis pintoi* regeneration *in vitro*. Soil Crop Sci. Soc. Fla. Proc. 59: 90-93.
Rey, H.Y., Scocchi, A. M., Gonzalez, A. M., and Mroginski, L. A.2000. Plant regeneration in *Arachis pintoi* (Leguminosae) through leaf culture. Plant Cell Reports 19: 856-862.

Genetic and molecular characterization of temperate and tropical forage maize inbred lines

B. Alarcón-Zúñiga, E. Valadez-Moctezuma, T. Cervantes-Martinez, T. Cervantes-Santana and M. Mendoza
Animal and Crop Science Depts. Universidad Autónoma Chapingo, Mexico, 56230, Email: camilaa@iastate.edu

Keywords: genetic similarity, maize races, forage quality, PIC, SSR

Introduction Livestock feeding in the Central highland of Mexico is based on harvest, grazing and annual forage conservation, with forage maize being the most important silage crop (Alarcón, 1995). Even though forage maize is extensively bred in Europe, USA and Asia since the 1900's, this started in Mexico only in the 1960's, and little is known about genetic diversity in both agronomic and nutritive value traits. Our breeding program goals are to analyze combining ability of biomass and quality predictors and to study the genetic relationship of inbred lines between lowland tropical and temperate races from Mesa Central, by genetic and molecular approaches.

Material and methods Fourteen inbred lines (IL) highly selected for forage biomass and quality value were used. 6 S_5 IL were collected from Mesa Central, Mexico, and 8 S_9 IL were originally single crosses from the tropical races: Tuxpeno, Vandeno, Olotillo, Naltel, Blandito, Reventador, Comiteco, Tepexintle, Celaya and Oloton. The single crossbred tropical races were recombined up to F_{13}, and selfpollinated up to S_9 (Cervantes *et al.*, 1978). In early 2004, the ILs and temperate x tropical crosses were field established in two locations, and agronomic and quality traits evaluated when the kernel 2/3 milklined. Traits included total and per component dry matter, plant height (PH), days at flowering; and soluble (SP) and insoluble protein (ISP), NDF, ADF & ADL, IVDMD, FAME, volatile fatty acids, sucrose and starch, were assayed. 27 out of 40 SSR markers were used to estimate genetic similarities among ILs, and to compute a discriminatory analysis by PCGA. The genetic components of variance, additive genotypic correlations and narrow sense heritabilities were estimated by MANOVA and standard errors computed by the delta method (Lynch and Walsh, 1998).

Results Genotypic effects were highly significant for all investigated traits (P<0.001), and much higher than genotype x environment interaction effects in all tropical ILs, but only higher in two of four temperate ILs. Transgressive segregations were observed in both tropical and temperate ILs for traits related to total DM, plant height, LSR, fiber predictors, free and volatile fatty acids. However, for each of the investigated digestibility traits, sucrose and starch, transgressive segregations were observed only on temperate Cacahuasintle and tropical ILs Tuxpeno y Vandeno. Narrow sense heritabilities on an entry basis ranged from low (~0.15; total and per component DM, IVDMD, FAME, ADL), medium (~0.3, PH, ISP, NDF, ADF, sucrose) and high (~0.6, days at flowering, volatile FA, starch). The average dry matter ear content was 51%, with a range of 2-3% among ILs, and a high heritability (0.5). High DM ear weight was found in ILs Cacahuasintle, Vandeno and Tuxpeno x Naltel. Positive additive genetic correlations were found between ear size, sucrose, starch and ADF content or IVDMD which had similar absolute values (0.55). So each of these traits was an important but not a unique determinant of silage maize quality. A low genetic correlation was found between ADL/NDF and IVDMD, suggesting that digestibility can be improved in both tropical and temperate ILs independent of lignin content. The 27 SSR marker primers detected 86 alleles, with a range per locus from 2 to 7 (average=3.19). The PIC values ranged from 0.09 to 0.75, with an average of 0.48. On the basis of the marker analysis the 14 inbred lines were classified into three distinct groups: temperate inbreds were included in group 1, with two distinct subgroups: Cacahuasintle and Chalqueno. The tropical inbreds were clustered into two groups: group 2 derived from dent grain germplasm and included Tuxpeno, Blandito, Comiteco, and Tepexintle; and group 3 (flint germplasm) included Vandeno, Naltel and Oloton.

Conclusions These results suggest that tropical ILs Tuxpeno and Vandeno can be used as top parental ILs for forage maize in the highland Mexico. The same was observed for Cacahuasintle as a temperate IL. The primer loci of SSR markers did not cover all genomes completely but enabled a determination of genetic similarities among ILs. More SSR primers and ILs are being used to associate with loci that positively determine heterotic groups, supported by a field diallel analysis in 2005.

References

Alarcon Z.B. 1995. Tropical and temperate forage crops for the Mexican Livestock. XII Livestock Nat. Cong. Mexico.
Cervantes S.T., M.M. Goodman, E.D. Casas and J.O. Rawlings. 1978. Use of genetic effects and genotype by environmental interactions for the classification of Mexican races of Maize. *Genetics*, 90, 339-348
Lynch M. and B. Walsh. 1998. Genetics and analysis of quantitative traits. Sinauer Assoc. Pub. USA.

Random amplified polymorphic DNA analysis in section *Pnigma* of the genus *Bromus* L.

M. Tuna[1], O. Barzani[2], K.P. Vogel[3] and A. Golan-Goldhirsh[2]

[1]*Trakya Univ., Agric. Faculty, Dept. of Field Crops, 59030, Tekirdag, TURKEY. Email:mtuna@tu.tzf.edu.tr*
[2]*The jacob Blaustein Institute for Desert Research, Ben-Gurion Univ. of the Negev, Sede Boker Campus, Israel.*
[3]*USDA-ARS, 344 Keim Hall, Univ. of Nebraska, P. O. Box 830937, Lincoln, NE 68507-0937, USA.*

Keywords: *Bromus,* genomic relation, phylogeny, RAPD

Introduction The section Pnigma consists of about 60 species. It was shown that Eurasian species of the section with the same ploidy level have similar nuclear DNA contents (Tuna *et al.*, 2001). A linear correlation between nuclear DNA content and ploidy level was also found (Tuna *et al.*, 2001). Furthermore, karyotypes of Eurasian species are similar (Tuna *et al.*, 2004a) but differ from the North American species (Tuna *et al.*, 2004b). Yet, the genetic relationship among species within the section is poorly known (Armstrong, 1991). The objective of this study was to assess the phylogenetic relationships among species of section *Pnigma* by using the RAPD technique.

Materials and methods Thirty one *Bromus* accessions representing 7 species from the USDA Regional Plant Introduction Station, Pullman, WA and Plant Gene Resources of Canada, Saskatoon, Saskatchewan S7N 0X2 were used in the study. DNA was extracted following the protocol described by Doyle and Doyle (1987). Eighty ten-mer primers (Operon Technologies, Inc.) were screened for polymorphism. A cluster dendogram was generated by the UPGMA clustering method using NTSYS-pc.

Results Thirteen primers used in the study generated 71 polymorphic RAPD markers. The size of the markers ranged from 491 bp to 3403bp. Four accessions (all diploids and *B.ciliatus*) could be distinguished from the other species by specific markers. Briefly, the dendogram includes two main clusters (Fig 1). One cluster includes only decaploids except PI 325235 (*B. riparius*) while the other cluster includes tetraploids and octaploids. Accessions of *B. ciliatus* and all of the diploids appeared as distinct species in the dendogram. Almost all accessions representing a species grouped most closely together.

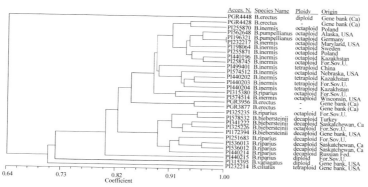

Figure 1 Dendogram of 31 *Bromus* accessions constructed from RAPD-based genetic distance

Conclusion It appears that diploid species are more distinct than their related polyploids. The results of this study confirm the previous reports (Armstrong, 1991) suggesting that Eurasian polyploid species of section *Pnigma* are closely related and carry different number of copies of the same genomes.

References

Armstrong, K. C. 1991. Chromosome evolution in *Bromus*. p. 363-377. In T. Tsuchiya, and T. K. Gupta (ed.) Chromosome engineering in plants. Part B. Elsevier, Amsterdam, The Netherlands.
Doyle, J. J. and J. L. Doyle. 1987. A rapid isolation procedure for small quantities of fresh leaf tissue. Phytochem. Bull. 19:11-15.
Tuna, M., K. P. Vogel, K. Arumunagathan and K. S. Gill. 2001. DNA Content and Ploidy Determination of the Bromegrass Germplasm Accessions by Flow Cytometer. Crop Sci. 41:1629-1634.
Tuna, M., K. P. Vogel, K. S. Gill and K. Arumuganathan. 2004a. C-Banding Analysis of *Bromus inermis* Genomes. Crop Science. 44:31 (2004)
Tuna, M., K. P. Vogel, and K. Arumuganathan. 2004b. Cytogenetic Characterization of Tetraploid *Bromus ciliatus* Genome. Proc. XVIIth EUCARPIA General Congress, Tulln, Austria.

Genetic characterization of prairie grass (*Bromus catharticus* Vahl.) natural populations

R. Sellaro[1], E.M. Pagano[1], B. Rosso[2], P. Rimieri[2] and R.D. Rios[1]

[1]*Instituto de Genética "Ewald A. Favret", CICVyA, INTA, cc 25 (1712) Castelar, Argentina.* [2] *EEA INTA Pergamino, cc 31 (2700), Argentina. Email: rrios@cnia.inta.gov.ar*

Keywords: genetic variability, RAPDs, *Bromus catharticus*

Introduction Prairie grass, *Bromus catharticus* Vahl., is a winter annual or biennial grass, native of South America which is widely distributed in the Pampeana area of Argentina and also cultivated in temperate regions of the world. Morphophysiological traits are currently used to assess the variability from natural populations and cultivars of this species. Molecular markers, which are not influenced by the environment, allow a more accurate assessment of genetic variability. Previous results from our group (Puecher *et al.*, 2001a) showed a narrow genetic basis for the prairie grass cultivars used in Argentina. On the other hand, we also observed that natural populations of this species collected in the typical area where prairie grass is cultivated in Argentina, showed a RAPD variability pattern similar to that previously observed for cultivars (Puecher *et al.*, 2001b). The objective of this work was to establish, using RAPDs, the genetic relationships among prairie grass natural populations including accessions from the margins of the cultivation area of this species in Argentina.

Materials and methods Twelve Argentinean *B. catharticus* natural populations collected from different geographical provenances were analyzed. Cultivars: Martín Fierro (*Bromus catharticus*), Zamba (*Bromus stamineus*), Pampera (*Bromus auleticus*) and Ona (*Bromus inermis*) were also included as controls. Genomic DNA was isolated of bulked samples derived from 10 plants per population using the protocol described by Puecher *et al.* (2001) and was further used as template in PCR with 8 selected primers (Operon) to produce RAPDs. Each band was scored as present or absent and low intensity bands were not considered. Genetic distances among populations were estimated using the Jaccard coefficient and UPGMA was used for cluster analysis and dendrogram plotting. All calculations were performed using the NTSYS-pc version 2.02g.

Results A total of 104 amplification products were scored from selected primers in the accessions and the control cultivars. The number of bands per primer showed a variation from 8 to 19. Considering the whole number of bands, 90% of them were polymorphic in at least one of the accessions. Cluster analysis clearly distinguished the different Bromus species (Figure 1). The genetic relationships among the controls were similar to those observed in our previous studies confirming the reproducibility of this method. The natural populations and the cv Martín Fierro of *B. catharticus* were grouped in five clusters with a genetic similarity coefficient ranging from 0.5 to 0.8. Two of these groups, one containing a

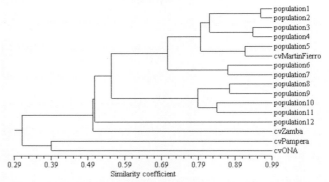

Figure 1 Dendogram for *Bromus* accession and cultivars, resulting from a UPGMA cluster analysis based on Jaccard similarity coefficient using RAPD markers

single population and the other one four populations, were clearly separate from the other groups, indicating that these accessions were genetically distinct from others in this study. This difference was coincident with significant differences in morphological characters (flag leaf width, growing habit, vegetative tillering).

Conclusions The results of the present study demonstrate that the inclusion of accessions collected at the marginal area of prairie grass cultivation in Argentina, contributed to broaden the genetic variability available for the breeding of this Argentinean native species.

References
Puecher D.I., C.G. Robredo, R.D. Rios & P. Rimieri (2001a). Genetic variability measures among Bromus catharticus Vahl. populations and cultivars with RAPD and AFLP markers. *Euphytica* 121: 229-236.
Puecher D., C. Robredo, E. Pagano, P. Rimieri & R.D.Rios (2001b). Evaluación de la variabilidad genética en poblaciones naturales de cebadilla criolla a través de RAPD. IV REDBIO 2001. Goiânia, Goiás, Brasil.

Analysis of *Bromus inermis* populations using Amplified Fragment Length Polymorphism markers to identify duplicate accessions

V.L. Bradley and T.J. Kisha
USDA,ARS Western Regional Plant Introduction Station, P.O. Box 646402, Pullman, Washington 99164-6402 USA Email: bradley@wsu.edu

Keywords: grass, AFLP, duplicate accessions

Introduction The temperate grass germplasm collection maintained at the USDA, ARS Western Regional Plant Introduction Station (WRPIS) in Pullman, Washington, consists of more than 18,000 accessions. Passport and collection data suggest that some of these accessions are duplicates, and their maintenance unnecessarily drains limited resources. The objective of this study was to use Amplified Fragment Length Polymorphism (AFLP) marker analysis on 4 populations of smooth bromegrass, *Bromus inermis* Leyss. subsp. *inermis* `Manchar´, a cross-pollinated perennial grass, to determine if the genetic variation among them was significant. If not, then maintaining separate populations would be unnecessary.

Materials and methods Total nucleic acids were extracted from new-leaf tissue of 24 plants from each population using the MagneSil$^{\circledR1}$ Kit by Promega and the concentration was adjusted to 250g/:l. Restriction enzyme digestion, ligation of adapter sequences, preliminary amplification, and selective amplification was performed using AFLP kits from Life Technologies. Selective amplification was performed using primers labeled with WELLRED$^{\circledR}$ dyes for analysis on the Beckman CEQ8000 capillary electrophoresis apparatus. Two primer pairs were analyzed providing a total of 780 polymorphic markers whose polymorphic information content (PIC) ranged from 0.02 – 0.50. Total and within population variances (PIC values) were calculated using the formula:

$$1 - \Sigma p_{ij}^{2}$$

where p is the frequency of the ith allele at the jth locus. Identification of markers was performed automatically on the CEQ8000 with ancillary software and exported to a spreadsheet where allele frequencies and variances were calculated. PIC values are measures of genetic variance within populations (Wright, 1951; Nei, 1973, Hamrick and Godt, 1989, Culley *et al.*, 2002). PIC values were calculated over all plants analyzed (total variance) and for each of the populations, and these values were used in the equation:

$$G_{ST} = 1-(H_S/H_T)$$

where H_S is the mean PIC value of each of the populations and H_T is the PIC value totaled over all individuals (combined from all populations). G_{ST} is defined as the proportion of genetic diversity that resides among populations (Nei, 1973). G_{ST} was calculated using the methods of both Nei (1973) and Hamrick and Godt (HG) (1989). The difference in the two methods is that the HG method averages G_{ST} values over all loci, while the Nei (1973) method calculates G_{ST} by first averaging H_T and H_S over all loci.

Results The methods produced G_{ST} values of 0.068(Nei) and 0.071(HG). Using the average of the PIC values obtained at each locus for each population as variances within populations, and the proportion of the total variance calculated using G_{ST} as variance among populations, an F-test (3 and 92df for among and within population variances, respectively) revealed no significant differences among populations. (F = 0.35).

Conclusions The results indicate that the 4 populations are very similar and it is not necessary to maintain them separately. Germplasm managers at the WRPIS will maintain one of the populations for distribution and the other populations will be inactivated.

References

Culley, T.M., L.E. Wallace, K.M. Gengler-Nowak, and D.J. Crawford. 2002. A comparison of two methods of calculating G_{ST}, a genetic measure of population differentiation. J. of Bot. 89(3):460-465

Hamrick, J.L. and M.J.W. Godt. 1989. Allozyme diversity on plant species. *In*: A.H.D. Brown, M.T. Clegg, A.L. Kahler, and B.S. Weir [eds.], Plant population genetics, breeding, and genetic resources, 43-63. Sinauer, Sunderland, Massachusetts, USA.

Nei, M. 1973. Analysis of gene diversity in subdivided populations. Proceedings of the National Academy of Sciences, USA. 70:3321-3233.

Wright, S. 1951. The genetic structure of populations. Annals of Eugenetics. 15:323-354.

[1]Mention of trademark or proprietary product does not constitute a guarantee or warranty by the USDA and does not imply its approval over other suitable products.

Characterisation of naturalised populations of *Thinopyrum ponticum* Podp through indexes obtained under saline stress

S.M. Pistorale[1], A.N. Andrés[2] and O. Bazzigalupi[2]
[1]*Depto. Ciencias Básicas, Genética. Universidad Nacional de Luján, Buenos Aires, Argentina, Email: genetica@mail.unlu.edu.ar;* [2]*INTA EEA Pergamino, cc 31 (2700) Pergamino. Buenos Aires, Argentina*

Keywords: saline stress, variability, relative tolerance index, Shannon-Wiener index, *Thinopyrum ponticum*

Introduction Argentina has 2.5 million hectares considered marginal for agriculture, with salinity problems that limit productivity. Among the adapted species, *Thinopyrum ponticum* is one of the perennial grasses worth mentioning because of its adaptation to lowland and salty soils. On soils with low osmotic potentials, available water is reduced, and toxicity due to ion concentration arises. Salt tolerance is genetically determined and its improvement is similar to any other character in that requires genetic variability and evaluation methods to identify superior genotypes. Due to the essentially low-intensity livestock production in Argentina, and the soil limitations in marginal environments, the characterisation and evaluation of germplasm is important in order to have a higher genetic diversity. The knowledge of parameters, procedures and methodologies for the evaluation of genetic diversity would help to plan future collection efforts aiming to add divergent populations. The objectives of this work, were to characterise *T. ponticum* germination under saline stress in ten naturalised populations through the Relative Index of salt Tolerance (RIT) (Pearen *et al.*, 1997) and to quantify population diversity through the Shannon-Wiener index (**H**)(Pielou, 1969).

Materials and methods The seeds were sown in plastic trays, paper substrate, and wetted with NaCl solutions, with conductivity at 0, 6, 12 y 18 decisiemens per meter (dS/m). The osmotic potential of these solutions were 0; -0,26; -0,6 y −1,0 MPa. The conditions for germination were: 20-30°C, 16L:8D), in a completely randomised block design, with 4 replicates and counts up to 25 days. For each population the RIT was calculated, values close to 1 indicate higher tolerance. The Hi were calculated as: $Hi = -\sum_{i=1}^{n} pi \cdot \log_2 pi$;   where $pi = nji/nj$.;

nji= germinated seeds of the j population at each salinity level (i), and nj = germinated seeds at the three levels. This H value indicates niche breadth regarding tolerance for each population. This breadth has both genetic and phenotypic components; the lower the value, the more restricted the response range. This index allows for comparing populations. The H by salinity level is, $Hsk = -\sum_{i=1}^{n} pk \cdot \log_2 pk$; where $pjk = nk/n.1$ indicates germination diversity for all the populations taken together at a given level (s). A low value indicates that there are few populations that germinate well at that level.

Results The results allowed the identification of more tolerant populations both in quantity and germination time. At the higher stress levels (12 y 18 dS/m) populations were more susceptible but three of them showed tolerance to all the salinity levels tested. The Hi ranged from 1.56 to 1.37, with some populations differing significantly (Table 1). The Hsk were lower at higher salinity levels ranging from 3.46 to 3.36. The contribution to the total diversity was 70% for salinity level, and 30% for niche breadth.

Table 1. Estimates of the diversity index (Hi) and Relative Index of salt Tolerance (RIT) for each population over salinities of 6, 12 and 18 dS/m. Based on germinated seeds data. Duncan Test, significant at 0.05 level.

Populations	Hi	RIT	Populations	Hi	RIT
1	1.45 bc	0.31	7	1.47 ab	0.34
2	1.44 c	0.27	8	1.43 bc	0.27
3	1.56 a	0.64	9	1.53 ab	0.48
4	1.49 bc	0.36	10	1.55 a	0.57
5	1.52 ab	0.47	11	1.42 c	0.27
6	1.37 c	0.21			

The RIT would be an effective tool for salinity tolerance evaluation. Along with the Shannon-Wiener index it could provide new tools for the collection, characterisation and evaluation of genetic resources.

References
Pearen, J.R.; Pahl, M.D.; Wolynetz, M.S.; Hermesh,R. 1997. Canadian Journal of Plant Science. 77:81-89.
Pielou, E. C. 1969. An Introduction to Mathematical Ecology. Wiley. New York.

Genetic structure of Mongolian Wheatgrass (*Agroypron mongolicum* Keng) in Inner Mongolia of China

Y. Jinfeng, Z. Mengli and X. Xinmin
College of Ecol. and Env. Sci., Inner Mongolia Agricultural University, Huhhot, Inner Mongolia 010018 P.R..China, Email: csgrass@public.hh.nm.cn

Keywords: genetic diversity, RAPD

Introduction Mongolia wheatgrass (*Agroypron mongolicum*) is a cross-pollinated, long-lived, cool-season and drought-resistant perennial bunchgrass, which plays an important role in arid and semi-arid grasslands of Inner Mongolia. Collections of *A. mongolicum* from different areas of Inner Mongolia are valuable sources of useful genes for its breeding. The genetic diversity of 8 accessions of *A. mongolicum* were examined in this study. A dendrogram was constructed to obtain information on the relationship between cultivated and wild *A. mongolicum* genotypes, which is basic information to explore the possibility of its use in intra- and inter-specific breeding programs.

Materials and methods A total of 8 accessions (6 wild and 2 cultivated) of *A. mongolicum* were collected from 7 areas in arid and semi-arid grasslands of Inner Mongolia. Fifteen plants (spaced at least 10 m apart) were randomly sampled from each site. Seeds were collected from each plant and planted in a greenhouse. Leaves were collected from one seedling from each plant. The collected leaves were prepared to extract genomic DNA and RAPD markers that were used to detect the genetic diversity. Seventeen arbitrary primers were used and their markers were analysed by using SPSS 8.0 to generate Jaccard's similarity coefficient (S_{ja}), genetic distance ($D=1-S_{ja}$), and diversity coefficient (DC). A UPGMA dendrogram was constructed using MEGA software.

Results The diversity coefficients (DC) ranged from 0.147 (Zhenlan) to 0.273 (Qingshuihe), with an average of 0.237 (Table 1) and the six wild populations showed a higher diversity than the cultivated populations. The diversity coefficient (DC) among the 8 populations was 0.222, among the six wild populations 0.250, while among the two cultivated varieties it was 0.181. This evidence was supported with a relatively small genetic distance between the two cultivated populations (average 0.290 among wild populations and 0.213 among cultivated populations). The UPGMA dendrogram showed that the eight populations could be divided into 3 groups, based on their geographic origin and on the soil type of their distribution areas.

Table 1 Diversity coefficients (DC) within populations

Site	Xiwu	Baiyinxi	Qingshuihe	Zhenlan	Yijihuole	Sunitezu	Var.1	Var.2	Average
DC	0.243	0.269	0.272	0.226	0.258	0.273	0.210	0.147	0.237

Conclusion The average diversity coefficient of 0.237 indicated considerable genetic diversity among populations of *A. mongolicum*. The open pollination and out-crossing system, as well as the strong gene flow led to a great diversity of *A. mongolicum* being retained in recessive genes in the heterozygotic state. As a geographically widely distributed species, *A. mongolicum* has a lot of ecotypes, each of them with their own physiological traits and adaptive capacity. It is noteworthy that there was a clear tendency for *A. mongolicum* that originated from the same location to be clustered together in the dendrogram.

References
P.Virk, B.,M. Ford-lloyd, Jackson & H. John, 1994, Use of RAPD for the study of diversity within plant germplasm collections, *Heredity*, 74:170-179
G. Yan, S. Zhang, 1985, Karyotype analysis of Mongolia wheat grass, *Grassland of China*, 2: 38-42

RFLP analyses of chloroplast DNA of the crested wheatgrasses

K.P. Vogel[1], D.J. Lee[2] and C.A. Caha[2]

[1]USDA-ARS, 344 Keim Hall, Univ. of Nebraska, PO Box 830937, Lincoln, NE 68583-0937; [2]Dept. Agronomy & Horticulture, Univ. of Nebraska, Lincoln, NE -68583-0915, USA. Email:kpv@unlserve.unl.edu

Keywords: Agropyron, wheatgrass, chloroplast DNA, RFLP

Introduction The crested wheatgrasses (*Agropyron* spp.) are widely distributed Eurasian grasses that have been used to re-vegetate millions of hectares of land in the northern plains and intermountain areas of North America. The genus consists of about 10 species that are based on the 'P' genome and includes diploids ($2n = 14$), tetraploids ($2n = 28$) and octoploids ($2n = 42$) (Dewey, 1984, Vogel *et al.*, 1999). The two main agronomic species used in North America are *A. cristatum* (L.) Gaertner and *A. desertorum* (Fischer ex Link) Schultes. Chloroplast genomes are predominately maternally inherited through the cytoplasm in angiosperms. Chloroplast DNA (cpDNA) variation can provide information on the evolutionary development of plant species. The objective of this study was to determine if chloroplast DNA polymorphisms occur within and among ploidy levels of these two main species of the *Agropyron* complex.

Materials and methods Cultivars, strains, or germplasms of *A. cristatum* diploids (Fairway, Ruff, NE AC1, NE AC2) and hexaploids (PI 222955, PI 401077) and *A. desertorum* tetraploids (Hycrest, NE AD1) and octaploid (E1 990) were grown in the greenhouse. Plants represented germplasm from the range of the species. Restriction fragment length polymorphisms (RFLP) procedures were used to assay for chloroplast DNA polymorhphism using the methods and probes described by Hultquist *et al.* (1996). Four restriction enzymes (*Bam*HI, *Eco*RI, *Eco*RV, and *Hind*III) and 20 chloroplast DNA probes were used in the RFLP analyses. Technical assistance of Risa Cohen and Sherry Hulquist is acknowledged with appreciation.

Results A total of 72 different bands were produced by the 80 probe-enzyme combinations of the RFLP analyses. All evaluated crested wheatgrass strains produced similar bands with the probe-enzyme combinations that were used. No chloroplast DNA polymorphisms were detected among the diploid, tetraploid, and octaploid *A. cristatum* and *A. desertorum* strains. See example in Figure 1.

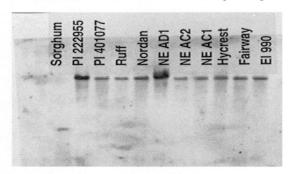

Figure 1 Crested wheatgrass and sorghum (control) restriction fragments produced by *Hind*III on a Southern blot hybridized with chloroplast DNA probe pLD 7

Conclusions The chloroplast DNA of the evaluated species of *Agropyron* complex appears to be relatively uniform which indicates that this DNA is conserved within the genus. The evaluated *Agropyron* species apparently have a common ancestral cytoplasm. This suggests that the primary genetic variation in the genus is due to nuclear DNA variation including variation due to polyploidy.

References

Hultquist, S.J., K.P. Vogel, D.J. Lee, K. Arumuganathan, and S. Kaeppler. 1996. Chloroplast DNA and nuclear DNA content variations among cultivars of switchgrass, *Panicum virgatum* L. Crop Sci.36:1049-1052.

Dewey, D.R.. 1984. The genomic system of classification as a guide to intergeneric hybridization with the perennial Triticeae. In J.P. Justafson (ed.). Gene manipulation in plant improvement. pp. 209-280. Plenum Press, New York.

Vogel, K.P., K. Arumuganathan, and K.B. Jensen. 1999. Nuclear DNA content of perennial grasses of the Tribe Triticeae. Crop Sci. 39:661-667.

Tracing the origins of Timothy species (*Phleum sp.*)

A.V. Stewart, A. Joachimiak[1] and N. Ellison[2]
PGG Seeds, PO Box 3100, Christchurch 8015, New Zealand
[1]*Department of Plant Cytology & Embryology, Institute of Botany, Jagiellonian University, Cracow, Poland*
[2]*AgResearch, Private Bag 11008 Palmerston North, New Zealand, Email: alan.stewart@pggseeds.com*

Keywords: *Phleum pratense, Phleum alpinum,* timothy

Introduction The section Phleum of the genus Phleum contains 3 species groups and, depending on the taxonomic classification used, these may be treated as 3 or 6 individual species (Joachimiak & Kula, 1997; Joachimiak, 2004). Firstly, there is the *P. pratense* group consisting of a series of diploid, tetraploid, hexaploid and octoploid forms. The diploid (2n=14) is usually known as *Phleum bertolonii* (syn. *P. pratense* spp *bertolonii*: syn *P. nodosum*) and/or *P. hubbardii*, while the tetraploid, hexaploid and octoploid are known as *P. pratense*. Secondly there is the *P. alpinum* group consisting of 2 contrasting diploid species, *P. rhaeticum* and *P. commutatum*, and a tetraploid *P. commutatum*. Finally there is the lesser known Mediterranean annual, *Phleum echinatum*, with a reduced chromosome number of 10.

Materials and methods A collection of natural populations of *P. bertolonii, P. hubbardii, P. pratense* 4x, 6x and 8x, *P. echinatum* and the 3 entities within the *P. alpinum* group. *P. commutatum* 2x and 4x and *P. rhaeticum* 2x, were studied using the sequences of the trnL (UAA) gene intron of chloroplast DNA (cpDNA) and the internal transcribed spacer (ITS) region of nuclear ribosomal DNA.

Results CpDNA sequences accurately discriminate between the 4 diploid species; *P. bertolonii; P. commutatum* and *P. rhaeticum,* as well as *P. echinatum.* They also provide an insight into the origin of some polyploids.

Both ITS and chloroplast sequences are able to discriminate between all of the diploid species. They also enable some of the genomes of the polyploids to be identified as well as revealing geographic patterns of variation. Although ITS sequences exhibit more variation than chloroplast DNA the patterns revealed are almost identical.

References

Joachimiak, A., Kula, A. 1997 Systematics and karyology of the section Phleum in the genus Phleum. *Journal of Applied Genetics* 38: 463-470
Joachimiak A. 2004 Heterochromatin and Microevolution in Phleum. Plant Genome: Biodiversity and Evolution. Vol. 1, Part B: Phanerogams. A.K. Sharma & A. Sharma (Eds.). Science Publishers Inc. Enfield, New Hampshire 03748, USA. pp 88-117

Genetic diversity and heterosis in perennial ryegrass

U.K. Posselt
University of Hohenheim, State Plant Breeding Institute(720), D-70593 Stuttgart, Germany
Email: posselt@uni-hohenheim.de

Keywords: molecular markers, diversity, heterosis

Introduction Plant breeders are concerned with the diversity among and within breeding populations, because it largely determines the future prospects of success in breeding programs. DNA markers provide a powerful tool for the assessment of genetic diversity. The relationship between genetic diversity and heterosis has been investigated in several species (Melchinger, 1999). In hybrid breeding programs divergent genepools have been established in the past, according to heterotic patterns based on testcross information. In the last decade it has been shown, that new breeding materials can be assigned to already existing gene pools using molecular markers. The phenomenon of heterosis has been of interest in grass breeding research for a long time, and the occurrence of heterosis has been demonstrated for particular crosses (review in Posselt, 2003). However, no attempts have been made to group perennial ryegrass populations according to diversity measures. So far, breeders have mostly combined diverse materials into a base population and applied intra-population improvement. Accidently, heterosis was captured in the new variety. For more reliable exploitation of heterosis in grasses, divergent genepools have to be established.

Materials and methods In a previous diversity study (Bolaric *et al.*, 2004), ecotypes and cultivars have been investigated using a RAPD marker. Genetic distance (GD) among all populations was calculated, and eight populations (6 ecotypes and 2 cultivars) with GDs from 0.21 to 0.42 were selected as parents for diallel mating. The 8 parents and their 28 hybrids were investigated in performance trials for 2 years in 2 testing sites.

Results On average the parents yielded 10.7 t ADMY (annual dry matter yield) while the hybrids performed only slightly better (10.9 t ADMY). Average heterosis (expressed as midparent heterosis MPH) across all crosses was only 2 %. Highest MPH was 7.6 % for hybrid P4 x P5 (Table 1). Compared to total ADMY, MPH in the first cut was much more pronounced (14.6 %).

Table 1 Annual dry matter yield (t/ha) of hybrids (above diagonal) and their parents (italics), and midparent heterosis in % (below diagonal)

P	1	2	3	4	5	6	7	8
1	*11.1*	11.1	10.9	10.6	11.8	10.5	10.9	11.6
2	0.5	*10.8*	11.1	10.1	11.2	10.2	10.4	11.1
3	2.7	5.6	*10.2*	10.7	11.5	10.2	10.9	11.4
4	0.3	-2.5	6.2	*9.9*	11.4	10.6	10.7	10.9
5	5.4	1.9	6.6	7.6	*11.3*	11.1	11.1	12.1
6	0.1	0.9	1.6	6.8	5.4	*9.8*	10.7	11.3
7	-1.0	-3.8	3.9	3.0	0.6	3.9	*10.8*	11.9
8	1.1	-2.2	2.9	0.7	4.6	3.9	5.1	*11.8*

The average association between GD and MPH was rather low (r = 0.10), but higher (Figure 1) for GD *vs.* Hybrid performance (r = 0.27). For in-dividual parents and their hybrids, much closer associations (r = 0.48) could be identified.

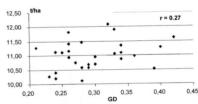

Figure 1 Hybrid performance (ADMY in t/ha) *vs.* genetic distance (GD)

Conclusions The results of the present study demonstrate, that parents can be pre-selected according to genetic similarity based on molecular markers. Thus, the number of cross combinations to be produced and tested can be reduced. The most promising cross combination to identify heterotic potential is hybrid 5 x 8. The former is the best performing ecotype, while the latter is the highest yielding cultivar among the parents. Thus, parents P5 and P8 could be assigned to divergent genepools. To broaden the respective pools, P7 could join the pool of P5, since it is also heterotic with P8.

References
Bolaric, S., S. Barth, A.E. Melchinger & U.K. Posselt (2004). Molecular genetic diversity within and among German ecotypes in comparisos to European perennial ryegrass cultivars. Plant Breeding (in press).
Melchinger, A.E. (1999). Genetic Diversity and Heterosis. In:J.G. Coors &S. Pandey (eds.) The Genetics and Exploitation of Heterosis in Crops. CSSA, Madison, WI., 99-118.
Posselt, U.K. (2003). Heterosis in Grasses. Czech. J. Genet. Plant Breed., 39, 48-53.

Population genetics of perennial ryegrass (*Lolium perenne* L.): differentiation of pasture and turf cultivars

M.P. Dobrowolski[1,3], N.R. Bannan[1,3], R.C. Ponting[2,3], J.W. Forster[2,3] and K.F. Smith[1,3]

[1]*Primary Industries Research Victoria, Hamilton Centre, Hamilton, Victoria 3300, Australia* [2]*Primary Industries Research Victoria, Plant Biotechnology Centre, La Trobe University, Bundoora, Victoria 3086, Australia* [3]*Molecular Plant Breeding Cooperative Research Centre, Australia. Email: mark.dobrowolski@dpi.vic.gov.au*

Keywords: perennial ryegrass, population genetics, simple sequence repeat markers, cultivar differentiation

Introduction Cultivar differentiation using molecular markers to assess genetic variation may be of value in obtaining or protecting plant breeders rights. A knowledge of genetic variation and how it is structured within perennial ryegrass (*Lolium perenne* L.) populations will also help us understand the consequences to fitness and adaptation when implementing molecular breeding strategies. In a study of the population genetic structure of a number of perennial ryegrass varieties we examined the cultivar differentiation potential of marker technology.

Materials and methods We used 26 genomic DNA-derived simple sequence repeat (SSR) marker loci. Based on genetic mapping in the p150/112 reference population, these loci are distributed across each of the 7 linkage groups (chromosomes) of the perennial ryegrass genome. Two groups of populations were compared to assess the capacity to distinguish them using the SSR marker loci: pasture varieties derived from the Kangaroo Valley ecotype from New South Wales, Australia, and turf varieties from a range of sources.

Results An analysis of molecular variance (AMOVA) showed that 90% of the variation is located within the cultivars, 8% among cultivars within the pasture and turf groups and 2% between the pasture and turf groups. Individuals in this analysis could not reliably be allocated to their source cultivar or group (pasture or turf) on the basis of their 26 loci SSR genotype (Figure 1). Due to the obligate outbreeding nature of perennial ryegrass, relatively high levels of diversity are maintained within cultivars that have undergone selection in their development.

Definition and identification of cultivars based on molecular markers may depend on careful selection of markers, informed by the population structure and breeding history of the cultivars to be distinguished. This data has important implications for the design and selection of metapopulations for association genetics analyses in outbreeding grass species.

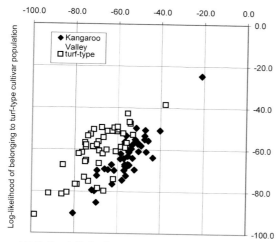

Figure 1 Log-log plot of the likelihood of a perennial ryegrass individual belonging to the population defined by the Kangaroo Valley-derived cultivars or the population defined by the turf-type cultivars. The probability that an individual is derived from the defined populations is calculated from the observed allele frequencies in the population excluding that individual if necessary, and assuming Hardy-Weinberg equilibrium and panmixia.

Analysis of genetic changes in single-variety ryegrass swards

C. Straub[1], G. Boutet[2] and C. Huyghe[1]
[1]Unité de Génétique et d'Amélioration des Plantes Fourragères, INRA, 86600 Lusignan, France
[2]Unité mixte de recherche Amélioration et santé des plantes, 63039 Clermont-Ferrand cedex 2, France Email: huyghe@lusignan.inra.fr

Keywords: genetic changes, ryegrass, synthetics, molecular markers

Introduction Ryegrass varieties are synthetics, with a wide within-variety genetic variance for most traits. Ryegrass swards are likely to experience genetic changes with seasons and years because of plant death, asymmetric vegetative reproduction or plant recruitment through reproduction or seed immigration. These changes may be related to or induce changes in agronomic traits, such as biomass production or dry matter composition. The present research was undertaken to measure genetic changes in swards obtained from sowings of a single variety of ryegrass. These changes were evaluated using both neutral molecular markers and morphological traits. The present paper deals with the molecular markers.

Material and methods A diploid ryegrass variety, cv Herbie, was cultivated under 12 different treatments combining different environmental conditions (5 locations in France), duration of cultivation (from 2 to 7 years) and exploitation regimes (grazing or cutting). In spring 2003, samples of 100 independent tillers were taken from each treatment to provide different populations. Two 100 seed lots of Herbie were used as controls. Every individual plant from the field samplings and the seed lots may be considered as a different genotype. DNA was extracted from each individual initial tiller. Each genotype was characterised with 13 molecular markers homogeneously distributed over the 7 chromosomes of the ryegrass haploid genome. The primers used for revealing marker polymorphism were obtained from INRA Lusignan or the Noble Foundation (Ardmore, OK, USA). PCR reactions were performed and markers were revealed using an ABI 3100 capillary sequencer on the Genotyping Laboratory in INRA Clermont-Ferrand. The software used for population genetic analysis included: Genetix, Genepop (Raymond & Rousset, 1995) and Geneclass (Piny *et al.*, 2004). Allelic frequency for each marker was analysed as well as the panmictic status of each population. Analysis of allelic polymorphism makes it possible to detect the recruitment of new individuals. Wright F-statistics (Wright, 1965) are being calculated for the whole set of populations. Distance among populations and the initial seed batches are also being calculated.

Results Data analysis is still in progress. As expected from the structure of the initial polycross of the variety Herbie, wide polymorphism was observed within the initial seed lots of the variety with the number of alleles varying from 2 to 15 depending on the markers, with a prevalence of a few alleles for each marker (Fig 1). Small but not significant differences among initial seed batches were observed. Panmictic equilibrium was reached for all markers. This is relevant to the four generations of multiplication in the commercial seed production of this variety. However, a small heterozygous deficiency was detected for most markers. New alleles were detected for 6 markers at a low frequency, in some cultivated swards. It is relevant to the low immigration rate observed in grassland swards. This extra genetic variation may originate from either the pollination of plants from the swards by foreign pollen (Nurminiemi *et al.*, 1998) or seed immigration (Mouquet *et al.*, 2004)

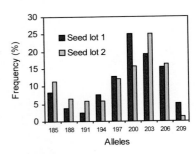

Figure 1 Distribution of allele frequency in the two seed batches for the microsatellite marker NFFA027

References

Mouquet, N., P. Leadley, J. Mériguet & M. Loreau (2004) Immigration and local competition in herbaceous plant communities: a three-year seed-sowing experiment. *Oikos*, 104, 77-90.

Nurminiemi, M., J. Tufto, N.O. Nilsson & O.A. Rognli (1998) Spatial models of pollen dispersal in the forage grass meadow fescue. *Evolutionary Ecology*, 12, 487-502.

Piny, S., A. Alapetite, J.M. Cornuet, D. Paetkau, L. Baudoin & A. Estap (2004). Geneclass 2 : a software for genetic assignment and first generation migrants detection. *Journal of Heredity*, 95, 536-539.

Raymond, R. & F. Rousset (1995) GENEPOP (Version 1.2) population genetics software for exact tests and ecumenicism. *Journal of Heredity*, 86, 248-249.

Wright, S. (1965) The interpretation of population structure by F-statistics with special regard to systems of mating. *Evolution*, 19, 395-420.

Genetic variability between adapted populations of annual ryegrass (*Lolium multiflorum* Lam) in Argentina

A. Andrés[1], B. Rosso[1], J. De Battista[2] and M. Acuña[1]

[1]INTA EEA Pergamino cc 31 (2700) Pergamino, Buenos Aires, Argentina aandrés@pergamino.inta.gov.ar;
[2]INTA EEA Concepción del Uruguay cc 6 (3260) C. del Uruguay, Entre Rios, Argentina

Keywords: genetic variability, breeding programme, annual ryegrass

Introduction Italian ryegrass (*Lolium multiflorum* Lam.) is one of the most important annual grasses used in Argentina because it adapts better to the intensive animal system of the Humid Pampas than other annual forage grass. Although much research has been done to study its productive potential and management technologies, little work has focused on breeding and selection. There is ample evidence that genetic variability occurs within grass species (Snaydon, 1987; Andrés and Barufaldi, 1997) both in morphology and physiology. As a result the variation of attributes related with yield potential, quality and adaptation to different management systems, is often used in plant breeding to develop new varieties. The objective of this work was to evaluate the genetic variability between 32 populations of annual ryegrass adapted to different grassland environments in the Humid Pampas Region of Argentina as an introductory part of a breeding programme at INTA. The final aim of this programme is to provide new varieties of annual ryegrass adapted to different management systems.

Materials and methods 32 adapted populations of annual ryegrass were grown from seeds in a cool greenhouse during summer 2004. At the stage of 8 tillers, forty five plants were randomly sampled from each population and transplanted 0.60 m apart in a randomized block design with three replicates, at the experimental grounds of INTA Pergamino. (Buenos Aires). All plants were measured or scored for a range of vegetative attributes related with winter dry matter production and herbage quality: tiller number (TN); growth habit (GH); leaf width (LW); vigour (V); dry matter yield (DMY) (24/8; 14/09; 25/10); and *in vitro* dry matter digestibility (IVDDM). Statistical analysis was performed on each attribute by using the SAS programme. The genetic parameters estimated were: genetic variance (GV); environmental variance (EV) and broad sense heritability (H).

Results and conclusions Annual ryegrass is naturalized and established in several environments of the "pampa" grasslands which have resulted in the formation of ecotypes with differences in their morphology and physiology. The results of this study showed that there were significant differences ($P<0.01$) between natural populations of annual ryegrass for the attributes related with winter growth (Table 1). There were also important differences between plants within populations. The evidence for high broad sense heritability values given by the present study may have an important applicability to the current breeding programme.

Table 1. Mean, range, variance components and broad sense heritability on phenotypic mean basis for attributes related with winter dry matter in annual ryegrass populations evaluated at Pergamino

Variable	Mean	Range	Genetic variance	Heritability
Growth habit (GH) (1...4)*	1.83	1.00-3.85	0.62	0.53
Leaf width (LW) (1...3)*	1.84	1.00-2.95	0.43	0.55
Vigor (V) (1...3)*	1.83	1.00-3.00	0.40	0.56
Dry matter yield (24/8) (g/pl)	13.08	8.87-16.65	49.65	0.65
Dry matter yield (14/9) (g/pl)	12.93	9.47-17.68	45.27	0.64
Dry matter yield (25/10) (g/pl)	28.07	17.81-33.29	91.59	0.52
Digestibility (IVDDM) (24/8)	71.47	61.71-79.78	14.39	0.41

*Scale: GH: 1 (postrate) to 4 (erect); LW: 1 (narrow) to 3 (broad); V: 1 (low) to 3 (high)

References

Andrés, A. and M. Barufaldi, M. (1997) Differences between adapted populations of *Dactylis glomerata* L. in Argentina. Proc. 18th Int. Grass. Cong. Winnipeg. Canadá, pp 103-104.

Snaydon, R.W. (1987). Population responses to environmental disturbance. Pg 15-31 in J.van Andel, J.P. Bakker and R.W. Snaydon. eds. Disturbance in grasslands-causes, effects and processes. W. Junk Publishers. Dordrecht. The Netherlands.

Does AFLP diversity reflect consanguinity within meadow fescue breeding material?

B. Boller and R. Kölliker
Swiss Federal Research Station for Agroecology and Agriculture, Agroscope FAL Reckenholz, Reckenholzstrasse 191, CH-8046 Zurich, Switzerland   Email: beat.boller@fal.admin.ch

Keywords: AFLP, molecular markers, diversity, *Festuca pratensis*, meadow fescue, breeding

Introduction Cultivars of perennial grass species are usually synthetics with a limited number of constituent parental clones, prone to inbreeding depression. Plant breeders aim at a balance between intensity of selection and maintenance of genetic diversity when making their choice of parent clones in an intuitive way, aided by fragmentary pedigree information. Molecular markers offer new opportunities for assessing genetic diversity among selected plants. The objective of the investigation presented here is to check if the genetic distance as measured by AFLP polymorphisms reflects consanguinity among *Festuca pratensis* individuals from our breeding programme.

Materials and methods 255 genotypes of *Festuca pratensis* originating from seven, partly interrelated gene pools (Table 1) were analyzed. AFLP templates were prepared by restriction digestion (*Eco*RI and *Mse*I) and adaptor ligation of 1 μg genomic DNA. For selective amplification, 11 combinations of 6 EcoRI and 2 MseI primers with 3 additional nucleotides each were used. Visual scoring resulted in 272 polymorphic markers. Genetic diversity among genotypes was calculated using Euclidean squared distances. The SAS v. 8.02 package was used for further statistical analysis.

Table 1 Origin of plant genotypes and mean Euclidean squared distances (E^2) within gene pools

Gene pool	no.	Programme	Origin	E^2
Ecotypes 99/03	31	FAL	Collection in 28 different permanent meadows	76.3
S1a	59	Schmidt	F1 progeny from random mating of 70 plants	65.9
S1b	46	Schmidt	F1 progeny from random mating of 65 plants	66.6
S2	62	Schmidt	F1 progeny from random mating of 60 plants	65.4
S3	11	Schmidt	various plants from recurrent selection schemes	66.2
cv. Preval	24	Badoux	Syn4 progeny from polycross with 16 clones	62.0
cv. Pradel	22	Badoux	Syn4 progeny from polycross with 15 clones	58.2

Results AFLP marker diversity within gene pools reflected the intensity of selection (Table 1). Non-related ecotypes showed the greatest diversity, followed by random mating breeding populations and narrow-based cultivars. Principal component (data not shown) and cluster analysis of AFLP data (Figure 1) clearly separated the gene pools and reflected known pedigree relationships among them. Half-sibs and half-cousins were significantly less distant to each other than to comparable non-related plants (Table 2). Average distances between half-sibs were 86.9 % and between half-cousins 93.5 % of the distances observed between pairs of unrelated plants. This corresponded well with the expected co-ancestry coefficients of 12.5 % for half-sibs and 6.25 % for half-cousins.

Table 2 Euclidean squared distances (E^2) between pairs of consanguineous and non-related plants. E^2 values within one row not followed by the same letter are significantly different.

Gene pool	half-sibs no.	E^2	half-cousins no.	E^2	non-related E^2
S1a+S1b	31	58.1c	44	64.4b	67.4a
S2	17	60.1b	21	61.0b	68.1a
both	48	58.8c	65	63.3b	67.7a

Figure 1 Relationship among gene pools as revealed by cluster analysis of AFLP data

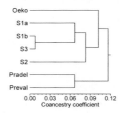

Conclusions There were two distinct lines of evidence suggesting that AFLP diversity reflects the degree of consanguinity within *Festuca pratensis* breeding material: grouping into gene pools by cluster analysis, and reduced distance between pairs of consanguineous plants. AFLP analysis can therefore be used to optimize genetic diversity among parent clones of synthetic cultivars.

Genetic diversity in *Festuca* species as shown by AFLP

X.Q. Zhang[1,2] and S.S. Bughrara[2]

[1]*Department of Grassland Science, Sichuan Agricultural University, Yaan, China, 625014.*
E-mail: zhangxq@sicau.edu.cn
[2]*Department of Crop and Soil Science, Michigan State University, Room 286 Plant and Soil Sciences Building,*
East Lansing, MI, USA 48824. Email: bughrara@msu.edu

Keywords: fescue, Festuca spp., genetic diversity, AFLP

Introduction   Fescues (*Festuca* spp.) are widely occurring temperate grasses with more than 450 species that represent a vast resource for genetic improvement of turfgrass and forage cultivars. Fescues are normally outcrossing species and exhibit many ploidy levels(x=7). Much of the work in classification of *Festuca* is predominantly based on morphological and cytological features. Difficulties in morphological characterization, which are largely subjected to environmental influences, have resulted in many synonymous species and uncertainties in phylogenetic relationships. DNA fingerprinting is considered a more stable and reliable technique to explore genetic diversity and relationships.

Materials and methods   To study the genetic diversity and relationships between fescue species, forty-six accessions from the USDA's germplasm collection representing thirty-seven species of *Festuca* from twenty countries were investigated using AFLP analysis. Selective amplifications for fescue were made using different primer combinations from the final products of EcoRI/Mse digestion, adapter ligation and pre-amplification. From the initial twenty selective primer combinations, nine combinations were chosen for clarity, repetitiveness in duplicated gel runs. Four hundred and forty-eight AFLP markers from 9 chosen primer combinations were used to differentiate between fescue accessions using a bulk of 25 genotypes per accession. Polymorphic bands were scored manually and analyzed using NTSYS v.2.1. Cluster analysis using the UPGMA was run to produce the dendrogram.

Results and discussion   Initial testing of AFLP techniques on fescue species demonstrated that selective amplification with the *Eco*RI + 3 and *Mse*I + 3 primer combinations produced too many fragments (≈200) to be resolved on polyacrylamide gels, whereas *Eco*RI + 3 and *Mse*I + 4 primer combinations generally produced the desired number of fragments (40~100). a total of 448 AFLP (418 polymorphic and 30 monomorphic) fragments were scored from nine primer combinations. The E-AGC and M-CGCG primer combination produced the greatest number of polymorphic fragments (61), while the E-ACA and M-CTCG primer combination yielded the fewest polymorphic fragments (33). The majority of scored fragments were in the 80- to 350-bp range, however, fragments were found across the entire size range of 60 to 500 bp.

Genetic similarities between accessions ranged from 0.36 to 0.87 showing no duplication in the collection and a high level of diversity in *Festuca*. According to both principal component analysis and UPGMA dendrogram, all accessions can be clearly divideded into seven groups. Genetic relationships between fine-leaved species and broad-leaved species were revealed in the cluster groupings. Specific markers were found for *F. occidentalis, F. pallens, F. arizonica, F. pungens, F. novae-zelandiae, F. heterophylla, F. gigantea, F. altaica, F. ovina*. Our results support earlier observations based on morphological characters that broad-leaved taxa are distinct from fine-leaved species but closely related.

Analyses of genetic change in grass-clover based systems over time

A. Ghesquiere, K. Mehdikhanlov, M. Malengier and J. De Riek
Department of Plant Genetics and Breeding, Caritasstraat 21, B 9090 Melle, Belgium.
Email: a.ghesquiere@clo.fgov.be

Keywords: genetic shift, molecular markers, white clover, red clover, grass-clover mixture

Introduction Since the use of nitrogen fertilisers is reduced, swards based on grass-clover mixtures regain importance in grassland production. Management of these swards is more complicated than the management of pure grass swards. The population structure will develop in response to abiotic and biotic stresses. In this study we will test the genetic change in the clover components of grass-clover mixtures.

Material and methods We have sown 15 varieties of white clover and 15 varieties of red clover in the greenhouse on the 18th of November 2004. We will screen these varieties by AFLP markers and select 5 varieties of each of the species which have narrow diversities within the variety and high genetic distances between the varieties. We extracted DNA from 30 plants of these varieties by the Modified CTAB DNA-isolation protocol. AFLP markers are being performed.

Based on the results of the AFLP study we will carry out a field experiment which will be sown in May 2005. (Table 1). The clover will be sown in association with perennial ryegrass. The plots will be cut 4 times per year and we will sample the plots at each cut to know which varieties are dominating the mixture or to test the genetic shift within the variety. AFLP and SSR markers will be used to analyse the genetic change due to selection or shift.

Table 1 Different types of grass clover mixtures used to analyse the genetic change. (RC: red clover variety; WC: white clover variety)

type	number of plots	description
A	5	each of the RC
B	5	each of the WC
C	1	mixture of the 5 RC
D	1	mixture of the 5 WC
E	1	mixture of the 5 RC and the 5 WC
F	5	each of the RC with a mixture of the 5 WC
G	5	each of the WC with a mixture of the 5 RC

Analyses In the plots of type A and B we will test the genetic shift of one white or red clover variety growing in association with perennial ryegrass. These studies will be performed with AFLP markers. These results can be compared to the AFLP results of the plots of type F and G where the clover variety is growing not only in association with grass but also in competition with another clover species. To analyse the genetic change in the plots with competition within the species (types C, D and E) we will use AFLP markers as well as SSR markers.

Keyword index

Author index

Printed in the United States
by Baker & Taylor Publisher Services